D1566810

ADVANCES IN Agronomy

VOLUME 48

Edited by

Donald L. Sparks
Department of Plant and Soil Sciences
University of Delaware
Newark, Delaware

ACADEMIC PRESS, INC.
Harcourt Brace Jovanovich, Publishers
San Diego New York Boston London Sydney Tokyo Toronto

This book is printed on acid-free paper. ∞

Academic Press, Inc.
1250 Sixth Avenue, San Diego, California 92101-4311

United Kingdom Edition published by
Academic Press Limited
24–28 Oval Road, London NW1 7DX

Library of Congress Catalog Number: 50-5598

International Standard Book Number: 0-12-000748-7

PRINTED IN THE UNITED STATES OF AMERICA
92 93 94 95 96 97 BC 9 8 7 6 5 4 3 2 1

Contents

Forms, Reactions, and Availability of Nickel in Soils

N. C. Uren

Electrochemical Techniques for Characterizing Soil Chemical Properties

T. R. Yu

History and Status of Host Plant Resistance in Cotton to Insects in the United States

C. Wayne Smith

Contributors

Numbers in parentheses indicate the pages on which the authors' contributions begin.

D. J. ANDREWS (89), *Department of Agronomy, University of Nebraska-Lincoln, Lincoln, Nebraska 68583*

C. O. GARDNER (1), *Department of Agronomy, University of Nebraska-Lincoln, Lincoln, Nebraska 68583*

K. A. KUMAR (89), *International Crops Research Institute for the Semi-Arid Tropics (ICRISAT), Sahelien Centre, BP 12.404, Niamey, Niger*

SHIVAJI PANDY (1), *CIMMYT c/o CIAT, A.A. 6713, Cali, Colombia*

C. WAYNE SMITH (251), *Department of Soil and Crop Sciences, Texas A&M University, College Station, Texas 77843*

N. C. UREN (141), *School of Agriculture, La Trobe University, Bundoora 3083, Victoria, Australia*

T. R. YU (205), *Institute of Soil Science, Academia Sinica, Nanjing, People's Republic of China*

Preface

Continuing in the tradition of previous volumes of *Advances in Agronomy*, Volume 48 contains contributions from authors from around the world. Scientists from the United States, China, Colombia, Niger, and Australia have contributed to this volume. These authors discuss important topics and advances in both the crop and the soil sciences.

The first article reviews advances in recurrent selection for population, variety, and hybrid improvement in tropical maize. Emphasis is placed on tropical maize germplasm and improvement, recurrent selection in maize, population formation and factors affecting choice of selection methods and progress from selection, recurrent selection methods used in the tropics, and contributions of recurrent selection and tropical maize germplasm to maize production and improvement. The second article is a review on various aspects of pearl millet and its use as food, feed, and forage. The third article is concerned with forms, reactions, and availability of nickel in soils. The fourth article presents advances in the use of electrochemical techniques for characterizing soil chemical properties. Fundamentals of electrochemical methods, potentiometry, problems with the liquid junction potential, voltammetry, and conductometry are covered in detail. The fifth and final article is an historical review of the status of host plant resistance in cotton to insects in the United States. Emphasis is given to evolution of cotton types, the boll weevil invasion of Texas, the pink bollworm, the cotton and tobacco bollworms, and other insect pests of cotton.

I am grateful to the authors for their first-rate contributions.

Donald L. Sparks

RECURRENT SELECTION FOR POPULATION, VARIETY, AND HYBRID IMPROVEMENT IN TROPICAL MAIZE

Shivaji Pandey[1] and C. O. Gardner[2]

[1]CIMMYT c/o CIAT,
A. A. 6713, Cali, Colombia

[2]Department of Agronomy,
University of Nebraska-Lincoln,
Nebraska 68583

Advances in Agronomy, Volume 48

I. INTRODUCTION

Maize (*Zea mays* L.) was first used as a source of food by ancient American Indian civilizations, and it also played an important role in their culture. They were responsible for its early domestication and cultivation. Today, maize has become one of the leading agricultural crops and is used for food, feed, fuel, and fiber in many parts of the world, not only in temperature regions but also in tropical and subtropical areas. Much has been written about maize research and production in temperature regions; far less has been written about tropical maize. Recurrent selection theory was developed primarily in the United States and many papers have been published on its application for the improvement of maize populations in the U.S. Corn Belt and other temperate regions. Much less has been written about its use for the improvement of tropical and subtropical maize.

The International Center for Maize and Wheat Improvement (CIMMYT) was established April 12, 1966, to increase quantity and quality of maize and wheat and alleviate hunger in developing countries of the world. Its center of operations is at El Batan, just outside of Mexico City, but its outreach spans the globe with major programs in Latin America, Africa, and Asia. Over 9 million ha of the world's maize crop are now seeded to CIMMYT-related varieties and there is solid evidence of increasing utilization of CIMMYT germplasm in national programs in tropical and subtropical countries, and even in temperate regions. The heart of the CIMMYT maize program has been recurrent selection at the population level. No other organization has done recurrent selection on such a large scale over such a long period of time. For this reason, CIMMYT maize breeding work will be highlighted in this chapter on recurrent selection in tropical maize.

The objectives of this paper are to (*a*) discuss tropical maize germplasm and environments in which maize is grown; (*b*) provide information on recurrent selection systems and how they are applied by the breeder; (*c*) review maize population improvement work in national programs in tropical regions; (*d*) highlight recurrent selection work of CIMMYT; and (*e*) relate population improvement work to hybrid programs for the tropics.

A. The Status of Maize

Maize is grown from 58° N to 40° S in latitude, from sea level to 3808 m above sea level (masl), and under 25.4 to 1016 cm rainfall (Hallauer and Miranda, 1988). The world produced around 450 million tons of maize in 1986. The developing countries produced 40% of the total in about 64% of the total maize area. The developed countries produced 47% of the total in 26% of the area. The remaining area and production occurred in Russia and Eastern Europe. Since the early 1960s, annual maize production in developing countries has been increasing faster (4% per year) than in developed countries (3% per year). Maize yields have increased by 2.8% per year (45 kg/ha/yr) in the developing countries and by 2.4% annually (104 kg/ha/yr) in the developed countries. The average maize yield for the developing countries is about 2.2 t/ha compared to 6.2 t/ha for the developed countries. Developing countries imported about 11 million tons of maize during the period 1986–1988, where maize use per capita grew by 1.8% per year, compared to a 1.2% increase in per capita consumption in the developed countries during the same years. Approximately 53% of the maize area in the developing countries is planted to improved maize—39% with hybrid seed and 14% with open-pollinated cultivars (OPVs) (CIMMYT, 1987, 1990).

B. Utilization of Maize

About 66% of the maize produced in the world is used as feed, 25% is used as food and as industrial products, and the remaining is used as seed or is lost. Whereas 20% of the world's maize is used as food in developing countries, only 7% of the maize is used as food in developed countries. Use of maize as feed generally increases with increases in per capita income in developing countries (CIMMYT, 1984).

Maize is a staple food for several hundred million people in Latin America, Asia, and Africa. It provides 15% of the world's annual protein produced from food crops and represents 19% of the world's food calories. Per capita consumption of maize in Mexico, Guatemala, Honduras, Kenya, Malawi, Zambia, and Zimbabwe is about 100 kg per year and in Nepal and the Phillippines about 40 kg per year (National Research Council, 1988).

Maize is consumed in many different ways around the world. Its principal use in Mexico and some Central American countries is for tortillas, an unleavened bread made from the dough of dehulled kernels. In the Andean highlands, maize is mainly used as green ears (choclos). Large seeded, soft (floury) and semisoft (morocho) maize types are used for this purpose, as well as for preparing "kancha," "mote," and a drink, "chicha." In the lowlands of the Andean countries, maize is generally used as a compact small bread, named "arepa." In Eastern and Southern Africa, maize is used as thick porridge and in West Africa as "kenkey." In Asia, cracked maize is usually cooked and eaten like rice or as bread. Maize is also used in soups and broths throughout the world. Maize-based products such as corn flour, thickenings and paste, syrups, soft drinks, candies, chewing gums, corn flakes, maize grits, and beer and other alcoholic beverages are available on store shelves throughout the developing and the developed world (CIMMYT, 1984).

C. Tropical Maize Environments

Approximately 60 million ha of maize are grown in the tropics with an average yield of 1.2 t/ha (Brewbaker, 1985). Approximately 42% of all maize grown in developing countries lies in the tropics (Edmeades *et al.*, 1989). Maize yields are primarily limited in the tropics by intercepted radiation to heat unit ratio. The ratio is much lower in the lowlands compared to high altitudes and is lower in the tropics compared to temperate latitudes (G. Edmeades, 1991, personal communication). Relatively less light is intercepted during the rainy season in the tropics, which coincides with the grain-filling period of the crop. Light interception is further reduced by lower plant densities. Extreme weather variations, erratic rainfalls, high temperatures, particularly during nights, and low temperatures at high altitudes also reduce yields.

Low fertility of most tropical soils reduces maize yields. Nitrogen (N) is the most limiting nutrient in the tropics and provides the major difference between low- and high-input agriculture. Phosphorus (P) is the second most limiting nutrient and is immobilized by aluminum (Al) and iron (Fe) in acid soils and by calcium (Ca) and magnesium (Mg) in the calcareous soils. The vast majority of tropical soils are acid and Al toxicity is the most important reason for crop failure in such soils (Brewbaker, 1985).

Maize cultivars in the tropics yield less than temperate cultivars in their environments due to their lower grain yield potential. The morphological characters, such as height and leafiness, that might have been important for the adaptation of maize to the tropical environment are responsible for a considerable portion of the increase in growth brought about by improved management going to nongrain plant parts (lower harvest index) in tropical maize. Extensive growth in height also contributes to yield losses through lodging.

High pest pressures and suboptimum moisture supply reduce maize yields in the tropics. Diseases frequently reduce production by 30–40% and insects can reduce yields by 100%. Weeds account for 50% of yield losses in the low-input agriculture of the tropics (Brewbaker, 1985). Edmeades *et al.* (1989) estimated that less than 5% of maize area in the tropics was irrigated. Average annual yield losses from moisture stress in the tropics were estimated to be 0–10% in rarely stressed areas, 10–25% in occasionally stressed areas, and 25–50% in the commonly stressed areas. Poor crop management procedures, limited resources, application of inadequate and improper inputs, and a lag in technology transfer contribute to reduced maize yields in the tropics.

II. TROPICAL MAIZE GERMPLASM

A. Extent of Genetic Diversity

Maize seems to have originated in Mexico and was domesticated some 7000 to 10,000 years ago in south-central or southwestern Mexico (Goodman, 1988). Four major hypotheses have been advanced over the years relating to the origin of maize (Hallauer and Miranda, 1988):

1. Maize evolved from a now extinct ancestor native to the highlands of Mexico or Guatemala (Weatherwax, 1955).
2. Maize evolved from a cross between *Coix* spp. and sorghum, perhaps in southwestern Asia (Anderson, 1945).
3. Maize evolved from teosinte directly as a result of human selection (Beadle, 1939).
4. Wild maize was a form of pod corn native to the lowlands of South America and modern maize arose from crosses between maize and teosinte or *Tripsacum* spp. (Mangelsdorf and Reeves, 1939).

Goodman (1988) thoroughly reviewed the available literature and concluded (*a*) that maize originated as a result of a single evolutionary event that occurred very long ago and involved only a few plants that caused the divergence between maize and annual Mexican teosintes; (*b*) that a single large population of toesinte was converted into maize; (*c*) that different populations of teosinte gave rise to various maize populations; or (*d*) that maize or annual Mexican teosinte arose as a single plant or a few plants and over time acquired its cytological, enzymatic, and morphological variability by mutations and backcrossing to the parental taxon. He stated, however, that these hypotheses were not supported by the available archeological and biosystematic evidences.

Anderson and Cutler (1942) proposed the need for a natural classification of the available variability present in maize germplasm and developed the concept of races of maize. They defined a race as a "group of related individuals with enough characteristics in common to permit their recognition as a group—in genetical terms it is a group with a significant number of genes in common." The classification of genetic variability in maize of Asia and Africa is not as advanced as that of Latin America (Goodman and Brown, 1988). A total of 285 races have been described in the Western Hemisphere. Approximately 265 are present in Latin America, the majority in South America. A closer and more critical look at the races, however, suggests that there may in fact be only about 130 distinct races in the Western Hemisphere (Hallauer and Miranda, 1988). Approximately 50% of these races are adapted to 0–1000 masl, 10% to 1000–2000 masl, and the rest to higher altitudes. About 40% of the races have floury endosperm, 30% are flints, 20% are dents, 10% are popcorns, and 3% are sweetcorns (Paterniani and Goodman, 1977). Direct selections from most of these races are currently grown on most of the area planted to maize in the developing world. Migration of original and later users of maize; isolation facilitated by geography, flowering differences, and gametophytic factors; mutations; cross-pollination; and natural and artificial selection for preferred grain and ear traits have contributed to the creation of a large amount of genetic variability in maize (Hallauer and Miranda, 1988).

Goodman and Brown (1988) summarized the most important characteristics and the heterotic patterns of major racial groups and races. Tuxpeno, a Mexican Dent, is reported to combine well with Coastal Tropical Flints, the race Tuson, and Cuban Flints. Tuxpeno has good stalk quality and tolerance to *Bipolaris* spp., but it has high ear placement, has a poor root system, and is susceptible to sugarcane mosaic virus. Tuson has excellent ear type, good grain quality, shorter plants, and tolerance to sugarcane mosaic virus and it combines well with Tuxpeno, Cuban Flints, and Chandelle. Coastal Tropical Flints possess excellent grain quality, good husk cover, stalk quality, good roots, and resistance to *Bipolaris* spp. and are included in such widely used maize cultivars as Suwan 1, ETO, Metro, Venezuela 1, and Mayorbela. The racial complex known as Cuban Flints in the Caribbean Islands, Cateto in Brazil, and Argentine Flint in Argentina combine well with Tuson, Tuxpeno, and Coastal Tropical Flints. Their greatest weakness is their susceptibility to viral diseases. Chandelle is an excellent source of prolificacy, low ear placement, and virus resistance but has poor roots. It combines well with Coastal Tropical Flints and Haitian Yellow. The race Haitian Yellow is similar to Coastal Tropical Flints but is late in maturity.

Crossa *et al.* (1990a) reported on evaluations of diallel crosses among 25 Mexican races, evaluated during 1963 and 1964 at high (2249 masl) and medium (1800 masl) elevations and during 1961 at a lower (1300 masl) elevation in Mexico. The races Conico, Conico Norteno, and Chalqueno had higher mean yields per se and in crosses at high elevations. Cacahuacintle and Maiz Dulce showed higher cross-performance but lower per se yields. Comiteco, Harinoso de Ocho, Celaya, Maiz Dulce, Tabloncillo, and Tuxpeno had higher per se yields and cross-performances at intermediate elevations. Highest yields were exhibited by Harinoso de Ocho, Celaya, Tabloncillo, and Pepitilla at lower elevations. Races Cacahuacintle, Maiz Dulce, and Harinoso de Ocho were the best general combiners across environments.

Estimates vary, but a relatively small proportion of genetic variability available in tropical maize is being utilized by maize breeders and serious efforts for promoting its use have only just begun (R. Sevilla, 1990, personal communication). Many national programs, International Board of Plant Genetic Resources (IBPGR), Latin American Maize Program (LAMP), North Carolina State University, CIMMYT (Vasal and Taba, 1988), and some other programs and groups are directing their efforts to collection, preservation, regeneration, evaluation, and utilization of maize germplasm. These efforts should prove to be of incalculable value to the present and future maize breeders of the world.

B. Utilization of Genetic Diversity

Most of the currently available superior maize cultivars and hybrids in the tropics trace their origin to selections from the collections now present in various germplasm banks. However, some of the collections in gene banks may not have been thoroughly evaluated, if at all. Available maize collections could be evaluated at several sites in replicated plots in the general areas of their adaptation. Choice of environments should be such that characterization of the germplasm for major traits of interest and importance to the breeders is possible. Currently grown cultivars should be included as checks in the evaluation trials to provide a measure of problems and potentials of the accessions. Most bank accessions are inferior to currently grown cultivars in most major agronomic traits and, therefore, are increasingly difficult to utilize.

The selected bank collections can be utilized in several ways. Crossa *et al.* (1990a) suggested use of collections to develop new heterotic populations for improvement through reciprocal recurrent selection (RRS), to develop lines and hybrids based on their combining ability, or to introgress them into heterotic populations. The collections or the populations developed by crossing them can also be improved through intrapopulation improvement procedures for development of superior OPVs.

Few accessions have been found usable directly. At CIMMYT, the selected collections are crossed to already agronomically superior populations, and after two cycles of intermating, the resulting populations are improved through recurrent selection. CIMMYT has also systematically introgressed many bank accessions into its broad-based gene pools, which were improved using a modification of the modified ear-to-row (MER) selection scheme (CIMMYT, 1979, 1981). The procedure involved planting preevaluated and selected accessions as females in the appropriate gene pools. This avoided the possibility of unproven materials contaminating the pool and provided the opportunity for the comparison of the accessions being crossed to the pool. The superior progenies were harvested and their seed was again planted as female rows in the next cycle to obtain an indication of combining ability of the accessions with the pool. Promising accessions thus identified were incorporated in the pool in the next cycle, using either remnant seed or seed of the crosses with the pool. Methodology for developing new populations using bank accessions and other materials is described later.

C. Morphological and Physiological Characteristics

Paterniani (1990) reviewed the plant characteristics that distinguish maize in the tropics from that in the temperate areas. Tropical maize generally has taller plants, larger tassels, tighter husks, larger leaves, and a greater density of leaf area than its temperate counterparts of equivalent maturity. Tropical maize has a lower harvest index (0.30–0.40) than temperate maize (0.50–0.55). This difference in harvest index (grain dry weight/total above-ground crop dry weight) is associated with differences in morphology between the two groups of germplasm. Tropical maize is also more sensitive to reduced photosynthesis per plant at flowering than temperate maize and is generally highly photoperiod sensitive (G. Edmeades, 1991, personal communication).

III. TROPICAL MAIZE IMPROVEMENT IN NATIONAL PROGRAMS

A. SURVEY OF NATIONAL PROGRAM OBJECTIVES

We conducted a survey of 48 maize scientists from the developing countries in Latin America (24), Asia (15), Africa (5), and the Middle East (4) to determine what traits they considered important in their research programs. Ninety-six percent of the scientists considered yield improvement to be an important objective and devoted about 46% of their resources to it (Table I). Sixty-seven percent of those sampled were involved in breeding for disease resistance and devoted nearly 13% of their resources to it. Other traits that were considered important by the breeders were reduced plant and ear height, early maturity, drought tolerance, and grain type, in that order. Lodging resistance, insect and cold tolerance, photosynthetic efficiency, tassel size, stability, protein quality, plant type, prolificacy, etc., were considered important by some but only 2% of the efforts were devoted to these traits.

Yield was the most important trait across the four continents. The next most important traits were maturity and grain type, in the Middle East; diseases and drought, in Africa; diseases and maturity, in Asia; and height and diseases, in Latin America. Some attention was devoted to quality traits in Asia and to insect resistance in Africa.

Sprague (1974) and Miranda (1985) advised maize breeders to use different strategies in developing superior cultivars for their farmers, depending

Table I

Percentage of Efforts That Tropical Maize Breeders Devote to Different Traits and Percentage of Breeders Involved in Improvement of Those Traits[a]

Traits	Efforts (%)	Breeders involved (%)
1. Yield	46.4	96
2. Disease resistance	12.7	67
3. Reduced plant/ear height	8.2	63
4. Early maturity	8.7	44
5. Drought tolerance	4.9	29
6. Grain type	3.8	42
7. Other	2.0	18

[a] Results based on survey of 48 maize scientists from Latin America, Asia, the Middle East, and Africa.

on the level of development of the national programs, human and other resources available for research, and capabilities of farmers for effectively using them. Development of OPVs instead of hybrids were suggested for the developing areas. For intermediate areas, breeding programs that introduce and evaluate germplasm sources, followed by population improvement and hybrid development, would be advisable. In the developed areas, breeding should involve development and improvement of base populations followed by development of inbred lines and hybrids from the improved populations.

B. Morphological and Physiological Traits

Paterniani (1990) reviewed research conducted in the tropics on the improvement of morphological traits. Reduced plant and ear height is desirable. Paterniani (1990) compared recurrent selection with use of the brachytic-2 gene to reduce plant height and found that both systems were effective. He noted, however, that in using the brachytic-2 approach, recurrent selection would also need to be practiced to adjust the genetic background represented by the modifiers.

Tropical maize cultivars have large tassel size (25–35 branches) compared to temperate cultivars (10–12 branches) and the number of tassel branches has been reported to be negatively correlated with yield, particularly in stress environments (Paterniani, 1990). In Brazil, the brachytic Piranao VD-2 population was mass selected for fewer tassel branches for two cycles. About 60% of the plants with large tassel were detasseled before pollenshed. The next cycle was planted with seed from non-detasseled plants. A 20% reduction in tassel branch number was obtained after two cycles of selection. While the cycles did not differ in yield under favorable environments, the selected population yielded 12% more than C0 under drought and had lower plant and ear height as well. Greater synchronization between male and female flowering is reported to contribute to greater drought tolerance.

C. Drought Tolerance

Palacios identified the trait "latente" in the line Michoacan 21 Comp 1-104 in 1957 (Molina, 1980; Munoz *et al.*, 1983). The plants with the trait stop or markedly slow down their development under water stress and resume growth when moisture is supplied. Miranda (1982) observed involvement of the latente gene in tolerance to heat, drought, and frost in maize. In crosses to Tuxpeno, the gene segregated 3:1 and in crosses to Cateto the gene segregated 15:1. He found the latente trait to be recessive and controlled by two genes. Brunini *et al.* (1985) reported that latente (*lt*) influences stomatal activity, markedly reducing evapotranspiration during water stress. No one, however, has reported successful use of these genes in developing drought-tolerant cultivars.

In Mexico, Molina (1980) practiced stratified visual mass selection (MS) for yield under drought to improve drought tolerance in Zacatecas 58, Criollo del Mezquital, and Cafime. The gains varied with germplasm, years of evaluation, and level of water stress in the trials and averaged 6.8% per cycle. He concluded that mass selection based on yield under drought was effective in improving drought tolerance in maize. Mashingaidze (1984) evaluated eight genetically diverse lines in diallel crosses for seed germination, seedling recovery, and grain yield under moisture stress. He found greater general combining ability (GCA) effects than specific combining ability (SCA) effects for the three traits. The high SCA effects generally occurred in higher yielding crosses where at least one of the parents had high GCA. Significant maternal effects were detected for seed germination. He concluded that simple selection techniques should be effective in increasing drought tolerance in maize.

D. Cold Tolerance

Improving cold tolerance is an important objective for breeders serving the needs of maize farmers in the tropical highlands as well as in winter maize grown in some subtropical areas. Miederna (1979) reported that CIMMYT's Pool 4 ("Highland Early Yellow Flint") and Pool 5 ("Highland Early Yellow Dent") were good sources of resistance to cold temperatures. Hardacre and Eagles (1980) reported on the capacity of autotrophic growth at 13°C in CIMMYT Pool 5 and Criollo de Toluca from Mexico and San Geronimo, Hurancavelicano, and Confite Puneno from Peru. Khera *et al.* (1981) evaluated 96 diverse materials for yellowing (chlorosis and bands induced by chilling), purpling, and early vigor in relation to grain yield. The high-yielding families had lower yellowing, higher purpling, and higher early vigor rating than the low-yielding families. In another study, Khera *et al.* (1988) reported that C3 of the population Pratap, selected for cold tolerance, yielded higher than C0 (8.23 versus 7.45 t/ha), had lower leaf yellowing rating (3.8 versus 4.5), and superior early vigor rating (3.2 versus 3.5), averaged over 10 trials.

Hardacre *et al.* (1990) evaluated three maize populations for frost tolerance under temperatures ranging between −1.5 and −5.0°C. Leaf damage and chlorophyll concentrations were assessed at 7 and 28 days after the treatments. Significant and consistent differences were observed for frost tolerance across the populations and the greatest level of tolerance was observed in the hybrid NZ1A × 5–113, derived from CIMMYT's Pool 5, based on materials from the highlands of Mexico. Paterniani (1990) reported on the effect of four cycles of MS in the population ESALQ-VD-2 for cold tolerance, after *ABPl* genes for purple color were incorporated into the population. The selected population yielded 7.7% higher than C0 in the winter and 2.4% lower during the summer. The selected population outyielded the commercial hybrids by 2.5–9.8% in the winter. Purple plants were reported to be superior to both green and dark purple plants under cold conditions. Purple plants have been shown to have 0.5 to 3.5°C higher temperature than green plants (Greenblat, 1968). Inside the stalk, purple plants had 1.2 and 1.8°C higher temperatures than green plants at 1.0- and 0.5-cm depth, respectively (Chong and Brawn, 1969).

E. Maize Diseases and Insects

At least 40 significant diseases and as many insects that affect maize are reported in the tropics (Brewbaker, 1984). Maize diseases reduce world's annual maize production by 7–17% (Payak and Sharma, 1985) and only rare maize fields lose less than 5% of their production to pests and diseases. The diseases that significantly suppress yield in the tropics are stunt, streak, downy mildew, and stalk rots (Renfro, 1985). Research conducted with the major maize diseases will be discussed.

1. Downy Mildews

Nine species of fungi are known to cause the disease in maize (Safeeulla, 1975). Resistance to Philippine downy mildew (*Peronosclerospora philippinensis* (Weston) Shaw) has been reported to be polygenic and of threshold-type in mature plants. Genes for resistance to this species have been shown to be additive with partial to complete dominance at approximately 50% infection; however, under severe disease pressure, this resistance breaks down and the plants are susceptible (Kaneko and Aday, 1980). Resistance to sugar cane downy mildew (*P. sacchari* (Miyake) Shirai and Hara) has been reported to be generally monogenic with possible existence of polygenic effects (Chang, 1972). Resistance to Java downy mildew (*P. maydis* (Racib.) Shaw) has been reported to be quantitative (Hakim and Dhalan, 1973). Brown stripe downy mildew (*Sclerophthora rayssiae* var. *zeae* Payak and Renfro) was reported to be under control of both additive and nonadditive gene effects (Handoo *et al.*, 1970). Resistance to sorghum downy mildew (*P. sorghi* (Weston and Uppal) Shaw) was found to be polygenic with additive genetic variance of major importance and dominance and additive × additive type of epistasis of minor importance (Jinahyon, 1973; Singburaudom and Renfro, 1982; Borges,1987). Resistance is the most efficient, effective, and economical means of control, and resistance to one species of *Peronosclerospora* has proven useful against other species as well (Frederiksen and Renfro, 1977).

2. Stunt Disease

Corn stunt is one of the most important biological stresses affecting maize productivity in Latin America (Bajet and Renfro, 1989). The disease is more prevalent in warm humid environments, but it is not restricted to such environments. The pathogen is transmitted by five leafhopper vectors and the geographical range of the disease coincides with that of its vectors (Renfro and Ullstrup, 1976). Two distinct types of corn stunt disease have been identified (Bascope, 1977; Nault, 1980):

1. The Rio Grande corn stunt, now simply known as corn stunt, caused by *Spiroplasma kunkelii*, or corn stunt spiroplasma, a helical, motile mycoplasma, and
2. The Mesa Central corn stunt, now simply called maize bushy stunt, caused by the maize bush stunt mycoplasma, a nonhelical or pleiomorphic mycoplasma.

3. Streak Virus

Except for a few reports from some states of India, maize streak virus (MSV) is restricted to Africa and surrounding islands (Seth and Singh, 1975). The disease is considered one of the most important maize diseases in sub-Saharan Africa. It occurs both in the forest and in the savanna zones and from sea level up to 1800 masl. The magnitude of yield loss varies from season to season and depends on the percentage of infected plants and growth stage at the time of infection. Yield losses of up to 100% have been measured in experiments conducted at the International Institute of Tropical Agriculture in Nigeria (IITA), under artificial streak epiphytotics.

Reliable methods for mass rearing of the vector (*Cicadulina* spp.) and for the screening of a large number of genotypes have been developed at IITA (Dabrowski, 1989). Kim *et al.* (1982) reported that 55% of the total variation for disease reaction was due to additive effects, 26% due to nonadditive effects, and 19% due to environmental effects. They calculated that two to three genes controlled resistance to the disease. The symptoms vary with the genetic background of the susceptible parent and the simple recurrent selection or the modified backcrossing method has been recommended for breeding for tolerance to the disease (Kim *et al.*, 1989). A single major gene with modifiers was reported by Storey and Howland (1967) and a system of three genes with modifiers was reported by Bjarnason (1986). Excellent sources of resistance to the disease have been developed through backcrossing by many national programs and S_1 selection is being used effectively by CIMMYT in improving resistance (Short *et al.*, 1990). Several lines resistant to MSV are available from IITA (Kim *et al.*, 1987).

4. Stalk Rots

Diplodia, Gibberella, and *Fusarium* stalk rots, caused by *D. maydis* (Berk) Sacc., *G. roseum* f. sp. *cerealis* (Cke.). & Hans, and *F. moniliforme*, respectively, cause premature death of plants and contribute to stalk lodging. Any form of stress makes the plants more susceptible to the pathogens. Plants infected by these and *Cephalosporium, Colletotrichum*, and *Macrophomina* charcoal stalk rots are conspicuous a couple of weeks before physiological maturity. The stalk rots caused by *Pythium aphanidermatum, Erwinia chrysanthemi* pv. *zeae*, and *Pseudomonas lapsa* occur during the active growing stage of the plant. Payak and Renfro (1966) reported that stalk rot caused by *Cephalosporium acremonium* was a serious disease in all the maize-growing areas of India. Diplodia and Gibbrella stalk rots are probably the most destructive seedborne stalk rots in the world (B. L. Renfro, 1985, personal communication). Bacterial stalk rots (*E. chrysanthemi* Burkholder, McFadden, and Dimock, and *P. lapsa* (Ark) Starr and Burkh) are important in the hot and humid tropics (Renfro and Ullstrup, 1976). Irrigation with sewage water, a practice common in the tropics, is known to increase the incidence of these diseases.

Khan and Paliwal (1980) observed that GCA effects were more important than SCA effects for resistance to *C. acremonium* and that resistance was partially dominant over susceptibility. While qualitative inheritance has been reported for the bacterial stalk rot (Singh, 1979), additive gene effects have been reported for Pythium stalk rots (Diwakar and Payak, 1975). Resistance to Diplodia stalk rots involves additive gene action and genes for resistance have also been located on various arms of several chromosomes. Resistance to Fusarium stalk rot appears to be qualitative (Hooker, 1978). Younis *et al.* (1969) reported that resistance to stalk rot caused by *F. moniliforme* was dominant over susceptibility. Presence of the "stay green" trait has been associated with tolerance to stalk rot (De Leon and Pandey, 1989).

5. Ear Rots

Ear rots of maize in the tropics are usually caused by *D. maydis, Rhizoctonia zeae* Voorhees, and *F. moniliforme*. Diplodia ear rot is more common in the warmer and more humid areas. Cultivars with loose husks, exposed ear tips, and upright ears show more susceptibility to the disease. Grains infected with *F. moniliforme* are reported to be toxic and cancerous to animals. Inheritance studies for resistance to most ear-rot-causing pathogens indicate it to be polygenic with generally additive effects. Resistance of inbred lines has been found to be a poor indicator of the tolerance in hybrids (Renfro, 1985).

Ear rots are also caused by *Aspergillus flavus* and *A. parasiticus* in maize, particularly in hot and humid environments. Although the disease is generally considered a storage disease, field infection does occur. *A. flavus* requires at least 17.5% moisture and between 28° and 35°C temperature for optimum growth (Diener and Davis, 1987). The fungi produce aflatoxins that have many effects on farm animals including malabsorption of nutrients, coagulopathy, decreased tissue integrity, poor growth, increased susceptibility to diseases, medicine failures, and reduced reproductive potential (Hamilton, 1987). In mammals, they cause hepatic necrosis, coagulopathy, and death: they are considered a hazard to public health (Pier, 1987). The same environmental factors that influence aflatoxin production in storage also affect levels of field infection. Insects, particularly *Sitophilus zeamais* Motschulsky, increase kernel infection with *A. flavus* (McMillian, 1987). Significant differences have been reported for resistance among genotypes and primarily additive genetic variation has been observed for resistance to *A. flavus*. Recurrent selection has been suggested to increase the level of resistance (Widstrom, 1987).

6. Leaf Diseases

Leaf diseases are of lesser importance now than 10 years ago because newer cultivars are more tolerant to the diseases (Renfro, 1985). Northern leaf blight (*Exserohilum turcicum* (Pass.) Leonard and Suggs) is usually present at higher altitudes in the tropics. Resistance is both polygenic and qualitative in nature. Southern leaf blight (*Bipolaris maydis*) is very destructive in the tropics. Three races (O, T, and C) of the pathogen have been identified (C. De Leon, 1991, personal communication). Both monogenic and polygenic resistances are known but even in the polygenic system, few genes are involved (Renfro, 1985). Faluyi and Olorode (1984) reported resistance to *B. maydis* to be monogenic recessive in their studies based on resistant varieties RbU-W and DIC. Curvularia leaf spot is widely distributed in the tropics and is caused by *Curvularia lunata* (Wakker) Boed. Although it causes only moderate damage to leaves, in certain areas the disease may become severe (Renfro and Ullstrup, 1976). Vasal *et al.* (1970) estimated that up to 15 genes might be involved in resistance and both additive and epistatic effects were important.

Two types of rusts, common (*Puccinia sorghi*) and southern (*P. polysora*), are known to be important. Both quantitative and qualitative systems are known to govern the resistance to the two diseases. Five loci have been identified that condition resistance to common rust and 11 loci, to southern rust (Renfro, 1985). A single complex locus (*Rpl*) identified in the "Cuzco" variety of Peru has provided resistance to many cultures of *P. sorghi* (Hooker and Le Roux, 1967).

7. Insect Resistance

Insects destroy approximately 12% of the world's maize production annually (Cramer, 1967). Stem borers, represented by different genera in tropical America, Asia, and Africa, the *Spodoptera* budworms, the *Heliothis* earworms, and the stored grain insects are considered to be the most important insect pests of maize. Four borers are known to cause significant yield losses in Africa: the pink stalk borer (*Sesamia calamistis* Hmps.), the African sugarcane borer (*Eldana saccharina* Walker), the maize stalk borer (*Busseola fusca* Fuller), and the spotted stalk borer (*Chilo partellus* Swinhoe). Whereas the first three are known to be present only in Africa, *Chilo* appears to have been recently introduced into East Africa from Asia. The borers can cause yield loss of 10–100% in Africa (Usua, 1986). Yield losses to *C. partellus* have been reported to be up to 18% in Kenya and 44% in Pakistan.

Breeding for tolerance to *S. calamistis* was less effective compared to that for *E. saccharina.* (Bosque-Perez *et al.*, 1989). Resistance to *Chilo* has been reported by Chatterji *et al.* (1971) and Ampofo *et al.* (1986). Sarup (1980) proposed half-sib (HS) recurrent selection for improving tolerance to *C. partellus* in maize and reported on use of modified full-sib (FS) selection scheme to develop *chilo*-tolerant experimental varieties. Recurrent selection under artificial infestation has been recommended by Ampofo and Saxena (1989) for development of cultivars resistant to *Chilo*. Omolo (1983) evaluated lines received from CIMMYT for their tolerance to this insect. He found several families from Populations 25 and 27, previously selected for tolerance to sugarcane borrer (SCB) (*Diatraea saccharalis*) (a few also were selected for tolerance to *Ostrinia nubilallis* and southwestern corborer (SWCB) (*D. grandiosella*)) were also resistant to *Chilo*. He suggested that crosses between local African maizes and exotic Mexican materials should offer multiple resistance or a multiline approach to pest management.

Whereas breeders in the past were largely unsuccessful in developing genotypes resistant to *B. fusca*, recent efforts seem more promising. Although no maize germplasm is immune to the insect, several materials have shown intermediate levels of resistance to the first-generation (whorl feeding) larvae. Resistance appears to be additive and efforts to transfer the trait through backcrossing have generally been unsuccessful, due mainly to the nature of the resistance (Barrow, 1989).

Williams and Davis (1989) reported on performance of eight maize lines under artificial infestation with 30 larvae of SWCB and *Spodoptera frugiperda* (FAW). Although the leaf feeding rating for SWCB for the susceptible check was 9.0 (rated on a scale of 1–9, where 1 = immune and 9 = highly susceptible), the lines ranged from 4.5 to 6.5. In another study, GCA was the only significant source of variation for leaf feeding by SWCB but both GCA and SCA were important for larval growth on maize calli in the same trial (Williams and Davis, 1985). Williams and Davis (1990) artificially infested a maize hybrid with 30 FAW and 30 SWCB larvae at the whorl stage to determine relative differences in the damage caused by the two insects. Whereas both insects caused extensive feeding damage, only SWCB caused 50% reduction in ear height. Yield reduction caused by FAW was 13% compared with 57% with SWCB infestation.

In the study of Williams and Davis (1989), the susceptible check averaged 9.0 for FAW leaf feeding and the lines ranged between 5.5 and 7.0. In a diallel trial of nine lines, Williams *et al.* (1978) reported that both GCA and SCA effects were important sources of variation for resistance to FAW. Most materials with higher levels of resistance to FAW contain Antigua germplasm. Two cycles of S_1 selection reduced leaf feeding by 3.5% per cycle while MS was ineffective in improving the trait. Heritability estimates for resistance range between 12 and 45% for leaf feeding, and leaf feeding is positively correlated with yield losses (Widstrom, 1989).

Ear tip invading insects are among the most destructive maize pests in the world. They are responsible for up to 4% losses in field maize, 14% in sweet corn in the United States, and higher losses in the tropics (Brewbaker and Kim, 1979). Tropical lowland maize races and composites have resistance to ear pests due mainly to high number of husks tightly covering the ear tip. Tropical highland germplasm, on the other hand, has a lower number of husks. The Mexican race Zapalote Chico has a high number of tight husks and provides resistance to ear worms (Brewbaker, 1974). Resistance to ear worms in this germplasm may also be due to the presence of an antibiotic factor in its silks (Straub and Fairchild, 1970). Waiss *et al.*

(1979) identified this factor to be Maysin, a flavone glycoside, that severely retarded growth of *Heliothis zea* (Boddie) larvae on maize silks. Highly significant GCA and SCA effects for tolerance, with GCA effects about twice as large as SCA effects, have been reported by Widstrom (1972) and Widstrom and Hamm (1969). Widstrom (1989) suggested use of recurrent selection to build up resistance to *H. zea* (Boddie) in maize populations. About 3.9% reduction per cycle in ear damage has been reported from HS selection (Widstrom *et al.*, 1970) and 3.5% reduction through S_1 selection based on an index involving husk tightness, husk extension, and maturity (Widstrom *et al.*, 1982).

The maize weevil (*S. zeamais* Motschulsky), often referred to as the rice weevil (*S. oryzae* L.), is an economic pest of maize. Pant *et al.* (1964) evaluated 11 cultivars for their tolerance to maize weevil and found variation among them. Ivbijaro (1981) reported greater tolerance to *S. zeamais* in the Nigerian material TZE 3. Widstrom (1989) suggested that improvement of resistance to maize weevil was seldom included as a breeding objective, although increased husk coverage and tightness reduced damage by the insect. He also reported variation for resistance to weevil damage to be largely additive and obtained a reduction of 3.6 and 5.1% per cycle in two populations involved in a RRS program after two cycles of selection.

F. Tolerance to Soil Acidity

Acid soils cover a significant part of 48 developing countries and involve 1660 million ha (Table II). About 64% of the land surface of tropical South America, 32% of tropical Asia, and 10% of Central America, Caribbean, and Mexico are considered acid soils (Sanchez, 1977). Over 820 million ha of this land are suitable for crop and pasture production and another 620 million ha for grazing (Overdal and Ackerson, 1972).

Table II

Oxisols and Ultisols in the Tropics[a]

	America	Africa	Asia	Total
		(million ha)		
Forests	555	455	190	1200
Savannas	205	195	60	460
Total	760	650	250	1600

[a] Adapted from Sanchez (1977).

Soils are acidic because their parent materials were acidic and initially low in the basic cations (Ca, Mg, potassium, and sodium) or because the elements have been removed from soil by leaching or by the harvested crops. Maize suffers in soils with pH lower than 5.6, due mainly to toxicity from Al and manganese and deficiency of Ca, Mg, P, molybdenum, and Fe. Maximum damage in acid soils is caused by Al toxicity. In mineral soils (<3–4% organic matter), the exchangeable acidity is mostly due to Al, whereas in soils with higher organic matter, soil acidity may be due to exchangeable hydrogen.

Aluminum is the most important cause of yield losses in acid surface soils and it can be precipitated and thus detoxified by liming. However, in areas remote from lime sources or lacking good road systems, on-farm costs of lime may be too high. Even in limed soils, excess soluble or exchangeable Al remains in the subsoils, as lime does not readily penetrate below the zone of application. Liming subsoils deeper than 30 cm is very difficult and prohibitively expensive. Even in limed soils, acid subsoils may restrict plant roots to surface layers inhibiting their normal formation, thereby reducing nutrient and water uptake and reducing yield. Also, lime application must be repeated every few years depending upon the soil conditions.

Genetic tolerance in maize to soil acidity would provide an ecologically clean, an energy conserving, and a more sustainable and cheaper solution to the problem than amending soils. Genetic differences for Al tolerance among inbreds, hybrids, and open-pollinated populations have been reported (Naspolini *et al.*, 1981; Magnavaca, 1983; Magnavaca *et al.*, 1987b, 1987c). Magnavaca *et al.* (1987a) showed that additive genetic variation was twice as important for Al tolerance as dominance. In a study in nutrient solution involving 11 Brazilian lines and 17 U.S. lines, Brazilian lines were reported to be more tolerant to Al and had more and longer seminal and adventitious roots than U.S. lines. Selection for high yield was reported to reduce root length in the population "Composto Amplo." When grown under acid soils, a reduction in frequency of highly tolerant plants was accompanied by an increase in intermediate tolerant plants.

Relative net seminal root lengths and the ratio of final seminal root lengths to initial seminal root lengths were the best indicators of Al tolerance among maize lines, particularly at relatively high Al concentrations. Evaluation of 50 cultivars in the field indicated that genetic variation existed among them for tolerance to Al and that yield, relative to the mean of the selected entries, would be the best measure of Al tolerance of maize genotypes. Field performance of genotypes was not highly correlated with responses under nutrient solution. However, the correlation was higher at higher levels of Al stress (Kasim, 1989).

IV. RECURRENT SELECTION IN MAIZE

A. Recurrent Selection Systems

The basic objectives of any recurrent selection system are: (*a*) to improve breeding populations in a cyclic manner by increasing the frequencies of favorable genes and gene combinations and (*b*) to maintain adequate genetic variability for further selection and improvement by intermating a sufficient number of superior genotypes each cycle of selection. Ultimately, these improved populations should be an excellent germplasm resource for direct release to farmers, for the extraction of superior OPVs, or for the development of inbred lines for use in new hybrids for farm production.

Recurrent selection systems include all those systems that repeat the same procedure cycle after cycle. The procedure generally involves (*a*) production of a set of genotypes (individuals or families of some type), (*b*) evaluation of those genotypes and identification of superior ones to be used as parents of the next generation, and (*c*) intermating the selected genotypes to produce the next generation (improved cycle).

The number of years required per cycle of recurrent selection may vary from 1 to 3 or more, depending on the system used, and whether two growing seasons in a year are available. The evaluation and selection phase should be conducted in environments representative of the target area in which a final product will be released for on-farm production. In the intermating of selected genotypes and/or in the development of new genotypes for evaluation, some visual selection can be practiced for plant type, disease, and insect resistance. Visual selection can be done most effectively in the target area, but this may lengthen the selection cycle. Use of second-season nurseries shortens the selection cycle and they can also be used for evaluating response to diseases and insects not always present in the target area.

Recurrent selection systems have been discussed extensively in the literature and summaries have been presented by Sprague and Eberhart (1977), Gardner (1978), Hallauer (1985), Hallauer and Miranda (1988), Hallauer *et al.* (1988), and Paterniani (1990).

Recurrent selection systems are generally divided into two types: (*a*) Intrapopulation Recurrent Selection Systems and (*b*) Interpopulation Reciprocal Recurrent Selection Systems. Within each system, there is (*a*) an evaluation unit and (*b*) a selection unit. In a few systems, the individual plant is the unit evaluated and the unit selected. In most systems, however, the evaluation unit is some kind of family. The selection unit may be the same family evaluated, but it may also be a related family.

Family evaluation normally requires replicated tests of families in environments representative of the area for which the population is being developed. Experiment station sites are most often used. Selection is based on performance of families evaluated. Superior families in the test or some family related to the selection unit are intermated to form the improved population used for the next cycle of selection.

In most intra- and interpopulation selection schemes, the intermating phase involves making crosses among selected families or their relatives (selection units). This may be done in several ways in a second-season nursery or in the main growing season. The simplest procedure is to grow the selected families in female rows to be detasseled in an isolated nursery, and to use a bulk composite of seed from all selected families to plant male rows for the pollen source. If isolations are not available, the diallel cross among all selections can be made but often involves too many crosses; hence, a partial diallel or chain crossing is used where each selection is crossed to the same number of other selections. A balanced composite involving the same number of crossed seeds from each selected family constitutes the improved population used to initiate the next cycle.

In addition to constituting the improved population, intermating selected families breaks gene linkages and permits genetic recombination to form new genotypes and maintain genetic variability. In some systems, intercrossing selected families and producing new families for testing can be done simultaneously (Compton and Lonnquist, 1982); however, genetic variability and selection limits may be reduced because of less opportunity for genetic recombination. For traits expressed prior to pollination, identification of superior families and intermating them to form new families for testing makes possible the completion of a cycle of selection per season. In short-term selection programs, loss of genetic variability may not be serious, and limiting recombination may be desired to retain favorable epistatic gene combinations. However, in long-term programs, maintenance of genetic variability and maximizing selection limits by separating the intermating and the new family production phases are desirable.

B. Intrapopulation Recurrent Selection Systems

1. Mass Selection

Simple mass selection (MS) is the oldest and simplest of all recurrent selection systems and was probably used by ancient native cultures for maize improvement. It was certainly used by early farmers, who saved seed ears from high-yielding plants in their fields at harvest time. Early maize breeders thought such a system was ineffective for yield improvement, although it was effective for other traits. Masking effects of soil heterogeneity and unequal competition among plants due to unequal spacing in the field contributed greatly to plant-to-plant variation and overshadowed the contribution of the genotype to individual plant yield. Keen observers, who evaluated each plant in relation to its surrounding neighbors, undoubtedly did make progress with this selection system.

Stratified mass selection (SMS) was suggested by Gardner (1961) as a simple procedure to control environmental influences on individual plant yield. The selection area is divided into small, approximately square plots within which the soil heterogeneity and other environmental variables are minimal. Selection is restricted to plants within each plot. Gardner also suggested use of overplanting and adjusting to perfect stands to provide equal competition to all plants, and use of an isolated field to avoid pollen contamination.

Stratified mass selection with pollen control (SMSP) can be done only when selecting for traits that can be evaluated prior to pollination, e.g., days to flower (Troyer and Brown, 1972, 1976), plant and/or ear height, number of leaves, number of shoots (Paterniani, 1978). In maize, all unselected plants are detasseled so that mating occurs only among selected plants. Additional selection for yield among female plants may be done at harvest time.

Mass selection with genetic stratification was discussed by Paterniani (1990) who credited Zinsly with the idea; however, the "Screening Honeycomb Design" suggested by Fasoulas (1973, 1977) and Fasoulas and Tsaftaris (1975) is basically the same idea. A constant check genotype is planted systematically throughout the field and individual plants of the selection population are compared to nearest check plants in making selections. To avoid contamination to a maize population undergoing selection, such a check must be (*a*) male sterile, (*b*) detasseled, or (*c*) possess a trait that makes outcrosses easily identifiable in the seed of selected plants. Since such a system is labor intensive and quite expensive to operate compared to SMS, it has not been widely used.

2. Half-Sib Family Selection

Ear-to-row (ER) selection involves open-pollinated seed of individual plants of a population planted and evaluated in a single HS progeny row with no pollen control (Hopkins, 1899; Montgomery, 1909). Ears from the best plants in the best families are chosen for ear-to-row planting and selection the following cycle. This system was used by plant breeders more than a century ago, but early maize breeders found it to be relatively ineffective for yield improvement but useful for improvement of other highly heritable traits.

Modified ear-to-row (MER) selection was suggested by Lonnquist (1964) and involves single replication testing in multiple environments with an isolated nursery used for intermating. In the nursery, each HS family is grown and detasseled. Male rows planted from a balanced composite of seed from all HS families are interspersed among the female HS families and provide the pollen source. Selection is done among HS families based on yield performance, and also among plants within HS families based on visual yield potential. Simultaneous yield testing and intermating permits advancement at the rate of a cycle per year.

Modified-modified ear-to-row (MMER) selection used by Paterniani (1967) for one cycle only and described in detail as a system by Compton and Comstock (1976), is based on testing and selection as in the above system. However, intermating is done in a second growing season, only selected HS families are planted as female rows to be detasseled in the isolated nursery, and only selected HS families are included in the balanced bulk composite used to plant the male rows. This system requires 2 years per cycle. An isolated second-season nursery can be used for the recombination phase to reduce cycle time to 1 year but if the environments are different, within-family selection may not be very effective.

Half-sib family evaluation and selection (HS) involves replicated yield testing, preferably in two or more environments as described above, of the progeny of individual plants that have been open-pollinated or hand-pollinated with a random sample of pollen from the same population (Jenkins, 1940). Selected HS families are intermated the following season. After intermating, a balanced composite of seed from crosses, with each selected family equally represented, is used to form new HS families. This system requires three growing seasons. The intermating phase and the production of new families phase are sometimes combined (Compton and Lonnquist, 1982) in the same season, which reduces cycle time requirement to two growing seasons. One advantage of using HS families is that an estimate of additive genetic variance and of heritability is available for each cycle of selection.

3. Full-Sib Family Selection

Full-sib family evaluation and selection (FS) involves yield testing of the progeny of intrapopulation paired-plant crosses. Testing and intermating are done using methods described above (Moll and Robinson, 1966). A balanced composite of seed from crosses (each selected parent represented equally) provides the improved cycle from which new FS families are developed for the next cycle. Cycle time is three growing seasons. This system has been used extensively by CIMMYT in the international maize-improvement program as well as by others. By intercrossing selected families and producing new FS families simultaneously, only two growing seasons are required.

4. S_1 Family Selection

S_1 family evaluation per se (S_1) involves yield testing of progenies produced by self-pollination of a number of plants in a population (Eberhart, 1970). Yield testing of progenies and intermating of selected ones is conducted in the same manner as described for all family systems. New families are produced by selfing plants in the balanced composite of the intermated population. Three growing seasons are required.

HS (topcross) family evaluation for GCA (S_1HS) involves simultaneous self-pollination of an individual plant in a population and topcrossing that same plant to a random sample of plants of a broad-based tester population (Jenkins, 1940). With prolific plants, the first ear can be topcrossed with a pollen-bulk of the tester and the second ear selfed. Topcrossed HS families are evaluated in replicated trials, but the S_1 families related to the best topcrosses in the yield trial are used in the intermating phase. Three growing seasons are required. A modification involves selfing S_0 plants and growing S_1 progenies as females in an isolated topcrossing block with the tester used as male to produce topcrosses. This would require four growing seasons but still 2 years.

FS (topcross) family evaluation for SCA (S_1FS) involves simultaneous self-pollination of an individual plant in a population and outcrossing that same plant to an inbred line tester (Hull, 1945, 1952). Topcrossed FS families are tested as described above, but S_1 families related to the best topcrosses are used for the intermating phase. Three growing seasons are required. Although designed originally to develop a source population from which lines could be extracted for use in hybrids with an available elite line, it has worked effectively for population improvement (Horner *et al.*, 1963, 1973; Russell *et al.*, 1973; Russell and Eberhart, 1975).

5. S_2 Family Selection

S_2 family evaluation per se (S_2) involves two generations of self-pollination, which can be done with or without selection in the S_1 generation (Horner *et al.*, 1973). S_2 families are tested and the superior ones selected and intermated as described for other family types. This system requires four growing seasons. Because of large genotype × environment (G × E) interactions, and difficulties of evaluating inbred maize progenies per se, this system has not been used extensively.

FS (topcross) family evaluation for SCA (S_2FS) involves simultaneously selfing S_1 progenies and topcrossing them to an inbred line and yield testing the topcrosses as described for other types of families (Horner *et al.*, 1963). Because the S_1 progenies trace back to a single plant in the population and are crossed to a homozygous line, the cross is basically the offspring of two parents, hence a FS family. For intermating, the S_2 families related to the best topcrosses in the test are used. Four growing seasons are required for each cycle. If selection within the S_1 families to be topcrossed is desired, this can best be done in the target environment and would require 3 years per cycle. This system can be used for extraction of lines for hybrid development as well as for population improvement.

6. S_n Family Selection (Single-Seed Descent)

Inbred family evaluation (S_n) involves development of inbred lines (S_n generation) through the single seed descent method (Brim, 1966). Use of two seasons per year greatly accelerates this process. Inbred families are tested in yield trials and the best families are intermated. The number of growing seasons required varies depending on the number of generations of inbreeding and time required for each.

Full-sib (topcross) family evaluation for SCA (S_nFS) involves selfing to near homozygosity before topcrossing to an elite inbred line. Topcrosses are yield tested and selected for SCA. S_n lines related to the superior topcrosses are intermated. The number of growing seasons required depends on generations of selfing. A variant of this system or of system S_2FS is used on F_2 populations of single crosses or other elite narrow-based populations by most U.S. commercial maize breeders in developing new lines for hybrids. Cycle length is so long that it is not very useful for broad-based population improvement.

C. Interpopulation Reciprocal Recurrent Selection Systems

1. Reciprocal HS Family Evaluation

S_1 family selection (RRS) was suggested by Comstock *et al.* (1949) as a breeding system that would permit utilization of all genetic effects in the improvement of the heterotic response in the cross of two populations. Individual plants of Population A are each self-pollinated and simultaneously crossed to a random sample of plants of the reciprocal Population B. Likewise, individual plants of Population B are simultaneously self-pollinated and crossed to a random sample of plants in Population A.

A balanced composite of the crossed ears of each plant constitutes a HS family. HS families of each population are evaluated in replicated yield trials in multiple environments. S_1 families related to the superior HS families within each population are used for the intermating process to produce the improved cycles of the two reciprocal populations. This system requires three growing seasons per cycle.

Because of high labor involvement and costs of manually making topcrosses plus inadequate sampling (6 to 10 plants) of the tester population in the original scheme, an alternative system described by Paterniani (1990) seems better. It involves selfing individual plants in each population and growing the two sets of S_1 families as females in separate isolation blocks with reciprocal populations used for male rows and border. S_1 families are detasseled and pollinated by the reciprocal population to produce the HS families for testing. Remnant seeds of S_1 families related to superior HS families are used for the recombination. This system takes an additional growing season.

HS family selection (RHS), suggested by Paterniani (1970), involves use of open-pollinated progenies (HS families) of each population grown in separate isolation blocks where they are detasseled and pollinated by the reciprocal population grown in male rows and in borders. Testcross families are grown in replicated trials and HS families related to the superior testcross families are used for recombination. Within each population, selected HS families are grown as female rows and detasseled in an isolated block and a bulk mixture of seed of selected HS families from the other population is planted in male rows and around the border to provide the pollen source to form new HS families to begin the next cycle and to constitute the improved reciprocal population. Thus the system requires only three growing seasons to complete a cycle. A cycle can be completed in two growing seasons in 1 year by growing the recombination block and yield trials simultaneously. However, selection will be on the female side only and gains per cycle will be reduced (E. Paterniani, 1991, personal communication).

HS family selection using prolific plants (RHSP) was also suggested by Paterniani (1970) to reduce the selection cycle to two growing seasons, which could be accomplished in 1 year with a second-season nursery. Two-eared plants are required in each of the two populations. Two isolation blocks are required. In Isolation 1, Population A is planted in female rows and Population B is planted in male rows. In Isolation 2, B is female and A is male. At flowering time, female rows are detasseled and the second shoot of prolific female plants is covered with a shoot bag. When bagged shoots are ready to be pollinated, bulk pollen is collected from 50 or more prolific plants being used as males in the other isolation (say,

Isolation 2) and are used to pollinate bagged shoots of that same population in Isolation 1. Likewise, males in Isolation 1 are used to pollinate bagged female shoots in Isolation 2. The open-pollinated top ear produces an interpopulation HS family, which is used in replicated yield trials; and the hand-pollinated second ear produces an intrapopulation HS family, which is available to be used simultaneously for recombination and for formation of new inter- and intra-HS families to begin the next cycle. With a second-season nursery, a cycle can be completed in 1 year.

2. Reciprocal Full-Sib Family Evaluation

S_1 family selection based on reciprocal plant-to-plant crosses (RFS) requires the use of two-eared plants (Villena, 1965; Lonnquist and Williams, 1967; Hallauer and Eberhart, 1970; Jones *et al.*, 1971). An individual plant in Population A is crossed to an individual plant in the reciprocal Population B using one of the ears. The reciprocal cross is also made if possible. At the same time the plants are each self-pollinated using the other ear. Selection is based on the cross performance in yield trials, and the S_1 families related to the selected crosses are used for intermating within each of the reciprocal populations. Intermating is often done in an isolated crossing block where S_1 families are detasseled for use as females, and a balanced bulk of seed of all S_1 families is used to plant male rows at intervals in the field and in the border. A balanced composite of seed with equal contributions from each selected family constitutes the completed selection cycle, and this seed is used to initiate the next cycle. Three growing seasons are required per cycle.

S_1 family selection based on reciprocal $S_1 \times S_1$ family crosses (FSRRS) is an alternative procedure in which S_1 families in each population are produced and paired-row crosses are made between S_1 families from opposing populations (Marquez-Sanchez, 1982). Such $S_1 \times S_1$ crosses are basically a cross between two noninbred plants as in the above scheme, but prolificacy is not required. These crosses, which might be called FS families, are evaluated, and S_1 families related to the best crosses are intermated. This lengthens the cycle to four growing seasons.

S_1 family selection based on reciprocal inbred line testcrosses (MRFS), suggested by Russell and Eberhart (1975), is a modified FS reciprocal recurrent selection procedure in which noninbred (S_0) plants of Population A are each self-pollinated and crossed to an inbred line of Population B derived from a previous cycle. Likewise, noninbred plants of B would be selfed and crossed to a previously derived line from A. Testcross performance would be the selection criteria and S_1 lines related to the best testcrosses would be the selection units. Two elite inbred lines from different

Table III

Recurrent Selection Systems Used for Intrapopulation Improvement and for the Improvement of Interpopulation Crosses

Population improvement systems	Value of c[a]	Unit evaluated	Unit selected and recombined
	Intrapopulation systems		
1. Mass selection of individual plants			
a. Simple mass selection (no pollen control)	0.5	Female plant in field	Female open-pollinated seed
b. Stratified mass selection (no pollen control)	0.5	Female plant in restricted area	Female open-pollinated seed
c. Stratified mass selection (pollen controlled)	1.0	Plant saved prior to pollination	Female open-pollinated seed
d. Mass selection with a constant genotype planted systematically for comparison and evaluation	0.5	Female plant relative to check	Female open-pollinated seed
2. Half-sib family selection			
a. Ear-to-row (no pollen control)	0.125	HS family, no replication	HS family
b. Modified ear-to-row (no pollen control)	0.125[b]	HS family replicated[c]	Plants within selected HS
c. Modified-modified ear-to-row (pollen controlled)	0.25[b]	HS family replicated[c] plants/family in isolated block	Plants within selected HS family with restricted crossing
d. Family evaluation	0.25	HS family replicated	HS family
3. Full-sib family selection			
a. Family evaluation	0.5	FS family replicated	FS family
4. S_1 family selection			
a. S_1 family evaluation per se	1.0[d]	S_1 family replicated	S_1 family
b. Half-sib family evaluation (GCA)	[c]	HS (TC) family replicated	S_1 family
c. Full-sib family evaluation (SCA)	[c]	FS (TC) family replicated	S_1 family

5. S_2 family selection			
a. S_2 family evaluation per se	1.5[d]	S_2 family replicated	S_2 family
b. Full-sib family evaluation (SCA)	[c]	FS (TC) family replicated	S_2 family
6. Single-seed descent			
a. Inbred family evaluation	2.0[d]	S_n family replicated	S_n family
b. Full-sib family evaluation (SCA)	[c]	FS (TC) family replicated	S_n family
	Interpopulation systems		
1. Reciprocal half-sib family evaluation			
a. S_1 family selection		HS testcross family	S_1 family
b. Half-sib family selection		HS testcross family	HS family
2. Reciprocal full-sib family evaluation			
a. S_1 family selection			
(i) Based on cross to a reciprocal plant		FS testcross family	S_2 family
(ii) Based on cross to a homozygous line		FS testcross family	S_1 family
b. Full-sib family selection		FS testcross family	FS family

[a] c is the proportion of additive genetic variance contained in the genetic component of variance among selection units.

[b] There is also an additional formula for expected additional gain from within-family selection in which $c = 3/8$ or 0.375.

[c] In topcross family evaluation, progress in population improvement depends on the distribution of gene frequencies in the tester (see Empig *et al.*, 1981).

[d] In inbred families, additive genetic variance among families increases to $(1 + F)\ \sigma^2_A$ and dominance variance is halved each generation.

heterotic groups, e.g., B73 and Mo17, that produce an exceptional hybrid might also be used as testers to improve two populations and their interpopulation cross. Testers for each population would be chosen according to heterotic responses noted in the population × tester crosses. Since the tester lines would be homozygous, greater genetic variability would exist among testcross families than when populations are used as testers. This suggests that greater progress from selection should be possible using inbred testers; however, Comstock (1979) presents theoretical arguments indicating that populations are slightly superior to inbred lines as testers in shifting gene frequency in directions favorable for increased cross-performance.

The units evaluated, selected, and recombined for most of the selection systems discussed are presented in Table III.

D. Expected Genetic Gains

For most of the recurrent selection systems discussed, equations to predict progress from selection have been developed. Detailed derivations are given for many systems by Empig *et al.* (1981). Sprague and Eberhart (1977), Gardner (1978), and Hallauer and Miranda (1988) summarize prediction equations. For intrapopulation selection, the general prediction equation for expected gain per year is

$G_s = SH/y = [(k\,\sigma_p)c\sigma_A^2]/[y\,\sigma_p^2].$

S = selection differential (superiority of the mean of selected units over the population mean).
= $k\,\sigma_p$ in normally distributed populations.

k = number of phenotypic standard deviations that selected units are expected to exceed population mean in normally distributed populations.

σ_p^2 = phenotypic variance among selection units.
= MS_f/re, where MS_f = the family mean square, r = number of replications, and e = number of environments.

σ_A^2 = additive genetic variance in the population.

c = proportion of additive genetic variance contained in the genetic component of variance among selected unit means. Values are given in Table III.

H = $c\sigma_A^2/\sigma_p^2$, heritability on a selection unit basis.

y = number of years per cycle.

Estimates of the magnitude of additive genetic variance in the population and the phenotypic variance among selection unit means are required.

Factors affecting gains from selection are the number of selection units evaluated, the proportion selected, the magnitude of additive genetic variance, and the precision of the experiment. Except for HS family evaluation and selection, substitution of the family component of variance for $c\sigma_A^2$ will give biased estimates of expected gain for traits where non-additive genetic effects are important. Where the units selected differ from those evaluated, estimates of additive genetic covariance between the two types of families are substituted for additive genetic variance.

For reciprocal recurrent selection systems, expected gains in the inter-population cross are calculated using components of variance among intercross family means (see references listed above).

V. POPULATION FORMATION AND FACTORS AFFECTING CHOICE OF SELECTION METHODS AND PROGRESS FROM SELECTION

In choosing germplasms and forming populations, the ultimate goal and time constraints are important. For intrapopulation selection, the population should possess the most desirable alleles at as many loci as possible. For interpopulation selection, heterotic patterns are of great significance in choosing germplasms to include in each of the two reciprocal populations. There is no lack of maize germplasm available for the formation of breeding populations. Locally adapted material and material from other countries with similar environments will be most useful; however, superior but less well-adapted germplasms from other areas no doubt contain favorable alleles not in adapted material. Such resources should not be overlooked.

Methodology for the formation of a population for use as a source of varieties or hybrids has been described by Pandey *et al.* (1984). Individual materials can be planted as female rows and detasseled in an evaluation-cum-HS crossing block where the pollinator is an adapted local variety or a balanced mixture of all the materials. A minimum of 50 plants of each germplasm should be planted in two- to four-row plots 5 m long and detasseled. Throughout the growing season, notes should be taken on various characteristics of the materials. Inferior entries should be rejected and each selected entry should be bulk-harvested.

Evaluation of these testcrosses in three to four environments in replicated yield trials should povide reliable information on the GCA of the materials to be intermated to form the population. Two-row plots 5 m long and two replications at each location would generally be adequate to obtain reliable estimates of genotype and environmental effects. Multi-location evaluation of entries in a single season has been suggested to provide sufficient information for selection of superior entries (Comstock and Moll, 1963). The materials with acceptable GCA should be selected for intermating.

Thorough recombination among the materials must precede any improvement of the population. Recombination can be accomplished by planting the materials in an arrangement of 2 or 3 females : 1 male, where individual components are the female and a balanced mixture (equal quantity of seed from each material mixed together) is the male. In the first cycle, each entry can be represented by up to 500 plants and detasseled. In each entry, 50–60 superior plants should be identified during the growing season and 20–30 ears from these plants selected at harvest. A large number of plants of each entry at this stage is desirable as it will permit more intense selection within the materials. When it is necessary to recombine among materials with different maturities, staggered plantings of male rows in time and space is suggested. Use of neutral environments has been advocated during recombination to overcome the photoperiod sensitivity of the accessions originating from varying latitudes.

Each selected ear should be planted ear-to-row next season and detasseled and a balanced mixture of all the selected ears is used as the pollinator. Within each entry, the inferior rows can be rejected and 2-3 ears selected at harvest from the superior rows so that each original entry is again represented by 20–30 ears in the population. Lower selection intensity in the recombination phase reduces the frequency of deleterious alleles in the population without causing significant genetic drift. The selection intensity can be manipulated so that the population is made up of approximately 500 ears for each cycle of recombination. Two to three additional cycles of intermating of the population should provide sufficient genetic recombination among the materials constituting the population and the identity of the components would largely cease to exist.

If one wishes to compare expected genetic progress using different intrapopulation recurrent selection systems applied to a specific population, one must have estimates of additive and nonadditive genetic variances and between- and within-plot environmental variances. For estimation of expected correlated responses in specific traits when single-trait or index selection is used, additive genetic covariances between the specific trait and the selection criterion are essential. Information about characteristics of many maize accessions held in germplasm banks has not been available in the past, but efforts to obtain such information are currently under way. Even so, there will continue to be a lack of information about the kind of gene effects operating and the magnitude of genetic variances controlling quantitative traits.

All of the recurrent selection systems have proven to be effective in maize improvement (Sprague and Eberhart, 1977; Gardner, 1978, 1986). The choice of which system and which populations to use must be made by each breeder and will depend upon (*a*) the ultimate goal—variety or hybrid; (*b*) traits to be improved—yield, disease resistance, drought tolerance, etc.; (*c*) time constraints—short-term vs long-term; (*d*) germplasms available—locally adapted varieties, improved populations, pools, and bank accessions; (*e*) knowledge about those germplasms; and (*f*) financial resources available.

If the ultimate goal is to develop and release an improved variety, one of the intrapopulation methods should be used with a population constructed to contain as many of the most favorable alleles at each locus as possible. If the ultimate goal is a hybrid of some kind, then one of the interpopulation reciprocal recurrent selection systems allows one to take advantage of all the genetic variability in two contrasting populations. When population improvement is integrated with a hybrid program, RRS and RFS provide new S_1 lines (units intermated) in each cycle. These can be advanced to S_2 lines during the testing phase, and those related to the best interpopulation crosses can be further inbred and/or topcrossed and evaluated in hybrids. RFS produces heterotic pairs of lines that may form a useful hybrid.

In early cycles of reciprocal selection involving two contrasting heterotic populations, intrapopulation selection in each has been as effective as reciprocal interpopulation selection in improving the interpopulation cross. In addition, intrapopulation selection has been more effective in improving the populations and derived inbreds (Odhiambo and Compton, 1989). This is not surprising since reciprocal selection in early cycles is likely to be more for GCA or additive effects, whereas, in the long run, additive genetic variance will be reduced and SCA involving dominance and epistasis effects will play a larger role in the interpopulation cross and in crosses between lines extracted from the improved populations. Recent evidence of Compton and Tragesser (Tragesser, 1991) indicates that improvement in the interpopulation crosses from intra- vs interpopulation selection do differ in the kind of gene effects operating.

If the trait to be improved is controlled primarily by additive genetic effects and heritability is high, any intrapopulation system will be effective, including mass selection. S_1 family evaluation and selection have been particularly effective in the improvement of disease and insect tolerances, as well as many other traits. Selfing during population improvement improves inbreeding tolerance in the populations and reduces genetic load and time required for inbred and hybrid development (Odhiambo and Compton, 1989; Tragesser, 1991). If the trait is controlled by overdominance and by dominance types of epistasis and heritability is low, e.g., yield, then interpopulation systems are required to maximize yields.

If there is a need to produce a new variety or hybrid in a short time, elite populations with high mean yields but perhaps limited additive genetic variability should be used. If there are no time constraints, more exotic germplasm can be used to add more favorable alleles, favorable interacting alleles, and favorable interacting nonalleles and linkage groups to the population. To bring in new favorable alleles not available in adapted material, one frequently must introgress less well-adapted germplasms, which brings in undesirable genes along with some good genes. Although one starts at a lower mean performance level, the greater additive genetic variability permits greater gains over a longer period of time to achieve a higher final level of performance in any variety released.

Financial resources available may dictate the choice of breeding system. Some systems require more hand-pollination and skilled personnel. Replicated testing in multiple environments is expensive and requires trained personnel to conduct the experiments properly, collect the data, and analyze the results. Reciprocal selection programs require two populations and twice the work of intrapopulation selection. Costs must be weighed against potential benefits in making decisions about which breeding system to use.

VI. RECURRENT SELECTION METHODS USED IN THE TROPICS

A. Survey of Methods Used by National Programs

According to our survey of 48 maize scientists in Latin America, Asia, Africa, and the Middle East, HS, S_1, and FS family selection schemes were the most widely used breeding methods (approximately 50% of breeders used them) and received 17.9, 14.1, and 12.4% of the research resources, respectively (Table IV). MER selection (Lonnquist, 1964), MER selection$^+$ (CIMMYT, 1974), and MS were used by 38, 27, and 44% of the breeders sampled and used 9.3, 11.6, and 7.2% of their research resources, respectively. All breeders practiced population improvement and this activity used 60.6% of their research resources. Hybrid development (line development, evaluation of line combinations, etc.) was practiced by 85% of the breeders and used 39.3% of their research resources.

Considerations of resource availability (27.3%), compatibility with program objectives (23.4%), and gains/resource use (20.1%) were major determinants of the method of selection breeders would use. Ease in implementation (11.0%) and the stage of development of the germplasm

Table IV

Population Improvement Methods Used (%) and Breeders (%) Using Them for the Improvement of Tropical Maize[a]

Selection method	Resources (%)	Breeders (%)
1. Half-sib family	17.9	52
2. S_1 selection	14.1	54
3. Full-sib	12.4	46
4. Modified ear-to-row (Lonnquist, 1964)	9.3	38
5. Modified ear-to-row (CIMMYT, 1974)[b]	11.6	27
6. Simple mass selection	7.2	44
7. Selection for SCA	4.3	31
8. Selection for GCA	4.3	31
9. Stratified mass selection	5.6	25
10. Others	1.7	8
Population improvement	60.6	100
Hybrid development	39.3	85

[a] Results based on survey of 48 breeders of tropical maize from Asia, Africa, the Middle East, and Latin America.
[b] Described in text under Backup Unit (Germplasm Development Unit).

being improved (9.1%) also affected choice of the breeding methods by the breeders. Other factors such as stage of development of the program and personal biases played relatively less important roles. Lack of financial and logistical resources, gains/resource used, and ease in implementation were cited as primary reasons for the use of HS, MS, and various MER selection schemes. Greater effectiveness at increasing frequency of desirable alleles and reducing the undesirable alleles and compatibility with hybrid development activities were cited as primary reasons for use of S_1 selection. Compatability with program objectives and seed availability for multilocation testing were cited as primary reasons for use of FS selection.

Half-sib, S_1, and FS selection schemes were the most widely used methods across the four continents. Half-sib selection was the most important method in Asia and the second most important method in the other three continents. Full-sib selection was the most important method in Latin America, the third most important method in Africa and the Middle East, and fifth in importance in Asia. S_1 selection was important in all continents except the Middle East. Only Latin America and Asia reported appreciable use of interpopulation selection schemes.

Population improvement was practiced by most of the scientists in the four continents and used 67.5% of the resources in Latin America, 54.7% in Asia, 52.5% in the Middle East, and 52% in Africa. Resources devoted to and scientists involved in hybrid development were 47.5 and 100% in the

Middle East, 45.3 and 93% in Asia, 48 and 80% in Africa, and 25.7 and 79% in Latin America, respectively.

B. Progress from Selection in National Programs

1. Mass Selection

From the published reports, MS seems to have been used most commonly. Using this procedure, Johnson (1963) reported a yield gain of 11% per cycle in Tuxpeno and Vencovsky *et al.* (1970) reported a gain of 3.8% per cycle after five cycles of selection in Paulista Dent and 1.7% per cycle in Cateto M. Gerais after three cycles of selection. Torregroza (1973) reported an increase of 48% in prolificacy and 35% in yield after 11 cycles of MS for increased prolificacy. Selection for single-ear plants for 11 cycles decreased prolificacy by 16% and yield by 7% in the same population. Nine cycles of MS for prolificacy in Populations MB-51 and MB-56 increased prolificacy by 2.0 and 3.38% and yield by 3.0 and 5.45% per cycle in the two populations, respectively (Torregroza *et al.*, 1976).

Arboleda-Rivera and Compton (1974) practiced MS for yield and prolificacy in a broad-based Colombian composite. Selection was practiced during dry season (A), rainy season (B), and during both dry and rainy seasons (AB). Selection in the B season increased yield and prolificacy by 10.5 and 8.8% per cycle, respectively, in the B season and 0.8 and 1% in the A season. Selection in the A season improved yield and prolificacy by 2.5 and 4.4% per cycle in the A season and by 7.6 and 11.4% in the B season, respectively. Selection in both seasons improved yield and prolificacy by 5.3 and 7.0% per cycle in the B season and by 1.1 and 3.3% per cycle in the A season. Vargas-Sanchez (1990) evaluated gains from selection and genetic variances in the same population after 10 cycles of selection in the A season, 10 cycles in the B season, and 20 cycles in A and B seasons, using Design I. Selection in season A produced higher gains than in B but highest gains were obtained with selection in both seasons. Gains from selection for A, B, and AB seasons for prolificacy were 2.3, 2.0, and 3.8% and for yield 3.8, 3.4, and 5.3%, respectively. The additive genetic variance for prolificacy increased in each of the selected populations. For yield the additive genetic variance was greater in AB than in the other two versions. Selection in A season reduced additive genetic x environmental effects for yield, whereas selection in the B season increased those effects.

Genter (1976) practiced 10 cycles of MS in a composite of Mexican races and obtained a gain of 19.1% per cycle in yield. He also reported that the advanced cycles of selection had lower plant and ear heights; reduced moisture, days to silk, and smut-infected plants; and improved synchronization of pollen-shed and silking. Darrah (1986) reported yield gains of 2.3% per year in Kitale Composite A (KCA), after 10 cycles of selection.

Delgado and Marquez (1984) used MS to improve adaptation of the maize variety Zacatecas 58. In one scheme, selection was practiced separately at each of three sites and the selections were recombined at one site (convergent–divergent MS). In another scheme, one cycle of selection was completed at each of the three sites; only one site was involved in any given cycle (rotative MS). The results indicated that rotative MS was superior to convergent–divergent selection for improving adaptation of the population.

2. Half-Sib Family Selection

Paterniani (1967) reported gains of 13.6% per cycle in the yield of population Paulista Dent after three cycles of MER selection. After 6 years of MER selection, Darrah *et al.* (1978) reported gains of 0.83, 2.59, and −0.43% per year in the yields of KII, Ec573, and H611, respectively. After 10 years of selection, they reported gains of 5.2% in the population KCA. Lima *et al.* (1974) obtained gains per cycle of 3% in Flint Composite and 2% in Dent Composite after two cycles of selection. After eight cycles of MER selection, Sevilla (1975) reported yield improvements of 9.5% per cycle in PMC-561. Segovia (1976) obtained 3.2% per cycle gains in the initial three cycles and no gains in the later three cycles of ER selection in Centralmex.

Gains per cycle from selection among and within HS families for yield have been reported to be 3.80% in the population Piramex (Paterniani, 1968), 10.8% in Tuxpeno (Lima and Paterniani, 1977), 1.90% in IAC-1 (Miranda *et al.*, 1977), 2.84% in IAC-Maya (Sawazaki, 1979), 2.70% in Dentado Composto Nordeste (Santos and Naspolini-Filho, 1986a), and 5.10% in Flint Composto Nordeste (Santos and Naspolini-Filho, 1986b).

Eberhart and Harrison (1973) reported results of two cycles of selection in the population Kitale, where the S_1 lines were topcrossed to the parental population (tester). They obtained a 15% increase in yield after the two cycles of selection. The selected population yielded higher than the original population in both poorer and better environments.

Santos and Naspolini-Filho (1986b) practiced 3 cycles of selection in the population Dentado Composto Nordeste and reported that additive genetic variance for ear weight decreased from C0 to C1 and then remained constant up to C3. Santos *et al.* (1988) reported 15.20 and 5.20%

yield improvement per cycle in the populations Dent Composite and Flint Composite, respectively, after 2 cycles of MER selection. Darrah (1986) employed MER selection to improve KCA. Gains per year for yield after 10 cycles of selection were 2.9%, with no change in the performance of the population cross when crossed to its heterotic counterpart.

3. Full-Sib Family Selection

Jinahyon and Moore (1973) reported results of four cycles of selection. Yield increased by 7.9% per cycle and plant and ear heights and lodging decreased in their experiments. Darrah (1986) practiced FS selection to improve KCA. Gains per year after five cycles of selection were 3.6% for yield. Four cycles of FS selection for prolificacy were conducted by Singh *et al.* (1986) in one variety. Linear response per cycle for prolificacy was greater under low plant density (5.5%) compared with that under high density (3.6%). They also reported a significant correlated response of 4.5% per cycle for yield and concluded that selection for prolificacy may be useful in developing high-yielding, early, and short maize genotypes.

4. S_1 Family Selection

Jinahyon and Moore (1973) reported a yield increase of 8.3% per cycle in Thai DMR Composite after two cycles of S_1 selection. The selected population had 17% stalk lodging compared to 53% in C0. In Thai Composite-1 (later called Suwan-1), improvement in yield and downy mildew resistance (DMR) were measured after nine and seven cycles of selection, respectively. Yields improved by about 5% per cycle and infection to downy mildew was reduced from 80% in C0 to less than 5% in C7 (S. Sriwatanapongse, 1990, personal communication). Darrah (1986) employed S_1 recurrent selection to improve KCA. Gains per year after five cycles of selection were 0.9%, with no change in the population cross when the population was crossed to its heterotic counterpart.

Gomez (1990) reported gains of 6.7**% per cycle in yield after two cycles, using a modification of S_1 selection scheme, where top ears of the plants were allowed to be open-pollinated and the second ears were selfed. The selfed ears were selected from higher yielding plants and planted next season ear-to-row as females and detasseled. A bulk mixture of all selfed ears served as the pollinator. Open-pollinated ears from the superior plants in visually selected S_1 families in the recombination block were used to initiate the next cycle.

5. Recurrent Selection Methods Used for Interpopulation Cross-Improvement

Vencovsky *et al.* (1970) practiced five cycles of SMS in Paulista Dent and three cycles in Cateto Minas Gerais. Although heterosis between the two populations increased in the first cycle of selection, it decreased in the third cycle. The cross involving the third cycle of selection of Paulista Dent and first cycle of Cateto Minas Gerais exhibited maximum heterosis and the highest yielding cross involved the fifth cycle of Paulista Dent and the third cycle of Cateto Minas Gerais without significant heterosis.

Torregroza *et al.* (1972) practiced two cycles of RRS in the populations Harinoso Mosquera and Rocamex V7 and reported increases of 4.5 and 15% per cycle in the yields of the two populations, respectively. The C1×C1 and C2×C2 yielded 32 and 34% higher than C0×C0. Gevers (1974) practiced three cycles of RRS and reported gains of 7.1, −0.5, and 3.3% per cycle in Teko Yellow, Natal Yellow Horsetooth, and the population cross, respectively, when the male parents were selected for agronomic traits. When random male parents were used, the yield improvements were 7.5, 7.4, and 5.8%, respectively. Darrah *et al.* (1978) practiced three cycles of RRS in the populations KII and Ec573 and reported gains of 4.96% per cycle in Ec573 and no gains in KII. The population cross showed an increase of 7.07% per cycle.

Paterniani and Vencovsky (1978) reported yield increases of 3.5% per cycle in the cross of the populations Dent Composite and Flint Composite after two cycles of RHS selection using prolific plants. Paterniani and Vencovsky (1977) also reported 7.5% gain in the population cross of Cateto and Piramex using RRS. They attributed 4.3% of the gain to improvement of the two populations and 3.2% to change in heterosis of the population cross. Darrah (1986) compared three methods (MER, S_1, and RRS) for the improvement of the cross KSII X Ec573. Only Ec573 showed a response per year of 4.6% for MER and 3.5% for S_1 selection. Neither population responded to RRS. Whereas 10 cycles of MER and S_1 selection did not improve the population cross, 5 cycles of RRS improved the population cross by 2.8% per year. Ochieng and Kamidi (1992) evaluated progress from 8 cycles of RRS in the same two populations at several sites over 3 years. Although no improvement was observed in either population, the varietal cross improved by 3.56% per cycle. Percentage midparent heterosis increased significantly ($b = 3.80$). Prolificacy increased in both the populations and the population cross. Although maturity did not change in either parent, the variety cross was earlier. Ear height was reduced in Ec573 but did not change in KSII.

C. Hybrid Development in National Programs

Generally, OPVs occupy 61% of the maize area and hybrids 39% in developing countries (CIMMYT, 1990). Exclusion of Argentina, Brazil, and China, the largest hybrid producers among the developing countries, increases percentage of area planted to OPVs to 84 and reduces percentage of area planted to hybrids to 16 (CIMMYT, 1987). The reasons for lower use of hybrids in the developed countries are insufficient yield advantage of hybrids under high stress and low yield conditions, limited financial resources for purchase of farm inputs, and lack of effective seed production and distribution systems (Paliwal and Sprague, 1981; CIMMYT, 1987). Nevertheless, when properly implemented, hybrid maize methodology has given good results in most tropical countries, particularly where a strong private seed industry is present (Paterniani, 1990).

One of the first hybrid development programs in the tropics was started in 1932 at the Agronomic Institute of Campinas, Brazil, where selfing was done in a local orange flint cultivar, Cateto, to develop a flint hybrid (Paterniani, 1990). The first superior semident hybrid was developed by crossing Cateto with a local yellow dent cultivar, Amarelao, at the School of Agriculture at Vicosa, Brazil. In a study involving 418 single crosses among 31 lines from Brazil, Colombia, and Mexico, Paterniani (1964) reported 55% superiority of yields in Colombian × Mexican lines over the best available double cross-hybrid in Brazil. He also reported greater increase in yield, standability, disease resistance, and vigor in hybrids of improved materials compared to the original cultivars, Cateto and Paulista Dent. The first successful double cross-hybrid developed in Brazil involved lines from Cateto and Tuxpan (based on a cross between Tuxpeno from Mexico and a yellow dent from Georgia). Most of the current maize hybrid development in Brazil is based on Tuxpeno and Caribbean Flints (R. Magnavaca, 1991, personal communication).

In Venezuela, heterosis between ETO and some other Caribbean flints with Tuxpeno is exploited (Paterniani, 1985). In Colombia, hybrids mainly based on Caribbean germplasm have been developed; lately some Tuxpeno materials have also been used. In Peru, Perla from Peru and flints from Cuba and Central America have been used for hybrid and composite development. Lately, Tuxpeno has played a more important role in hybrid development in Brazil due mainly to its heterotic potential with Cateto.

In a study involving 12 races of maize—four dents, five flints, and three floury types—Paterniani and Lonnquist (1963) reported that F_1 crosses within endosperm type were as productive as those between endosperm types. Castellanos *et al.* (1987) studied the use of S_3 lines in hybrid production in Guatemala. In one experiment, diallel crosses among 8 yellow S_3 lines and in another experiment diallel crosses among 10 white S_3 lines were evaluated at four locations. Up to 242 and 277% high-parent heterosis was detected in the two experiments, respectively. Superior lines had high GCA and their seed production potential was considered sufficient for their commercial use. This indicated that enough genetic variability exists within endosperm types and that not all hybrids need be between endosperm types.

Cordova (1984) reported on the performance and use of family hybrids developed in Guatemala during 1977–1983 (Table V). Methodologies for development of hybrids using FS families were elaborated, as such families were more productive as seed parents and were relatively easier to maintain in successive seed increases. He reported high-parent heterosis of up to 54% among families derived from La Posta, Mezcla Tropical Blanca, Tuxpeno-1, and Tuxpeno Caribe. On the basis of trials conducted at 33 locations during 1982, HB-83 was released in 1983. The superiority of HB-83 during 1982 averaged 37% over the local check. By 1984, approximately 2500 tons of seed of all nonconventional hybrids were produced, which reduced maize seed importation by 90%. Such nonconventional hybrids are superior in yield and stability to many of the commercial hybrids as well (Cordova, 1990; Azofeira and Jimenez, 1988; Perez *et al.*, 1988).

The search for heterosis between introduced germplasm and local maize strains was initiated in 1959 in Kenya. Crosses of KSII with Ecuador 573 and Costa Rica 76 yielded 40% more than the best parent, KSII. Based on

Table V

Yield of White Seeded Maize Varieties and Hybrids Evaluated at 183 Locations in Guatemala during 1978–1981

Material	Yield (t/ha)
HB-33	4.43
HB-19	4.34
ICTA T-101	4.33
H-5 (commercial check)	4.21
HB-67	4.21
HB-69	4.03
La Maquina	3.98
ICTA B-1	3.81
CRIOLLO (check)	3.33

this information a varietal cross, H611 (KSII × Ecuador 573), was released in 1964. A double cross, H622, and a three-way hybrid 632 were released for commercial production in 1965. Other hybrids released to date include H612, H613, H614, and H625. Hybrid maize in Kenya has become so popular that most farmers in high- and medium-yield potential areas will not grow OPVs. However, OPVs yielding 4–24% higher than the commercial hybrid H-625 have recently been identified (Ochieng, 1986; Mwenda, 1986). The maize breeding program of Zimbabwe is totally oriented toward hybrid development. Twenty-eight composites have been developed based on their heterotic patterns. Since 1932, 14 double cross hybrids, 4 three-way hybrids, 6 single crosses, and 4 modified single crosses have been developed. The seed production and promotion of the hybrids are efficiently carried out by the Seed Association of Zimbabwe (Oliver, 1986).

Singh (1986) reported that hybrids based on local flint germplasms were inferior to those developed from local × exotic (U.S.) or flint × dent crosses in India. Based on local × U.S. or Central American materials, 4 double cross hybrids were identified in 1961. By 1985, 18 double cross or double topcross hybrids of medium to full-season maturity had been released to the farmers.

Kim *et al.* (1990) conducted a diallel trial involving five tropical and five tropical × temperate maize lines to determine their combining ability. The GCA effects accounted for approximately 80% of the total genetic variance. In Nigeria, the five tropical inbreds showed positive GCA effects in forest region, and three tropical and two nontropical inbreds showed positive GCA effects in the savanna. The GCA effects were more important in crosses among tropical lines and the SCA effects were more important in crosses involving tropical and nontropical lines.

D. Recurrent Selection at CIMMYT

The activities of the various units in CIMMYT's maize program and their experiences and information are briefly summarized here. The results reported apply largely to selection methods used and progress made prior to 1985. Data on the effectiveness of the new methodologies will be available in the near future. A list of CIMMYT's maize pools and populations developed and improved for tropical lowlands is presented in Table VI. A more comprehensive list and a description of materials and the germplasm involved in their formation are given by CIMMYT (1982).

Table VI

Maize Gene Pools and Populations Developed and Improved by CIMMYT for Tropical Lowlands

Name of material	Other name
Gene pools	
Tropical Early White Flint (TEWF)	Pool 15
Tropical Early White Dent (TEWD)	Pool 16
Tropical Early Yellow Flint (TEYF)	Pool 17
Tropical Early Yellow Dent (TEYD)	Pool 18
Tropical Intermediate White Flint (TIWF)	Pool 19
Tropical Intermediate White Dent (TIWD)	Pool 20
Tropical Intermediate Yellow Flint (TIYF)	Pool 21
Tropical Intermediate Yellow Dent (TIYD)	Pool 22
Tropical Late White Flint (TLWF)	Pool 23
Tropical Late White Dent (TLWD)	Pool 24
Tropical Late Yellow Flint (TLYF)	Pool 25
Tropical Late Yellow Dent (TLYD)	Pool 26
Tropical White Flint QPM	
Tropical White Dent QPM	
Tropical Yellow Flint QPM	
Tropical Yellow Dent QPM	
Maize populations	
Tuxpeno-1	Population 21
Mezcla Tropical Blanco	Population 22
Blanco Cristalino-1	Population 23
Antigua Veracruz-181	Population 24
Blanco Cristalino-3	Population 25
Mezcla Amarilla	Population 26
Amarillo Cristalino-1	Population 27
Amarillo Dentado	Population 28
Tuxpeno Caribe	Population 29
Blanco Cristalino-2	Population 30
Amarillo Cristalino-2	Population 31
ETO Blanco	Population 32
Antigua Republica Dominicana	Population 35
Cogollero	Population 36
Tuxpeno o_2	Population 37
Yellow QPM	Population 39
White QPM	Population 40
Composite K o_2	Population 41
La Posta	Population 43
Early Yellow Flint QPM	Population 61
White Flint QPM	Population 62
Blanco Dentado-1 QPM	Population 63
Blanco Dentado-2 QPM	Population 64
Yellow Flint QPM	Population 65
Yellow Dent QPM	Population 66

1. Backup Unit (Germplasm Development Unit)

The Backup unit, more recently named Germplasm Development Unit, develops and improves germplasm with a broad genetic base that can be used either as new populations for international testing or for providing superior families for introgression into the existing populations. Several gene pools for different climatic conditions and with different maturity, grain color, and texture have been developed over the years. The methodology involved in their development and improvement has been described by Vasal *et al.* (1982), Pandey *et al.* (1984), and De Leon and Pandey (1989).

Pools were improved by means of the MER selection of Lonnquist (1964) with some alterations. In each cycle of selection, approximately 500 HS families making up a gene pool were planted in isolation in an arrangement of two females to one male. Each ear selected in the previous cycle was planted in an individual female row. Male rows were planted with seed prepared by mixing approximately 60 seeds of only those ears that came from superior families in the previous cycle. For each family, generally only one replication was grown in a 5-m row that included 16 to 21 plants. All female rows and undesirable male plants were detasseled prior to pollen shedding. For selection of families, data on yield, plant and ear height, days to silking, lodging, disease and insect reaction, and plant and ear uniformity were recorded at appropriate stages of plant development. The male rows closest to a female row served as a check in the selection. From the superior families identified (about half of the total), one to three ears were selected from good plants and planted as new families in the next cycle of selection. One cycle of selection was completed in a season. New germplasm was added to the gene pools regularly, as described previously. Pools were also improved for specific diseases and insects under artificial inoculations and infestations. The choice of stresses to be imposed in the selection process in a given pool was based on its susceptibility and on the importance of those stresses in the area of the world where the pool would be useful.

In a comparison of cycles of selection for eight tropical pools conducted at two sites, progress (linear response) averaged 2.50 (significant at the 0.01 probability level), −0.15, −0.35, −1.66, and −0.90% per cycle for yield, days to silking, plant height, stalk-rot score, and percentage ear rot, respectively (De Leon and Pandey, 1989). Yield improvement was greater in pools selected for stalk-rot resistance (3.02% per cycle) than in those selected for ear-rot resistance (1.38% per cycle). In the reduction of days to silk and plant height, however, greater progress was made in pools selected for ear-rot resistance (−0.50 and −0.75% per cycle, respectively) than in those selected for stalk-rot resistance (−0.04 and −0.25% per cycle, respectively). Results indicated that the selection scheme was effective in increasing grain yield and ear- and stalk-rot resistance and reducing days to silking and plant height (Table VII).

2. Advanced Unit

CIMMYT maize populations that showed potential in preliminary evaluations in the cooperating countries and that had an acceptable level of phenotypic uniformity were further improved through international testing and selection of the superior FS families based on those tests. Because the main growing season in the Northern Hemisphere was from April to September and in the Southern Hemisphere from September to April, one cycle of selection was completed in four seasons (2 yr). The FS families were developed during October to April in Mexico. Data reported were based on evaluation of 250 FS families of each population along with six local checks at six lowland tropical locations. The progeny trials were grown in the northern hemisphere during April to September (Season 1). One of the trials from each population was usually grown in Mexico. A simple 16×16 lattice design was employed. Each plot consisted of 22 plants in a single row at a density of approximately 53,500 plants/ha. While the trials were grown in the southern hemisphere, the 250 FS families wre planted in 10-m rows in the breeding nursery in Mexico (Season 2). Approximately 50% of the desirable plants of each family were selfed. For selfing, plants were selected for one or more of the following major traits: shorter plant height, earlier maturity, resistance to SCB, resistance to FAW, resistance to ear rot, and resistance to stalk rot. Selections for insect and disease resistance were practiced under artificial infestation and inoculation (Vasal *et al.*, 1982; Pandey *et al.*, 1986).

At harvest, two to three S_1 ears were selected from each FS family. Each S_1 ear was identified by its parental FS family and planted ear-to-row in Mexico in a 16-plant plot the next season (Season 3). Based on the international progeny test data across locations, about 100 FS families were selected. Better S_1 families from the selected FS families were identified, and approximately 50% of the plants within them selected and pollinated with a mixture of pollen from selected plants of most of the selected S_1 families. Selection was again practiced for the major trait. At harvest, two to three HS ears were selected from each selected FS family. Sixty-four plants of each HS ear were grown ear-to-row in Mexico in the next season (Season 4). Superior HS families in each FS family were identified and reciprocal plant-to-plant crosses made between them ensuring that the two HS families involved in a cross came from different FS families. It should

Table VII

Means for Different Cycles of Selection in Eight Tropical Gene Pools Evaluated in Two Tropical Environments

Pool	Cycle	Yield (t/ha)	Days to silk (No.)	Plant height (cm)	Stalk rot rating (1–5)[a]	Ear rot (%)
TEWF	C0	3.62	65	172	3.9	—
	C5	4.40	66	158	3.7	—
	C8	4.36	64	158	3.3	—
	C11	4.33	62	155	3.2	—
LSD ($P<0.05$)		0.57	1.4	7.9	0.2	—
TEWD	C0	3.48	65	171	3.8	—
	C5	4.52	65	161	3.5	—
	C8	4.84	65	162	3.1	—
	C11	5.14	63	162	3.1	—
LSD ($P<0.05$)		0.38	0.8	9.0	0.2	—
TEYF	C0	2.69	65	152	4.1	—
	C5	3.73	66	157	3.9	—
	C8	3.93	64	148	3.6	—
	C11	4.36	62	151	3.3	—
LSD ($P<0.05$)		0.44	1.6	4.8	0.3	—
TEYD	C0	2.72	65	146	3.8	—
	C5	4.42	66	149	3.4	—
	C8	4.12	65	147	3.2	—
	C11	4.62	62	148	2.9	—
LSD ($P<0.05$)		0.35	1.0	8.5	0.3	—
TIYD	C0	5.46	72	199	3.5	—
	C7	5.89	69	175	3.2	—
	C12	5.83	71	177	3.1	—
	C16	6.40	70	182	2.8	—
LSD ($P<0.05$)		0.63	2.2	6.6	0.2	—
TLWF	C0	5.98	73	201	3.4	—
	C7	5.55	71	182	3.4	—
	C11	5.92	70	169	2.9	—
	C15	6.18	70	171	2.7	—
LSD ($P<0.05$)		0.46	2.0	7.5	0.2	—
TIWD	C0	4.43	73	190	—	31
	C7	4.31	70	169	—	35
	C10	4.62	69	167	—	31
	C15	5.19	67	168	—	24
LSD ($P<0.05$)		0.37	2.1	4.3	—	—
TLYF	C0	4.04	74	201	—	26
	C5	3.89	73	192	—	23
	C9	4.33	71	183	—	22
	C14	4.86	70	186	—	15
LSD ($P<0.05$)		0.44	1.0	13.4	—	—
Gains/cycle (%)[b]		2.50	−0.15	−0.35	−1.66	−0.90

[a] 1 = good; 5 = poor.
[b] All gains were significant at the 0.01 level of probability.

be noted that the FS families were selected on the basis of international progeny test data, and selection for the major trait was emphasized during the selfing and bulk sibbing generations.

Introgression of germplasm from the backup pools was made regularly into the populations. When the progeny trials were growing in the southern hemisphere, approximately 50 HS families of the appropriate backup pool were planted alongside the FS families of the population in a nursery in Mexico. During this and subsequent seasons, pool families were handled exactly like population families, but no crosses were made between population families and the pool families. At harvest, about 220 FS pairs from the population and 30 FS pairs from the backup pool were selected to make up the 250 FS families for the international trials. The backup pool families equal to or superior in performance to the means of the population were used along with families selected from the population for the next cycle of selection.

Results of progress made in four medium-maturity populations for some important agronomic traits are presented in Table VIII. Gains per cycle averaged over the four populations in seven environments for yield, days to silk, ear height, and ears per plant were 2.11 (significant at the 0.01 probability level), −0.31, −0.47, and 1.05%, respectively (Pandey *et al.*, 1987).

Table VIII

Performance of Different Cycles of Selection of Four Medium-Maturity Tropical Maize Populations Grown in Seven Environments in 1983–1984

Population	Cycle	Yield (t/ha)	Days to silk (No.)	Ear height (cm)	Ears/plant (No.)
Blanco Cristalino-1	C0	5.11	64	113	0.91
	C2	5.64	62	103	0.98
	C5	5.29	63	108	0.97
Mezcla Amarilla	C0	4.97	63	108	0.96
	C2	4.64	62	93	0.96
	C5	5.40	60	101	0.98
ETO Blanco	C0	4.84	64	108	0.88
	C1	5.04	65	111	0.92
	C4	5.46	64	104	0.96
Ant. Rep. Dominic.	C0	4.66	59	94	0.95
	C1	4.53	59	87	0.92
	C4	5.13	59	92	0.98
LSD ($P < 0.05$) within population		0.28	0.7	5	0.04
Gains/cycle (%)[a]		2.11	−0.31	−0.47	1.05

[a] All gains are significant at the 0.01 level of probability.

Another trial was conducted in six environments involving different cycles of selection of eight late-maturing tropical populations (Pandey *et al.*, 1986). Gains per cycle for yield, days to silk, ear height, and ears per plant averaged 1.31, −0.59, −1.77, and 0.87%, respectively, across the populations and locations (Table IX).

Villena (1974) practiced two cycles of FS selection in 3 maize populations, where evaluation of progenies was conducted in Central America and Mexico. The yield superiority of C2 over C0 was 33%, with a range of 29–36%. Paliwal and Sprague (1981) reported gains per cycle for yield that ranged between 0.8 and 9.8% in 13 maize populations with an average of

Table IX

Performance of Different Cycles of Selection of Eight Full-Season Tropical Maize Populations Grown in Six Environments in 1983–1984

Population	Cycle	Yield (t/ha)	Days to silk (No.)	Ear height (cm)	Ears/plant (No.)
Tuxpeno-1	C0	5.98	66	112	0.96
	C2	6.04	65	110	0.97
	C5	6.34	66	114	0.96
Mezcla Trop. Blanco	C0	6.09	68	127	0.94
	C2	6.50	66	124	0.95
	C4	6.55	65	119	0.97
Ant. Veracruz-181	C0	5.68	67	127	0.96
	C2	5.33	67	117	0.94
	C5	5.76	66	117	0.96
Amarillo Cristalino-1	C0	5.27	68	120	0.94
	C2	5.28	67	123	0.95
	C5	5.75	66	117	1.00
Amarillo Dentado	C0	6.02	68	133	0.99
	C2	6.03	66	126	0.96
	C4	6.39	65	115	1.04
Tuxpeno Caribe	C0	6.20	67	120	0.93
	C2	6.01	66	122	0.92
	C5	6.21	65	111	0.95
Cogollero	C0	5.71	66	136	0.93
	C2	5.57	66	131	0.98
	C5	6.25	63	114	0.99
La Posta	C0	6.12	70	146	0.91
	C2	6.26	69	142	0.96
	C4	6.58	67	130	0.98
LSD ($P < 0.05$) within population		0.31	0.8	5	0.04
Gains/cycle (%)[a]		1.31	−0.59	−1.77	0.87

[a] All gains are significant at the 0.01 level of the probability.

3.4%. The gains depended on relative yield and genetic variability of the populations.

3. Development of Experimental Varieties (EVs) and Their Performance

On the basis of data of progeny trials at each site, approximately 10 FS families were selected and intermated (plant-to-plant crosses among families) using remnant seed to form an EV. The name of the EV was derived from the name of the site where the progeny test was conducted, followed by two digits indicating year of test. The last two digits indicated the population number. Thus, the EV Poza Rica 8121 was developed by intermating 10 FS families of the population Tuxpeno-1 (Population 21) whose 250 FS families were evaluated at Poza Rica, Mexico, during 1981. Data from all sites where a population was tested were used to select 10 FS families for the formation of the "across experimental variety," e.g., Across 8121. At harvest, clean ears from each family resulting from crosses with other families were saved and shelled in bulk. An equal quantity of seed was taken from each family bulk and mixed together to make an F_1 bulk. The F_1 bulk was advanced to F_2 and the ears saved were shelled in bulk to provide seed for experimental variety trials (EVTs), elite variety trials (ELVTs) (best from the EVTs), and for distribution to CIMMYT's collaborators upon request. National programs grew EVTs and ELVTs, along with their own checks, to identify potentially superior maize germplasm for use in their programs.

Forty Poza Rica and Across EVs developed from C0 and C3/C4 of 10 populations were evaluated in four environments in Mexico to measure performance and stability. Poza Rica EVs were derived on the basis of Poza Rica summer data and the Across varieties were developed on the basis of data from five to six locations. Percentage superiority of EVs derived from C3/C4 over those from C0 was 9.5 (significant at the 0.01 probablity level) for yield, −2.6 for days to silk, −5.1 for plant height, −9.6 for ear height, 5.6 for prolificacy, and −11.7 for ear rating (Table X). Poza Rica EVs were lower ($P < 0.01$) in plant and ear height (−6.1 and −10.3%, respectively) than Across EVs during summer at Poza Rica and Across EVs had fewer (−1.19%) days to silk than Poza Rica EVs at the other three sites. The results indicate that EVs developed from advanced cycles of populations undergoing recurrent selection have greater yield ($P < 0.01$) and are superior in other agronomic traits to those derived from original cycles (Pandey *et al.*, 1991).

Table X

Performance of Poza Rica and across EVs Derived from Different Cycles of Selection of 10 Tropical Maize Populations[a]

Type of variety	Yield (t/ha)	Yield $(b)^b$	Days to silk (No.)	Plant height (cm)
Poza Rica EVs	4.92	1.05	76	189
Across EVs	5.09	0.95	76	193
LSD ($P<0.05$)	0.19	0.15	0.4	2.7
EVs C0	4.78	1.06	78	196
EVs C3/C4	5.23	0.94	76	186
LSD ($P<0.05$)	0.13	0.08	0.3	2.3

[a] EVs were grown at four Mexican sites in 1983–1984.
[b] Stability parameter of Eberhart and Russell (1966).

4. Selection for Reduced Plant Height, Foliage, and Tassel Size

A modified FS selection scheme was employed for reducing plant he in an extremely tall population Tuxpeno (Johnson *et al.*, 1986). The cı rion for family and plant selection was a visual assessment of the sho genotypes at the time of pollination. Plant-to-plant crosses betw selected families were made using approximately five plants per fam each plant being crossed with a plant in one of the other selected famil Two or three ears were saved from each selected family at harvest to re about 300 families in each generation. Some families selected at the tim pollination were rejected at harvest for either their poor yield performa (visual) or their reaction to diseases. Selection and recombination w achieved in the same season, allowing two cycles of improvement per y This process was repeated from C1 to C12, at Poza Rica, Mexico, at elevation of 60 masl and a latitude of 19° N. From C12 onward, fami were visually evaluated in observation nurseries at Obregon (10 m 28° N latitude) in summer, and Tlaltizapan (900 masl, 19° N latitude winter, in addition to the location used for crossing (Poza Rica). observation nurseries were planted 1 to 2 weeks in advance of the cross block so that information on family height could be used in selection pı to pollination at Poza Rica. In the earlier generations, the plant populat of the crossing block was 53,000 plants/ha. From C12 onward, the cross block was grown at a higher population density of 106,000 plants/ Families with poor performance at higher density were eliminated. Polli tions were made among shorter plants with good synchronization of pol shed and silking under density stress.

Cycles 0, 6, 9, 12, and 15 were evaluated at two or three locations in Mexico during 1978–1979 to measure direct and correlated responses to selection. Selection was effective in reducing plant height from 282 to 179 cm (2.4% per cycle). The reduction in plant height was accompanied by 4.4% increase in yield (from 3.17 to 5.4 t/ha), 4.14% increase in kernal No./m^2 (from 1592 to 2667 kernel/m^2), and 3.1% increase (from 0.30 to 0.45) in harvest index per cycle, at the optimum plant density (Table XI). Optimum plant density for yield also increased by 2.1% per cycle (from 48,000 to 64,000 plants/ha). No change was noted in leaf area index, total dry matter production per square meter, effective filling period, or kernel weight at the optimum plant density (Johnson *et al.*, 1986).

Much of the improvement in yield potential of the shorter plant selections during the initial six cycles was due to their ability to respond to higher plant populations without lodging, while maintaining or reducing the level of barrenness. Further increases in the density for optimum grain yield (density tolerance) were due to a reduction in the number of barren plants particularly for the cycles following C12 which were selected under a higher plant population. Reducing the number of barren plants was associated with a shortening in the interval between anthesis and silking.

If the relationship between grain yield and population density for the shorter plant selection is extrapolated to those densities (approximately 30,000–35,000 plants/ha) currently being achieved by farmers in the tropics, yield would be greater for the C0. The higher yield potential of the shorter varieties is expressed only at higher density. This improvement in tropical maize is similar to that which occurred between 1930 and 1970 in

Table XI

Plant Height, Lodging, Grain yield, and Harvest Index for Selections for Reduced Plant Height in Tuxpeno Grown at Two to Three Locations during 1978 and 1979 in Mexico

Cycle of selection	Plant height (cm)	Grain yield (t/ha) 64,000 plants/ha	Grain yield (t/ha) 50,000 plants/ha	Lodging (%)	Harvest index
0	282	3.17	3.13	43	0.30
6	219	4.29	4.24	12	0.40
9	211	4.48	4.31	14	0.40
12	202	4.93	4.71	9	0.41
15	179	5.40	5.03	5	0.45
LSD ($P<0.05$)	22	0.30	0.37	7	0.04
Gains/cycle (%)[a]	−2.39	4.43	3.74	—	3.10

[a] All gains were significant at the 0.01 level of probability.

the U.S. Corn Belt. For farming conditions in the tropics in which weed management and crop nutrition are being improved, it is important to have a responsive and efficient genotype in order to maximize grain yield from those inputs. Selection for optimum plant height, yield, and density tolerance would provide management-responsive and stable maize varieties for the farmers in the tropics (Johnson *et al.*, 1986).

The large and almost linear response in plant height to selection indicates considerable additive variance for this trait. Even after 15 generations of selection, there still existed considerable variation for the trait in the population. When the short plant selections are crossed with other materials, the F_1 is approximately intermediate in plant height (Cordova *et al.*, 1979). The short plant selections are used intensively in national programs, where they are crossed with taller materials of other backgrounds in order to develop OPVs and hybrids with good plant stature (Cordova *et al.*, 1980).

There is evidence that careful removal of the male inflorescence prior to flowering increases grain yield (Poey *et al.*, 1977). Most tropical maize cultivars have a large tassel that negatively affects yield and tends to be associated with higher ear placement (Paterniani, 1981). Also, yield per unit of leaf area is low for tropical materials. Reducing individual plant leaf area while maintaining or increasing grain yield per plant would increase the efficiency of grain production per unit leaf area. Subsequent increases in leaf area index through density management might further increase grain yield per unit area (Paterniani, 1990).

Fischer *et al.* (1987) reported results of six cycles of FS selection for reduced tassel branch number and for reduced leaf area density above the ear (LADAE) in the populations Tuxpeno-1 and Antigua-Republica Dominicana. Selection for each trait was practiced separately. In another population, ETO Blanco, simultaneous selection was practiced for both traits. Evaluation of C0 and C6 of single-trait selections and C0, C2, C4, and C6 of combined trait selection was made at several densities at three sites in 2 years in Mexico. Gains per cycle for tassel branch number, LADAE, grain yield, and harvest index, respectively, were (*a*) Tuxpeno-1: selection for reduced tassel branch number: −7.4*, not significant (NS), 2.2*, and 2.4*%; selection for reduced LADAE: −1.2*, −3.8*, NS, and 3.89*%; (*b*) Antigua-Republica Dominicana: selection for reduced tassel branch number: −8.6*, −1.6*, 2.4*, and 2.2%*; selection for reduced LADAE: −5.7*, −2.6*, 2.0*, and 2.2%*; and (*c*) ETO Blanco: selection for reduced tassel branch number and LADAE: −6.8, −1.8*, 2.5*, and 2.9*%, respectively. In all populations, selection increased yields of the genotypes under higher density and decreased anthesis to silking interval (ASI), and days to 50% silking. Increase in grain yield potential (yield at optimum density) was due to an increase in harvest index. The number of grains/leaf area increased in these selections, which represents an increase in grain sink (grain number) relative to source (leaf area surface). The proportion of dry matter produced after flowering that was stored as grain (partitioning index) did not change in these selections. The increased grain yield, brought about by selection for reduced plant height, foliage or tassel, suggests that a decrease in tassel branch number and leafiness would provide further improvement in the yield and stability of tropical maize (Fischer *et al.*, 1987).

5. Selection for Drought Resistance

Tropical maize often suffers significant yield losses due to drought, which can occur any time during the crop cycle. A program was initiated during 1975 in Tuxpeno-1 (referred to as Tuxpeno Seleccion Sequia) to improve its drought tolerance (Fischer *et al.*, 1983). Full-sib progenies of C1 of Tuxpeno-1 were grown at Tlaltizapan in the dry season. Selection criteria, developed on the basis of their relationship to yield among various varieties in previous trials, under irrigated and stress conditions were (*a*) relative leaf and stem elongation rate; (*b*) anthesis-to-silking interval; (*c*) leaf senescence score during grain filling; (*d*) canopy temperature from C3 onward; and (*e*) yield.

Approximately 80 families were selected each cycle based on an index developed using the above traits. Fischer *et al.* (1983, 1989) reported results of evaluations conducted on the first three cycles of selection under both stress and nonstress conditions. Yields increased by 9.5% per cycle under drought, whereas little yield improvement occurred under nonstress conditions. Selection increased total dry matter under stress, root activity at 120- to 150-cm depth, ears per plant, and harvest index. Grain yield was correlated with ASI (−0.71*), relative leaf elongation (0.65*), and canopy temperature (−0.35*) under stress conditions and with leaf area index (−0.64*) under nonstress conditions.

Bolanos and Edmeades (1987) and Bolanos *et al.* (1990) evaluated cycles 0, 2, 4, 6, and 8 of Tuxpeno Seleccion Sequia during the dry season of 1987–1988 at Tlaltizapan. Cycle 8 yielded 0.80–0.90 t/ha more than C0 under environments with mean yields of 0.50 to 8.00 t/ha. Cycle 6 did not differ from C8, suggesting that perhaps genetic variability for the trait was reduced. The average gains at yield levels of 2 t/ha induced by drought stress at flowering were similar to those reported by Fischer *et al.* (1989). Selection increased ears per plant and harvest index by 1% per cycle without affecting total biomass, and reduced ASI. Number of grains and yield were reduced 10% for each day of delay in silking relative to anthesis, and after 10 days to delay the yield was practically zero. The latter cycles of selection had a greater accumulation of dry matter in the ear prior to flowering in all environments. The increase in dry matter in the ear increased by 18.2% under drought and 9% under optimum conditions. With an increase in the ear weight, there was a reduction in the tassel weight, perhaps because they competed for the same available photosynthates (Bolanos *et al.*, 1990).

Byrne *et al.* (1990) compared cycles 0, 2, 4, 6, and 8 of Tuxpeno Seleccion Sequia with cycles 0, 2, 4, and 6 of Tuxpeno-1 improved through international testing, in 12 environments with varying levels of moisture. A 1.6% yield gain per cycle was obtained for the drought selected population vs 1.2% per cycle for the same population improved through international testing.

Bolanos *et al.* (1990) concluded that:

1. There is sufficient genetic variability in CIMMYT maize germplasm for yield and ASI for improvement in drought tolerance. Stress should be manipulated to maximize expression of this variability.
2. Drought reduces number of grains and number of ears due mainly to silk delay and loss of silk viability.
3. The genetic variability for ASI under drought is due to differential capacity for accumulation of dry matter in the ear. Genotypes with reduced ASI are more drought tolerant.
4. Gains made for drought tolerance to date are largely due to better distribution of dry matter to the ear and selection for ASI under stress may increase yield potential in all environments.

6. Selection for Early Maturity

Early maturing varieties offer flexibility in planting dates under rainfed conditions, opportunity for increasing cropping intensity, and the flexibility for escaping drought that may occur at the beginning or end of the growing season in the tropics. Early cultivars facilitate early land clearing and help conserve soil moisture for the next crop. They also permit an early harvest of the crop, when scarcity of food is common, as occurs among subsistence farmers in Africa, Asia, and Latin America.

Early maturity is generally associated with reduced yield. Therefore, in 1975, CIMMYT undertook a program aimed at developing earlier maize varieties without sacrificing grain yield in the tropical environments. Several approaches were tried:

1. One approach was the development of early populations by crossing low-yielding early materials from the tropics and subsequent selection for

higher yield while maintaining earliness. This approach was not successful because the recovery of yield potential proved difficult.

2. A second approach was introgression of different levels of temperate germplasm into tropical maize and subsequent selection for disease resistance, higher yield, and early maturity. This approach was also dropped after a few cycles of selection because of extreme disease and insect susceptibilities of temperate maize in the tropical environment.

3. Development of populations by crossing early, lower yielding tropical maize types with intermediate maturity and higher yielding tropical materials and subsequent selection for early maturity and yield was another approach. This methodology was used in the development of early tropical gene pools. The pools were improved using the procedure described below and progress was made for early maturity, yield, and other traits (De Leon and Pandey, 1989).

4. A population, Compuesto Seleccion Precoz, was formed by recombining intermediate and early-maturing families of late maize populations. A HS selection program was initiated in the population in 1976, and two cycles of selection were completed each year in two locations, Poza Rica and Tlaltizapan, Mexico, with a total of 15 cycles of selection.

The selection procedure involved in Approaches 3 and 4 above involved planting of 500 HS families in a structure of 2 females : 1 male. The male was the bulk of the superior families of the population. Each family was planted in a 5-m row with standard row spacing of 0.75 m and plant-to-plant spacing of 0.33 m. Thus a density of 43,600 plants/ha was maintained for the females and the density in male rows was 87,200. The higher density in the male rows was used to select for tolerance to density pressure by eliminating weak, lodged plants and plants with poor synchronization of anthesis and silking. These undesirable plants, together with plants in the female rows, were detasseled before anthesis. In the male rows, when 70% of the plants had silked, all plants were detasseled and, at anthesis, early plants in the female rows were marked. Final selection was made from these plants. At both locations, materials were harvested relatively early to aid in the visual separation of drier ears from the superior plants previously selected for earliness. These ears were used for initiating the next cycle.

Islam, Pandey, and Diallo (unpublished data) evaluated cycles 0, 4, 8, and 12 of Compuesto Seleccion Precoz in four tropical environments in Mexico at densities of 53,000, 80,000, and 106,000 plants/ha. A reduction of 0.5 day per cycle in days to 50% silking (69 days for C0 vs 61 days for C12) with little reduction in grain yield (4.8 t/ha for C0 vs 4.3 t/ha for C12) was observed.

Narro (1988) evaluated cycles 0, 3, 6, 9, 12, and 15 of Compuesto Seleccion Precoz at Ames (Iowa), Cali (Colombia), and Chiclayo (Peru) during 1987, at two densities of 53,000 and 75,000 plants/ha. The results indicate that the method employed was useful in reducing days to maturity, plant and ear height, and leaf area and in increasing prolificacy (Table XII). The changes in yield, averaged over the three locations, varied with cycles and were nonsignificant. The yields increased at Ames, decreased at Chiclayo, and remained the same at Cali, with increasing cycles of selection. The changes in the traits were linearly related to cycles of selection. The later cycles yielded particularly better at Ames, due mainly to shorter growing cycle. No cycle × density interaction was observed. Narro (1988) concluded that development of early-maturing and high-yielding cultivars for a given site would be more effective, if practiced at that site.

7. Selection for Nitrogen Use Efficiency

Maize yields vary with the supply of N, and N is frequently deficient in tropical soils, where supply of chemical fertilizers is also limited (Edmeades and Lafitte, 1987). Most breeding programs develop cultivars under high-input levels under the assumption that a variety superior under such conditions would also be superior under low-input levels, though by a reduced margin. In 1985, CIMMYT initiated a study to show whether it was possible to select for improved performance of maize under N stress

Table XII

Progress from Selection for Early Maturity and Other Traits in the Population Compuesto Selection Precoz at Three Locations (Ames, Iowa; Cali, Colombia; and Chiclayo, Peru) during 1987

Cycle of selection	Yield (t/ha)	Days to silk (No.)	Plant height (cm)	Ear height (cm)	Leaf area (cm^2/plt)	Ears/plant (No.)
0	3.2	71	209	112	6300	0.74
3	2.8	67	193	96	5816	0.73
6	3.4	65	192	94	5738	0.79
9	3.1	64	188	91	5530	0.77
12	3.4	60	176	82	5340	0.81
15	3.2	58	172	75	5097	0.81
LSD ($P<0.05$)	0.40	2.3	5.6	3.3	219	0.05
Gains/cycle (%)[a]	ns	−1.0	−1.2	−2.0	−1.3	0.72

[a] Gains for all traits, except yield, were significant at the 0.01 level of probability.

without sacrificing its responsiveness to high N levels. Preliminary results obtained by evaluating 18 entries (16 improved materials and 2 landraces; 12 late and 6 early maturing) indicated that there was no correlation between grain yield and chlorophyll concentration at flowering and grain filling at high N levels. Yield and chlorophyll content in the ear leaf, 2 weeks after silking, however, were closely correlated under low N levels. The entries × N interaction was significant at the 8% level of probability, indicating that selection under N stress might be justified. Two hundred fifty FS progenies of variety Across 8328 were evaluated both under 0 and 200 kg N/ha levels in 5-m row plots. A 20% selection intensity was applied and selection was based on an index that comprised grain yield under N stress and without N stress, and high ear-leaf chlorophyll content during grain filling, short ASI, and large leaf area under N stress. During the first cycle of selection, families that yielded well under limited N also retained their lower leaves in a functional state for a longer time and had a shorter ASI.

While further results on progress from selection in this study are awaited, Machado and Paterniani (1988) have reported satisfactory improvement in the N use efficiency of a population after one cycle of MS under low N levels. At CNPMS, Sete Lagoas, Brazil, efforts are under way to identify traits that would aid in selection of genotypes more efficient in N absorption and utilization. Traits receiving attention are root length, capacity for assimilation of free ammonium ions, and glutamine synthetase activity. Genetic variation for the traits has been observed and selection has been practiced for three cycles in Nitroflint and Nitrodent populations. Preliminary results suggest excellent progress from selection in N uptake and utilization capacity and yield (Machado *et al.*, 1990).

8. CIMMYT's Quality Protein Maize (QPM) Program

QPM maize possesses a recessive allele, o_2, that confers significantly higher levels of lysine and tryptophan, two essential amino acids that are important in the diets of humans and monogastric animals. At the beginning of 1970, CIMMYT initiated a breeding program to solve problems associated with the QPM maize that hindered use of such cultivars by the farmers. These problems included reduced grain yield, soft and chalky kernel appearance, greater vulnerability to ear-rot organisms, more severe damage by stored-grain insects, and slower drying following physiological maturity. Donor stocks were developed with vitreous endosperm and protein quality similar to that of soft endosperm types, exploiting modifiers of the o_2 locus. A modified backcrossing scheme was used that permitted continuous accumulation of modifiers and use of the most advanced versions of the recurrent parents. The maize germplasm with o_2 locus and modified kernel types was later termed quality protein maize. Thirteen gene pools, 10 populations, and hundreds of OPVs have been developed using modified backcrossing and recurrent selection for distribution and use by the national programs. During 1985, CIMMYT initiated development of QPM inbred lines and hybrids to exploit heterosis, ensure seed purity, and reduce effect of contamination (CIMMYT, 1981; Vasal *et al.*, 1984; Bjarnason, 1990).

Various QPM variety trials were conducted by CIMMYT during 1988, at 80 sites in Africa, Asia, and Latin America. In 15 of the 80 sites where the trials were conducted, the yield of QPM varieties was higher ($P < 0.05$), in 54 cases equal, and in 11 cases lower ($P < 0.05$) than those of normal hybrids and varieties used as checks.

During 1988, CIMMYT personnel evaluated single cross QPM hybrids developed from S_4, S_5, and S_6 lines in 11 different diallel trials at various sites. In all trials, the superior QPM single crosses out-yielded the best normal checks. The superior five white QPM single crosses yielded 37–49% higher than QPM varieties, 5–14% higher than the normal hybrids, and 32–43% higher than local checks included in the trials (means of four sites). The yellow QPM single crosses yielded 18–25% higher than QPM varieties, 1–7% higher than normal hybrids, and 11–17% higher than local checks included in the tests (means of three sites).

A trial involving late white tropical QPM materials was conducted at five sites to compare QPM materials with local normal checks (Bjarnason, 1990). Included were three QPM double crosses, three QPM three-way crosses, one QPM family hybrid, and one QPM EV, along with a local normal check. One QPM hybrid yielded mroe than the best local check at each site and three QPM hybrids out-yielded the best local checks across sites. The QPM materials were generally earlier than normal materials (Table XIII).

Another trial was conducted with late tropical yellow materials at five sites. One QPM hybrid yielded equal to or more than the best local normal check at each site. No difference was observed in the performance of QPM and normal hybrids (checks) (Table XIV). These results indicate that QPM hybrids that have been developed are fully competitive with normal hybrids.

CIMMYT has discontinued improvement of QPM populations but will continue to develop QPM lines tolerant to ear rots and leaf diseases to be used in hybrids and synthetics by countries that demonstrate interest in such materials. In the future, use of restriction fragment length polymorphisms (RFLPs) to identify modifiers may enhance the efficiency of QPM breeding programs (Bjarnason, 1990).

Table XIII

Comparison of White Tropical QPM Materials to a Local Normal Check at Five International Sites in 1989

Material	Yield (t/ha)	Days to silk[a] (No.)	Ear height[b] (cm)	Ear rot (%)
Double cross 2 Q	7.95	58	127	5.6
Triple cross 2 Q	7.74	60	114	2.1
Double cross 1 Q	7.72	66	121	3.5
Triple cross 3 Q	7.67	64	134	6.9
Family hybrid Q	7.53	59	125	6.3
Triple cross 1 Q	7.44	61	138	4.4
Capinopolis 8563	7.07	61	123	8.8
Double cross 3 Q	6.75	62	129	9.2
Las Acacias 8562	6.36	65	110	6.1
Best local check	7.26	71	135	3.9
LSD ($P<0.05$)	0.58			

[a] Days to 50% silking from three sites.
[b] Ear height from four sites.

9. Selection for Disease Resistance

A collaborative program of disease resistance was initiated in 1974 to concentrate on downy mildew, corn stunt, and streak virus. Six national maize programs entered this collaborative project: Thailand and the Philippines for downy mildew resistance, El Salvador and Nicaragua for corn stunt, and Tanzania and Zaire for maize streak. Recurrent selection for improved disease resistance was practiced in three broad-based tropical

Table XIV

Comparison of Yellow Tropical QPM Materials to a Local Normal Check at Five International Sites in 1989

Material	Yield (t/ha)	Days to silk (No.)	Ear height (cm)	Ear rot (%)
Double cross 2 Q	5.54	49	113	8.5
Double cross 2 Q	5.47	53	111	10.5
Double cross 1 Q	5.36	51	116	7.2
Double cross 3 Q	5.26	50	141	11.2
Triple cross 3 Q	5.23	53	116	6.5
Triple cross 2 Q	5.14	51	108	6.6
Across 8565	4.57	52	102	12.7
S8766 Q	4.39	51	99	8.0
Poza Rica 8666	4.35	52	104	10.1
Best local check	5.46	56	121	9.2
LSD ($P<0.05$)	0.32			

populations, tropical late white dent (TLWD), tropical yellow flint-dent (TYFD), and tropical intermediate white flint (TIWF). Alternate cycles of selection were carried out in disease-prone areas in the collaborating countries to identify resistant families, and in Mexico to improve agronomic characters of the resistant selections. Plantings were made at the time of highest disease incidence in the collaborating countries to provide data for selection and recombination of resistant families in Mexico.

Four cycles of S_1 selection and recombination for downy mildew and corn stunt resistance had been completed by 1980. Trials for corn stunt were planted in Guatemala, El Salvador, Nicaragua, Panama, the Dominican Republic, and Mexico and those for downy mildew were planted in Honduras, Venezuela, Nepal, Thailand, the Philippines, and Mexico. Results showed that the advanced cycles of selection of these populations had improved disease resistance and higher yield under both disease-free and disease-pressure situations. Other agronomic characters such as plant and ear height and maturity have also improved (Table XV) (De Leon, 1982).

S_1 lines from the last cycle of selection of the three populations were screened for their reaction to streak virus at IITA and several lines showed good to excellent resistance (M. Bjarnason, 1991, personal communication). Improvement of tolerance to stunt was reinitiated in a population formed by intermating eight EVs from Population 73, using S_1 recurrent selection. The synthetic formed during C3 was superior to previously stunt-tolerant EVs across seven locations during 1990 (H. Cordova, 1991, personal communication).

Nearly 20 varieties released and grown in Nicaragua, El Salvador, Honduras, Mexico, Venezuela, Philippines, Nepal, Vietnam, and Indonesia

Table XV

Progress in Improving Resistance to Downy Mildew and Corn Stunt Diseases in Collaborative Research, 1980

Disease and population tested	Cycle (No.)	Infection (%)	Yield (t/ha)	Plant height (cm)	Days to silk (No.)
Downy mildew (Thailand)					
Trop. Interm. White Flint	C0	41.4	3.84	185	68
	C4	0.0	4.40	160	66
Trop. Yellow Flint-Dent	C0	67.4	2.68	199	68
	C4	2.6	4.49	183	66
Stunt desease (El Salvador)					
Trop. Interm. White Flint	C0	30.6	3.34	206	49
	C4	21.8	4.12	200	49

trace their origin to this germplasm (Cordova *et al.*, 1989; C. De Leon, 1990, personal communication).

Selection for DMR was also practiced and progress made in three maize populations (Populations 22, 28, and 31) improved through international testing, by the CIMMYT's Asian Regional Maize Program (ARMP) based in Thailand. Four new maize populations (EW-DMR, EY-DMR, LW-DMR, and LY-DMR) have been developed and are being improved for DMR, using S_1-S_2 recurrent selection. Bulks of C0, C1, C2, and C3 of all four populations were evaluated for DM incidence in disease nurseries and for agronomic traits under disease-free conditions in Thailand and the Philippines during 1990. Results indicated that the recurrent selection scheme resulted in simultaneous improvement of levels of DMR ($-11^{**}\%$ per cycle) and grain yield (507^{**} kg per cycle) over the four populations (De Leon and Granados, 1991).

Techniques have been refined at CIMMYT for inoculation with (*a*) *F. moniliforme* and *D. maydis* ear and stalk rots; (*b*) *E. turcicum*, and *B. maydis* leaf diseases; and (*c*) stunt disease (CIMMYT, 1979, 1981). Studies have been conducted to determine progress from selection for ear- and stalk-rot resistance (De Leon and Pandey, 1989). They reported that the MER selection scheme was effective in increasing grain yield ($2.50^{**}\%$) and resistance to ear rot ($-0.90^{**}\%$) and stalk rot ($-1.66^{**}\%$), and in reducing days to silking ($-0.15^{**}\%$) and plant height ($-0.35^{**}\%$) (Table VII). Ceballos *et al.* (1991) reported 19 and 6% improvement in the resistance to *E. turcicum* (Pass) and *P. sorghi* per cycle after four cycles of combination of FS and S_1 selection in eight subtropical materials. They concluded that resistance to *E. turcicum* was highly heritable and that breeding methodology employed was effective in improving resistance to the two diseases. Significant GCA and SCA effects have been observed for resistance to *P. maydis* (Ceballos and Deutsch, 1989). The trait is under control of one to two dominant genes in the most resistant lines. They also reported additive gene effects for resistance in a related study.

10. Selection for Insect Resistance

At CIMMYT, four species of insects, SWCB, SCB, FAW, and corn earworm (CEW) are mass reared for artificial infestation of over 100,000 plants per year. Resistance breeding is done in the fields with emphasis on reducing damage from larvae. All artificial infestations are carried out with bazooka at the appropriate stage of plant development (Mihm, 1989).

A multiple-borer-resistance population, 590 (MBR), has been developed with the objective of recombining sources of resistance to temperate and subtropical stalk borers such as *O. nubilalis* (ECB), SWCB, SCB,

C. partellus, S calamistis, E. saccharina, and *B. fusca*. Multiple resistance provides the poor farmers in the tropics protection against multiple insects present in their fields. Inclusion of germplasm with diverse resistances ensures a more durable and higher level of resistance.

Two hundred FS families of MBR were tested at six locations for tolerance to six insect species. All families were either resistant or intermediate in their reaction to ECB and *C. partellus*, all but 1% of the families were either intermediate or resistant to *Diatraea* spp., and 77% of the families were either intermediate or resistant in their reaction to *B. fusca*. Between 18 and 52% of the families also rated resistant to FAW. Approximately 39% of the families were resistant to one species, 24% to two species, 9% to three species, 2.5% to four species, and 0.5% to five species. None was resistant to all six species (Smith *et al.*, 1989).

Studies have also been conducted that relate to the genetics of insect resistance. Sosa *et al.* (1990) evaluated eight cycles of modified FS selection in the population Antigua-Veracruz 181 for leaf feeding and tolerance to FAW at several sites. They reported 5.5% yield increase per cycle under infestation. Since there was no change in leaf feeding, and selection of progenies in each cycle was largely based on yield under infestation, they concluded that the progress was due to a change in tolerance of the population to FAW. Thome *et al.* (1990) observed high GCA effects for resistance to ECB, SWCB, and SCB in three resistant lines derived from the MBR pool and concluded that resistance in these lines was largely dominant in nature. De Leon *et al.* (1990) have reported on RFLP analysis of six resistant lines from the MBR pool and one susceptible line. Six chromosome regions of the genome were similar to that of the resistant parent, particularly within chromosome 2; and several regions in chromosomes 4 and 6 were similar to those in the susceptible parent.

A multiple-insect-resistant tropical population, 390 (MIRT), has also been developed with some of the sources of resistance included in the MBR pool and with insect-resistant selections from CIMMYT's tropical populations and pools. It is assumed that the MBR population would probably never be the best source of resistance to tropical insect species. MIRT is expected to possess disease resistance and agronomic characters needed in maize germplasm for the lowland tropics. Resistance to SCB and FAW are the traits under selection in the population. Breeding for insect resistance also focuses on evaluation of germplasm bank accessions for tolerance to various insect species, on improving resistance to CEW in highland maize germplasm, and on transferring genes for borer resistance from *Tripsacum* spp. (CIMMYT, 1988).

EVs derived from MBR were recently compared with other insect-tolerant materials in a 10-entry trial in paired infested/protected row plots. Six replications were used at each of several locations in the United States,

Table XVI

Performance of EVs Derived from MBR, When Artificially Infested with ECB, FAW, SCB, and SWCB Larvae at Different Sites in the United States, Canada, and Mexico in 1990

Material	Leaf-feeding rating (1–9)[a]				Yield reduction Inf. vs prot.[e] (%)
	ECB[b]	FAW[c]	SCB[d]	SWCB[e]	
Font 6230	6.5	8.2	7.5	8.1	39.7
Ki3/Tx601	6.2	8.0	4.3	7.2	26.6
B73/Mo17	5.9	8.1	7.2	8.1	43.8
Pioneer 3184	5.0	7.0	6.9	7.6	31.8
MBR-ECB	4.0	5.8	3.5	5.7	11.5
MBR-Chilo	3.6	6.0	3.6	5.6	17.9
MBR-SWCB	3.6	6.0	3.9	5.6	23.5
MBR-SCB	3.5	6.0	3.9	5.7	3.3
MBR-All bugs	3.5	5.4	3.9	5.7	9.2
P47s3/Mp78:518	3.3	5.9	4.1	4.9	12.8
Susc. check	7.3	7.0	—	7.4	26.6
Res. check	3.8	5.5	—	4.6	18.1

[a] 1 = highly tolerant; 9 = highly susceptible.
[b] Means of four environments.
[c] Means of three environments.
[d] Means of one environment.
[e] Means of two environments.

Canada, and Mexico. Results indicated that resistant materials had lower leaf-feeding ratings and suffered lower yield reductions than the susceptible materials, MBR EVs were equal or superior to the resistant checks in performance, and resistant materials could be used as sources of resistance to multiple-borer and armyworm pests (Table XVI) (J. Mihm, 1991, personal communication).

11. Selection for Tolerance to Soil Acidity

Because Al and P stresses tend to occur together, development of improved cultivars should involve simultaneous selection for tolerance to the two stresses. Fortunately, genetic variation has been reported for tolerance to both stresses. CIMMYT at Cali, Colombia, has been improving a maize population, SA-3, for tolerance to soil acidity for about 20 cycles. The population was developed in the late 1970s at Santander Quilichao, Colombia. During 1977, 192 materials of diverse origins were planted in single-row plots at 45 and 75% aluminum saturation. At harvest, 70 ears were selected, and 45 materials were observed to have good plant development. In 1978, remnant seed of these 45 materials was planted as females, and a bulk mixture of seed from the 70 selected ears as males, in a

HS recombination block. The resulting population was improved using the MER selection scheme at 45 and 75% Al saturation and 6-7 ppm P (G. Granados, 1991, personal communication). The population and the varieties derived from it show higher levels of tolerance and good agronomic attributes (S. Pandey, unpublished).

Research was initiated in 1985 to develop two white (SA-6 and SA-7) and two yellow populations (SA-4 and SA-5) tolerant to acid soils. National program materials with tolerance to acid soils and CIMMYT materials that performed well in Latin America, Asia, and Africa were obtained. In 1986, 36 materials were selfed and sibbed at Cali, Colombia, and 613 yellow and 618 white S_1 lines were selected. During 1987–1988, the materials were evaluated in acid soil trials in Brazil, Colombia, Indonesia, and the Philippines to determine whether some of them might be of immediate use to cooperators and to get a better idea of their characteristics. During 1987 at Cali, the S_1 yellow lines were topcrossed to two yellow heterotic lines from the Colombian national program, and the white S_1 lines with two white heterotic lines from this program.

On the basis of the performance of the lines in the topcross formation block, topcrosses of 250 lines from each of the two groups were harvested for further evaluation. In the second season of 1987, the 613 yellow and 618 white lines were evaluated under different levels of Al and P concentrations and under nonacid soil conditions. The topcrosses were evaluated at Santander Quilichao in acid soils and at Cali in nonacid soils during 1987. In 1988, the four populations were developed on the basis of the lines' performance and their heterotic patterns. The two white populations form a heterotic pair as do the two yellow ones.

The populations are evaluated in the field under Al saturations ranging from 40 to 75% and P levels ranging from 5 to 12 ppm at Carimagua, and Santander Quilichao in Colombia. In addition, the materials are evaluated under acid soil conditions in Brazil, Indonesia, Peru, the Philippines, Thailand, and Venezuela, among others. Testing also takes place under normal soil conditions to facilitate the accumulation of genes for yield potential at the same time that acid soil tolerance is being improved. Where acid soils occur, this germplasm should perform significantly better than susceptible genotypes, but where they do not, they should still be highly responsive to favorable soil conditions. Two cycles of FS selection have been completed in SA-4, SA-5, SA-6, and SA-7.

E. Hybrid Development at CIMMYT

Many national programs have initiated hybrid development activities to satisfy needs of their farmers. To serve such national programs better,

CIMMYT initiated a program in 1985 to develop new germplasm products for hybrid development and to accumulate and publish information about the utility for hybrid development of the tropical and subtropical materials of CIMMYT (Vasal *et al.*, 1987). Eight diallel trials were developed based on adaptation, maturity, grain color, and protein quality of materials to study combining ability of CIMMYT materials and to determine their heterotic patterns. The evaluations were made at several locations in Mexico, the United States, and various countries in Central and South America and Asia. The five highest yielding crosses in each diallel trial were selected. The high parent heterosis in these five crosses across the diallel trials ranged from 2 to 19%.

CIMMYT has also developed and evaluated several types of hybrids during the last several years. Many of these hybrids are superior to the checks (OPVs and hybrids) included in the trials by the national programs (Table XVII).

Crossa *et al.* (1990b) evaluated diallel crosses among seven tropical late yellow populations in seven environments and found Suwan-1 and Population 24 to have the highest mean yield in crosses and high heterosis. Populations 28 and 36 had a high cross mean but relatively lower heterosis. They recommended interpopulation improvement involving populations 24 and 36.

Beck *et al.* (1990) evaluated diallel crosses among 10 CIMMYT populations and pools with early and intermediate maturity at five locations. General combining ability was significant for all traits and SCA was significant for ear height only. Population 49 × Population 26 and Population

Table XVII

Performance of Different Types of Hybrids Developed and Evaluated by CIMMYT in Several Environments during 1986–1990[a]

Type of hybrids	Yield (t/ha)	Superiority over checks (%)	High-parent heterosis (%)
Single crosses	7.2	29[b]	—
Double crosses	7.1	20[c]	—
Three-way crosses	7.6	28[b]	—
Double topcrosses	7.6	—	25
Topcrosses	7.9	—	41
Interfamily hybrids	7.2	—	33
Intersynthetic hybrids	6.6	—	13

[a] S. K. Vasal, 1991, personal communication.
[b] Checks were OPVs.
[c] Checks were hybrids.

Table XVIII

Possible Heterotic Combinations for Some CIMMYT Populations

Population	Possible heterotic partner(s)
21	Pops. 25 and 32; Pool 23
22	Pops. 25 and 32
23	Pop. 49; Pool 20
24	Pops. 27, 28, and 36; Suwan-1
25	Pops. 21, 22, 27, 29, 43, and 44; Pool 24
26	Pop. 31; Pool 21
27	Pops. 24, 25, 28, and 36; Suwan-1
28	Pops. 24 and 27; Suwan-1
29	Pops. 25 and 32
31	Pop. 26; Pool 22
32	Pops. 21, 22, 29, and 44
33	Pop. 45
34	Pops. 42 and 47; Pool 32
36	Pops. 24 and 27
42	Pops. 34, 43, 44, and 47
43	Pops. 25, 42, and 44
44	Pops. 25, 32, 42, 43, and 47; Pool 32
45	Pop. 33; Pool 33
46	Pop. 48; Pool 30
47	Pops. 34, 42, and 44
48	Pop. 46
49	Pop. 23; Pool 19

26 × Pool 21 provided high-parent heterosis of 9.6 and 7.3%, respectively. Pool 22 showed the highest GCA effects (0.37 t/ha) and the only cross with significant SCA effect (6.7%) involved Population 23 and Pool 20. They recommended hybrid development involving Population 23 and Pool 20 for white materials and Population 26 with either Pool 21 or 22 for yellow materials. CIMMYT maize program is currently involved in development and improvement of heterotic populations tolerant to inbreeding depression and identification of new heterotic patterns (S. K. Vasal, 1991, personal communication).

Heterotic patterns of various CIMMYT materials are summarized in Table XVIII.

F. Recurrent Selection and Genotype and Genotype x Environment Variances

Gardner *et al.* (1990) calculated genotype, genotype × environment, and error variance components of CIMMYT's 11 tropical maize populations, using data from international progeny trials (Tables XIX and XX). Test

Table XIX

Genetic Components of Variance for Each Cycle of Selection in 11 CIMMYT Populations Undergoing Full-Sib Family Recurrent Selection

	Cycle						
Population	0	1	2	3	4	5	6
				($\times 10^{-4}$)			
21	1350	664	1,504	2414	2164	−1113	
22	−4479	1806	1,184	1194	2402	1032	2968
24	3461	2311	961	2579	2393	2180	2409
27	2106	1844	1,223	1534	2562	1760	3167
28	2393	1445	747	2362	2142	870	
29	1131	1011	807	1333	1938	1442	
32	2665	1400	60	1979	1148		
35	1440	867	23,210	1643	1183		
36	1810	1324	1,636	852	1738	1439	
43	−116	2009	2,637	1810	1073		
49	1970	1979	986				

sites and their numbers were generally different for different cycles of selection. The results indicated no reduction in either genotypic or genotype × environment interaction variances in any of the populations. The broad genetic base of CIMMYT populations, low initial frequency of desirable alleles, low selection intensity, introgression of additional

Table XX

Genotype × Environment Interaction Components of Variance for Each Cycle of Selection in 11 CIMMYT Populations Undergoing Full-Sib Family Recurrent Selection

	Cycle						
Population	0	1	2	3	4	5	6
				($\times 10^{-4}$)			
21	6,763	1954	364	2103	2973	38,075	
22	56,269	3247	774	1862	9319	4,116	2882
24	800	2529	1698	2147	669	2,148	1006
27	843	1958	1426	963	3387	995	849
28	2,604	1396	3232	1828	2127	2,819	
29	17,417	8230	1629	3798	1650	10,064	
32	4,291	919	9472	1480	2600		
35	1,228	839	803	3231	1266		
36	1,196	2637	2059	1685	93	1193	
43	22,593	1248	620	2140	2924		
49	1,153	1494	1152				

germplasm, extremely diverse environments used during different selection cycles, etc., were cited as possible explanations for the results obtained.

Weighted mean estimates (weights used were inverse of the variance of each estimate) were used to estimate mean yield, and genotype, genotype × environment, and error variances for the 12 populations by Gardner *et al.* (1990), using data pooled over cycles of selection (Table XXI). The results indicate that good data from two replications at six sites would result in satisfactory progress from selection. More replications could not be justified. With data from only three locations, expected gains would be greatly reduced. They emphasized the need for classification of environments based on the use of the resulting products and for collecting good quality data for maximizing progress from selection.

Pandey *et al.* (1986, 1987, 1991) used the stability analysis of Eberhart and Russell (1966) to analyze the data from cycles of selection trials of late and medium-maturity CIMMYT populations and from the trial of EVs derived from the original and the latter cycles of selection from most of the

Table XXI

Estimates of Means, Components of Variance, and Expected Progress from Full-Sib Family Recurrent Selection in 12 CIMMYT Populations

		Component of variance[a]			Expect gain[b] (%)	
Population	Mean[a]	Error ($\times 10^{-4}$)	G × E ($\times 10^{-4}$)	Genotypic ($\times 10^{-4}$)	6 env.	3 env.
21	5.09	8515	3712	1422	3.49	2.86
22	5.38	8111	2287	1375	3.39	2.83
24	4.97	6736	1760	1955	4.86	4.26
27	4.54	5997	1297	1771	5.13	4.52
28	4.75	7259	2138	1433	4.04	3.41
29	5.24	7449	3102	1108	2.92	2.38
32	4.08	6578	1704	1611	5.24	4.42
35	4.92	5525	1150	1286	3.89	3.37
36	5.19	6598	1486	1481	3.92	3.37
43	5.56	9555	2194	1705	3.75	3.17
48	6.09	5176	1282	1399	3.33	2.90
49	5.05	5783	1269	1440	4.04	3.51

[a] Weighted means where weights were the inverses of the variances of the parameters estimated each cycle of selection.

[b] Expected gains from selection calculated assuming the ratio of dominance variance to additive genetic variance to be 0.9377 (average of 99 maize genetic studies reported by Hallauer and Miranda, 1988) and a selection intensity of 20%.

same populations. The deviations from regression were generally nonsignificant (except in the case of medium maturity materials where they were significant at $P = 0.05$) and are not reported here. The *b* values for C4/C5 of the late materials were significantly lower than those for CO (Table XXII). The most recent cycles of selection had lower *b* values than C0 in the case of six of the eight populations. The last cycles had higher yields across the environments as well. The advanced cycles of selection yielded higher than C0 in 15 cases of 16 in low-yielding environments and in 21 cases out of 32 in high-yielding environments.

In the medium maturity materials, the *b* values decreased significantly from C0 to C2 and then increased again (Table XXII). In two of the four populations the advanced cycles of selection averaged higher yields and lower *b*'s than C0. In this study as well, the advanced cycles of selection yielded more than C0 in 11 of 12 cases in low-yielding environments and 12 of 16 cases in high-yielding environments.

The regression values of Across EVs derived from C3/C4 were lower than those from Poza Rica and Across EVs derived from C0 and Poza Rica EVs derived from the last cycle ($P < 0.01$). The results indicate that EVs developed from advanced cycles of selection of populations undergoing recurrent selection yield more ($P < 0.01$) and are more stable ($P < 0.01$) than those developed from the original cycle (Table X) (Pandey *et al.*, 1991). Gardner *et al.* (1990) attributed the improvement in stability to elimination of deleterious genes, shorter plant height, higher harvest index, and improvement in resistance to biological and environmental stresses.

Table XXII

Yield and Stability of Different Cycles of Selections, Averaged over Several Tropical Maize Populations

Late-maturity materials[a]			Medium-maturity materials[b]		
Cycle	Yield (t/ha)	b^c (t/ha)	Cycle	Yield (t/ha)	b^c (t/ha)
C0	5.88	1.04	C0	4.89	1.04
C2	5.88	1.01	C1/C2	4.96	0.92
C4/C5	6.23	0.96	C4/C5	5.32	1.04
LSD ($P < 0.05$)	0.11	0.04		0.14	0.06

[a] Means of eight populations evaluated at six sites.
[b] Means of four populations evaluated at seven sites.
[c] Stability parameter of Eberhart and Russell (1966).

VII. SOME CONSIDERATIONS IN ORGANIZING AN EFFICIENT MAIZE IMPROVEMENT PROGRAM

Recurrent selection should be an integral part of any applied maize breeding program whether the ultimate product is an improved variety or some knd of hybrid. The systematic accumulation of favorable alleles and favorable combinations of alleles in advanced cycles makes them superior sources of new inbred lines for use in hybrids as well as superior sources of new varieties for release to farmers.

Eberhart *et al.* (1967) suggested a comprehensive breeding system in maize based on experiences in Kenya, which was further discussed by Sprague and Eberhart (1977). Such a comprehensive system has three distinct phases: (1) the development of two or more breeding populations from diverse sources so that the population per se and the population-cross mean(s) will be at the highest level possible and the populations will have maximum additive genetic variation within each; (2) continuous population improvement by an effective recurrent selection program; and (3) the development of superior hybrids from each cycle of selection by an efficient and systematic procedure. Gardner (1986) discussed the development of breeding populations and the integration of population improvement with hybrid development activities to develop superior maize cultivars for the tropics.

In developing countries where the release of a new improved variety may be the most useful end product, intrapopulation recurrent selection provides a steady source of new varieties at each cycle of selection. Many such new varieties have been developed from superior FS families extracted from improved CIMMYT populations (Pandey *et al.*, 1986, 1987). These populations have been involved in FS family recurrent selection programs (or modification thereof) for several cycles. Unfortunately, most CIMMYT populations appear to have been developed from very broad germplasm resources without regard for heterotic patterns. Hence, there is little interpopulation heterosis observed in crosses among them (Crossa *et al.*, 1990b; Beck *et al.*, 1990). Nevertheless, considerable heterosis has been reported between lines derived from the same and different CIMMYT maize populations (Han *et al.*, 1991).

As developing nations advance and become more interested in hybrid programs, heterotic patterns among populations available becomes increasingly more important. Heterotic patterns among maize races, varieties, or other open-pollinated populations from tropical and subtropical countries have been observed (Crossa *et al.*, 1990a; Gardner and Paterniani, 1967; Castro *et al.*, 1968; Paterniani and Lonnquist, 1963).

Heterosis estimates of 100 to over 200% above the better parent are common. Such information is useful in forming two or more breeding populations to which reciprocal selection might eventually be applied. Such reciprocally developed populations can be improved to serve as useful resources for new varieties and will be far more useful for developing inbreds and hybrids. The main reason that reciprocal selection systems have not been used by CIMMYT is related to the primary objective, which has been improved varieties, and the cost of simultaneously improving two populations compared to one.

Comparisons between intrapopulation and reciprocal interpopulation selection methods indicate that selection in early cycles is for additive genetic effects with partial to complete dominance in both systems. They seem to be equally effective in increasing the interpopulation cross mean (Moll and Stuber, 1971; Odhiambo and Compton, 1989). At the same time, intrapopulation selection is more effective in improving population means, and S_1 family selection is definitely more effective in reducing inbreeding depression and improving means of selfed progenies (Odhiambo and Compton, 1989). S_1 family selection is also very effective for improving insect and disease resistances and environmental stress tolerances. Therefore, since reciprocal selection programs are more difficult and costly to run, it seems wise to use S_1 family selection in each of the two populations in the early cycles and switch to reciprocal FS selection after about five cycles when gains due to overdominance and dominance types of epistasis are likely to become relatively more important. For traits that are highly heritable, a few cycles of intrapopulation selection will be highly effective.

Jones *et al.* (1971) provide evidence to support the choice of reciprocal FS over reciprocal HS selection. An additional advantage is that new reciprocal pairs of S_1 lines for further development and use in hybrids are a natural spinoff of each cycle of selection. However, no one has compared the systems on the same population.

In international work, where it is sometimes difficult to conduct good evaluation trials of S_1 families in early cycles of selection, use of FS family evaluation and selection could be substituted for the S_1 family system provided that a generation of inbreeding with selection among plants within S_1 families is used as described by Pandey *et al.* (1986). This should result in improved tolerance to inbreeding where recurrent selection is fully integrated with a hybrid program.

In basic genetic studies designed to obtain information about population parameters and how they are influenced by selection, it is essential to work with closed populations. However, in any maize improvement program where the ultimate objective is release of a new cultivar, there is no reason to keep a breeding population closed. New promising germplasm can be introgressed continually as has been done at CIMMYT. In reciprocal programs, such additions must take into account heterotic patterns.

VIII. CONTRIBUTION OF RECURRENT SELECTION TO MAIZE PRODUCTION IN THE TROPICS

Studies to determine contribution of genetics and plant breeding in increasing maize yields in the tropics are limited. As has been previously mentioned, approximately 53% of the total maize area in the developing countries is planted to improved materials—39% to hybrids and 14% to open-pollinated varieties (CIMMYT, 1990). While hybrids have been reported to yield about 11% more than varieties in Mexico and Central America, the yields of the two types of germplasm are indistinguishable in many environments. In fact, in high-stress areas, varieties may outyield hybrids (CIMMYT, 1987). Some studies have compared improved varieties and hybrids with those of the farmers' cultivars both in farmers' fields and at experiment stations. They may provide an indication of the contribution of superior germplasm to increased maize yields in the tropics.

Maize scientists of ICA, Colombia, compared a local maize cultivar (Native) with the local cultivar contaminated with improved cultivars (Regional), the Regional improved through recurrent selection (Improved Regional), an improved OPV (Improved Variety), and a hybrid, at 10 sites in different seasons, with a total of 27 environments (F. Arboleda, personal communication, 1990). Five of the sites were in the lowlands, 3 in the medium altitude areas, and 2 in the highlands. The materials evaluated at each site were adapted to that ecosystem.

Averaged over environments, the Native yielded significantly less than the other types (Table XXIII). Although, no differences were observed in the performance of the Improved Variety and the hybrid, both were superior to Native, Regional, and Improved Regional cultivars. The Improved Regional was superior to the Regional in yield. This study also demonstrated the improvement of farmers' cultivars through chance introgression of improved germplasm in them, an indirect contribution of superior germplasm to yield improvements. Although the trials were conducted at experiment stations, the yields ranged from 1.42 to 6.00 t/ha across environments, indicating that a range of environments was sampled. No clear differences were detected in yield patterns of the genotypes for the different groups of altitudes sampled.

Table XXIII

Performance of Native, Regional, Improved Regional, Improved Variety, and Hybrid in 27 Environments at 10 Sites during 1986–1988 in Colombia

	Yield at altitudes (t/ha)			
Type of material	Low[a]	Medium[b]	High[c]	Mean
Hybrid	3.86	5.53	5.21	4.81
Improved variety	4.00	4.77	5.04	4.64
Improved regional	2.88	4.33	4.40	3.85
Regional	2.81	2.95	3.74	3.32
Native	2.22	2.24	2.47	2.36
LSD ($P < 0.05$)	0.18	0.32	0.25	0.18

[a] Means of 13 environments.
[b] Means of six environments.
[c] Means of eight environments.

In Venezuela, scientists have examined maximum yields of varieties and hybrids from different decades at experiment stations, showing approximately 50 kg/ha/yr (Table XXIV) increases in yields of the genotypes during 1930–1980 (A. Bejarano, personal communication, 1990). The national maize yields ranged from 34 to 39% of the experiment station yields, averaging about 35%, during the period 1957 through 1985.

Sanchez (1990) summarized results of about 200 trials conducted largely in the farmers fields in the lowlands of Peru and Bolivia during the 1970s and 1980s and compared performance of different types of cultivars (Table XXV). Unfortunately, it is not possible to determine the relative contribution of germplasm and improved management in the yield improvements reported in the study. The superiority of maize types during the 1980s

Table XXIV

Yields of Maize Varieties and Hybrids Developed in Different Periods in Venezuela

Years	Varieties (t/ha)	Hybrids (t/ha)
1930 (locals)	2.5	—
1940	3.0	—
1950	3.5	4.5
1960	4.0	5.0
1970	4.5	5.5
1980	5.5	6.0

Table XXV

Yields of Different Types of Maize Cultivars during the 1970s and 1980s in the Lowland Areas of Peru and Bolivia[a]

	Peru		Bolivia	
Type of germplasms	1970s	1980s	1970s	1980s
		(t/ha)		
Commercial hybrids	4.3	7.5	7.5	7.2
Improved varieties	4.6	8.2	—	6.1
CIMMYT varieties	4.7	7.2	8.0	7.8
Varietal crosses	5.0	8.2	5.0	5.0
Nonconventional hybrids[b]	—	11.2	—	—
Topcrosses	—	11.3	8.5	9.0
Conventional hybrids	—	12.0	—	—
Checks	3.0	6.5	4.0	5.2

[a] The results are based on 200 trials conducted in Peru and 150 trials in Bolivia, largely in farmers' fields.
[b] One or more parents are noninbreds.

ranged from 53 to 98% over that in the 1970s in Peru and only slight superiority was detected in Bolivia. The most dramatic improvement occurred in the performance of local checks, indicating a high adoption rate of improved germplasm, particularly in the coastal and jungle areas of Peru. The data demonstrate the differences in the efficiencies of different national programs in developing superior germplasm for their farmers, in addition to showing superiority of improved materials over farmers' cultivars.

An on-farm maize trial conducted at 10 locations in Ghana showed that improved germplasm plus recommended management practices yielded 4.20 t/ha compared to 1.45 t/ha yield of farmers' varieties under their own practices (Akposoe and Edmeades, 1981). In their study, improved germplasm contributed 21%, fertilizer use (90:60:60 kg $N:P_2O_5:K_2O$) 54%, and improved weed control 7% of the yield increases. In a similar study over a period of 10 years, Chutkaew and Thiraporn (1987) reported that improved germplasm contributed 37%, increased fertilizer use 43%, and improved weed control 20% of the yield increases in Thailand.

The current values accepted for maize yield increases in the tropics might be biased if we accept the idea that more maize is grown in marginal environments (environments where maize yields are reduced to 10–40% of their potential productivity) each year. More productive areas are used for higher return crops and activities (Edmeades and Tollenaar, 1990). An example of this phenomenon is found in Colombia. During 1972, 64,500 ha of maize were grown in Valle de Cauca, with an average yield of 2.5 t/ha (Londono and Andersen, 1975). During the same year, 68,500 ha were planted in Cordoba with an average yield of 1.14 t/ha. Due to expansion of the sugarcane industry and proliferation of social problems in Valle de Cauca, the maize area has gradually declined to 13,900 ha during 1990, with an average yield of 3.56 t/ha. Meanwhile, the maize area in Cordoba has expanded by 101%, to 137,800 ha, in 1990. Average yields in Cordoba, while increasing by 45%, were nevertheless only 1.65 t/ha in 1990 (F. Polania, 1991, personal communication). In Brazil, sugarcane for alcohol production and soybeans have displaced maize from better soils, as well (E. Paterniani, 1991, personal communication). This displacement of maize from higher yielding to lower yielding environments has been observed in several other tropical countries, lowering yield improvements rates. Fortunately, however, yield improvements have been positive in most developing countries.

Edmeades and Tollenaar (1990) estimated that about 70–80% of the yield increases reported in the developing countries are due to changes in crop management practices, principally through increased N use, higher plant densities, and better weed control. Maize yield increases in the developing countries have shown positive correlation ($r = 0.89^{**}$) with the use of improved germplasm and with N ($r = 0.76^{**}$), on a regional basis. The maize production increases through yield improvement in the tropics (45 kg/ha per year) amount to 3.7 million tons of additional maize annually. If only 30% of the yield improvement can be assumed to be contributed by superior germplasm, maize breeding/genetics would be contributing an additional 110-million plus dollars to the economies of the developing countries annually.

IX. CONTRIBUTION OF TROPICAL MAIZE GERMPLASM TO TEMPERATE MAIZE IMPROVEMENT

Although for many years, introgression of tropical maize germplasm into temperate maize has been suggested as a way to broaden the germplasm base, few successful commercial hybrids in the temperate areas have been reported to include tropical germplasm (Wellhausen, 1965; Goodman, 1985; Hallauer and Miranda, 1988; Castillo and Goodman, 1989). Castillo and Goodman (1989) concluded that factors limiting the use of tropical germplasm in temperate areas include (*a*) superiority of existing temperate germplasms and presence of adequate genetic variability in them; (*b*) poor adaptation of tropical materials in the temperate climate, due mainly to photoperiod sensitivity of tropical materials; (*c*) more extensive improvement of temperate materials through conventional breeding methods compared to use of tropical germplasm; (*d*) difficulty in overcoming unfavorable linkages in tropical materials; and (*e*) limited evaluations of exotic materials in temperate climates. In sorghum, tropical materials have been systematically converted to useful short-statured, photoperiod-insensitive forms, but no such effort has been made in maize.

Some successful use of tropical maize to improve temperate populations has been reported, but more efforts have probably been made that were not reported. Negative results are seldom published, and successful efforts by commercial hybrid companies are not likely to be revealed. Heterosis among temperate and exotic races has been studied (Wellhausen, 1965). Greater genetic variability has been observed in populations containing exotic germplasm. Selection has usually been practiced in populations with 25% or 50% tropical germplasm and with 75% or 50% temperate materials; however, Hallauer and Sears (1972) reported progress from direct selection in ETO Composite, and Gardner released to private and public breeders a population derived from CIMMYT's Compuesto L-Mc C2 (Antigua Gpo. 2 × Tuxpeno). Nelson (1972) reported on commercial use of inbreds developed using some exotic germplasm in the southern part of the United States. Oyervides *et al.* (1985) and Gutierrez *et al.* (1986) have reported lower photoperiod sensitivity in tropical × temperate crosses, compared with tropical materials. Castillo and Goodman (1989) suggested that evaluation of tropical materials under short-day conditions would be superior to that under long-day conditions. Hallauer and Miranda (1988) reported results of a survey of the north central region of the United States and found that 25.2% of the public sector populations undergoing improvement contained some exotic germplasm. They suggested that use of exotic germplasm in the research programs of the private sector was perhaps even greater. Several populations with different levels of several tropical materials have been developed and are being improved for adaptability (A. R. Hallauer, 1991, personal communication).

Integration of tropical maize germplasm into adapted maize populations at the University of Nebraska was initiated by Lonnquist in collaboration with tropical maize breeders in the late 1950s and early 1960s. Compton *et al.* (1979) used recurrent selection for adaptability to improve eight such populations and reported average gains of 5.4% per cycle. Gardner's well-adapted population, Compuesto L-Mc C2, abbreviated as TA, was used in a diallel cross involving several improved Nebraska populations and the best of the Iowa Stiff Stalk Synthetic and the Nebraska B Synthetic (Crossa *et al.*, 1987). TA was the highest yielding population per se and ranked second in GCA. It combined well with both Lancaster and Reid types and was considered to have good insect and disease tolerance. Extensive use has been made of CIMMYT germplasm in other populations developed by Gardner at Nebraska. They exhibit improved stalk quality, good standability, and insect and disease tolerance.

X. CONCLUSIONS

During the period 1985 to 2000, total annual maize demands are expected to rise by 3.5% in the developing countries—1.6% for food and 4.9% for feed. The production in developing countries is expected to rise by 3.6% per year. While this growth is lower than 4.0% annual increases during 1970–1985, it will still surpass the annual population growth rate of 2% during the same period. Despite impressive yield improvements and even some increases in maize area during the past decades, developing countries imported over 11 million tons of maize annually during the 1986–1988 and are expected to be by far the most important source of growth in demand for maize in the world (CIMMYT, 1989).

Most developing countries cannot afford to import maize; hence, getting improved germplasm and improved technologies into use on farms would play a major role in increasing maize production and reducing the need for maize imports. It is recognized that maize production practices in the tropics are generally poor. Improved production practices are bound to contribute more to increased yields initially, but as production practices improve, greater breeding effort is required to increase yields and adapt materials to improved practices. Yield potential of tropical maize should be increased by improving grain production efficiency, measured in time and space. Selection for desirable morphological traits, higher harvest index, reduced ASI, increased efficiency of nutrient uptake and utilization, improved stress tolerances, and superior quality traits should all be emphasized in the tropics. Growth in demand for food has encouraged agriculture to expand in marginal area, forcing more and more maize to be grown under stress. This trend is not expected to change soon. Breeding and crop management research should focus on reducing crop losses from weeds, diseases, insects, drought, and toxicities and deficiencies of mineral nutrients. More efficient technologies for germplasm conservation, regeneration, and evaluation must also be developed.

Increased technical capacity of maize scientists in the developing world would enable them to make more efficient use of the available limited resources, by allowing them to choose their germplasm and breeding and testing procedures more effectively. However, little advanced technology development is expected to take place in the developing countries, due mainly to declining budgets for agricultural research. Developing countries would generally be importers and modifiers of advanced technology (computing, biotechnology, etc.). Although current breeding methodologies will continue to play the major role in maize improvement in the foreseeable future, biotechnology is expected to aid in genetic improvement of stress tolerance in maize.

Lack of good seed production, processing, and storage and distribution facilities in many developing countries has greatly impeded progress in getting improved OPVs and hybrids into the hands of farmers. Thus the impact of superior germplasm developed by maize scientist has been greatly reduced. In developing countries in 1988, the area planted to improved OPVs was 14%, to hybrids 39%, and to local varieties 47% (CIMMYT, 1990). A commercial seed industry will not develop in countries where a profit is not ensured. Hence, initially, the public sector must assume a greater role in seed production, storage, and distribution, while at the same time encouraging development of a seed industry. Developing more realistic national policies, pricing the seed of improved cultivars at a level fair to both the farmer and the seed industry, and making superior germplasm available to the private sector should increase goodwill and pave the way for greater private sector involvement in the seed industry. In countries where the private sector has become involved, greater progress has been made in increasing maize production.

Acknowledgments

We thank Dr. R. P. Cantrell for suggesting this manuscript. We also thank all maize scientists from developing and developed countries and from CIMMYT who have contributed greatly to the information presented in this article. The authors express sincere appreciation to M. Bjarnason, R. Cantrell, H. Cordova, C. De Leon, J. Deutsch, G. Edmeades, M. Goodman, G. Granados, A. Hallauer, R. Magnavaca, J. Miles, E. Paterniani, S. Vasal, W. Villena, R. Wedderburn, and D. Winkelmann for careful and constructive review of the manuscript.

Dedication

This chapter is dedicated to the many maize scientists in national programs who have contributed greatly to increased maize yields in the tropics, and particularly to those who have collaborated with CIMMYT maize scientists in the international testing program, which has been mutually beneficial. It is also dedicated to CIMMYT maize scientists for their many contributions to tropical maize improvement and for their dedication to strengthening national programs and training leaders and scientists, thus improving the lot of poor farmers and consumers in developing countries.

References

Akposoe, M. K., and Edmeades, G. O. (1981). "Second Annual Report 1980: Part 2. Research Results." Unpublished report to the Canadian International Development Agency, Ottawa.

Ampofo, J. K. O., and Saxena, K. N. (1989). *In* "Toward Insect Resistant Maize for the Third World" (CIMMYT, ed.), pp. 170–177. CIMMYT, El Batan, Mexico.

Ampofo, J. K. O., Saxena, K. N., Kibuka, J. G., and Nyangiri, E. O. (1986). *Maydica* **31,** 379–389.

Anderson, E. (1945). *Chron. Bot.* **9,** 88–92.

Anderson, E., and Cutler, H. C. (1942). *Ann. Missouri Bot. Gard.* **29,** 69–89.

Arboleda-Rivera, F., and Compton, W. A. (1974). *Theor. Appl. Genet.* **44,** 77–81.

Azofeira, J. G., and Jimenez, K. M. (1988). "Evaluacion de hibridos dobles y triples de maiz en ocho localidades de Costa Rica." Paper presented at the XXXV Reunion PCCMCA, 2–7 April 1988, San Pedro Sula, Honduras.

Bajet, N.B., and Renfro, B. L. (1989). *Plant Disease* **73,** 926–930.

Barrow, M. R. (1989). *In* "Toward Insect Resistant Maize for the Third World" (CIMMYT, ed.), pp. 184–191. CIMMYT, El Batan, Mexico.

Bascope, B. (1977). "Agente causal de la llamada "raza Mesa Central" del achaparramiento del maiz." M. S. thesis, Esc. Nal. Agric., Chapingo, Mexico.

Beadle, G. W. (1939). *J. Hered.* **30,** 245–247.

Beck, D. L., Vasal, S. K., and Crossa, J. (1990). *Maydica* **35,** 279–285.

Bjarnason, M. (1986). *In* "To Feed Ourselves" (B. Gelaw, ed.), pp. 197–207. Proceedings, First Eastern, Central, and Southern Africa Regional Maize Workshop, Lusaka, Zambia, 1–17 March 1985. CIMMYT, El Batan, Mexico.

Bjarnason, M. (1990). "El programa de maiz con alta calidad de proteina del CIMMYT: Estado presente y estrategias futuras." Paper presented at the XVIII Congreso Nacional de Maiz y Sorgo of Brazil, July 29–August 3, Vitoria-ES, Brazil.

Bolanos, J., and Edmeades, G. O. (1987). *Agron. Abstr.*, 88.

Bolanos, J., Edmeades, G. O., and Martinez, L. (1990). "Mejoramiento para tolerancia a sequia en maiz tropical: La experiencia del CIMMYT." Paper presented at the XVIII Congreso Nacional de Maiz y Sorgo of Brazil, July 29–August 3, Vitoria-ES, Brazil.

Borges, O. L. F. (1987). *Crop Sci.* **27,** 178–180.

Bosquez-Perez, N. A., Mareck, J. H., Dabrowski, Z. T., Everett, L., Kim, S. K., and Efron, Y. (1989). *In* "Toward Insect Resistant Maize for the Third World" (CIMMYT, ed.), pp. 163–169. CIMMYT, El Batan, Mexico.

Brewbaker, J. L. (1974). *In* "Proceedings of the 29th Annual Corn and Sorghum Res. Conf," Vol 29, pp. 118–130.

Brewbaker, J. L. (1984). *In* "A Multidisciplinary Approach to Agrotechnology Transfer" (G. Uehara, ed.), Hawaii Inst. Trop. Human Resources Res. Ext. Ser. 26, March 1984.

Brewbaker, J. L. (1985). *In* "Breeding Strategies for Maize Production Improvement in the Tropics" (A. Brandolini and F. Salamini, eds.), pp. 47–77. FAO and Instituto Agronomico per L'Oltremare, Firenze.

Brewbaker, J. L., and Kim, S.K. (1979). *Crop. Sci.* **19,** 32–36.

Brim, C. A. (1966). *Crop Sci.* **6,** 220.

Brunini, O., Miranda, L. T., and Sawazaki, F. (1985). "Les Besoins en Eau des Cultures." INRS, Paris.

Byrne, P. F., Bolanos, J., Eaton, D. L., and Edmeades, G. O. (1990). *Agron. Abstr.* **82,** 82.

Castellanos, J. S., Cordova, H. S., Queme, J. L., Larios, L., and Perez, C. (1987). *In* "Proceedings of the XXXIII Reunion Annual del PCCMCA, 30 March–4 April 1987," Guatemala City, Guatemala.

Castillo, G. F., and Goodman, M. M. (1989). *Crop Sci.* **29,** 853–861.
Castro, M., Gardner, C. O., and Lonnquist, J. H. (1968). *Crop Sci.* **8,** 97–101.
Ceballos, H., and Deutsch, J. A. (1989). "Estudio de la herencia de la resistancis a *Phyllachora maydis* en maiz." Paper presented at the X Congreso ASCOLFI-V Reunion ALF-XXIV Reunion APS-Caribbean Division, 10–14 July 1989. Cali, Colombia.
Ceballos, H., Deutsch, J. A., and Gutierrez H. (1991). *Crop Sci.* **31,** 964–971.
Chang, Shin-Chi (1972). *In* "Proceedings of the Asian Corn Improvement Workshop," Vol. 8, pp. 114–115.
Chatterji, S. M., Bhamburkar, M. W., Marwaha, K. K., Panwar, V. P. S., Siddiqui, K. H., and Young, W. R. (1971). *Indian J. Ent.* **33,** 209–213.
Chong, C., and Brawn, R. I. (1969). *Can. J. Plant Sci.* **49,** 513.
Chutkaew, C., and Thiraporn, R. (1987). *In* "Proceedings of Asian Regional Maize Workshop, April 27–May 3, 1986," pp. 104–114. Jakarta and East Java.
Compton, W. A., and Comstock, R. E. (1976). *Crop Sci.* **16,** 122.
Compton, W. A., and Lonnquist, J. H. (1982) *Crop Sci.* **22,** 981–983.
Compton, W. A., Mumm, R. F., and Mathema, B. (1979). *Crop Sci.* **19,** 531–533.
Comstock, R. E. (1979). *Crop Sci.* **19,** 881–886.
Comstock, R. E., and Moll, R. H. (1963). *In* "Statistical Genetics and Plant Breeding" (W. D. Hanson and H. F. Robinson, eds.), pp. 164–196. Pub. 982. NAS–NRC, U.S.A.
Comstock, R. E., Robinson, H. F., and Harvey, P. H. (1949). *Agron. J.* **41,** 360–367.
Cordova, H. S. (1984). "Formacion de hibridos dobles y triples de maiz en base a familias de hermanos completos y sus implicaciones en la produccion de semilla comercial." Paper presented at the XV National Maize and Sorghum Meetings, Brazil, 2–6 July, 1984, Maceio, Brazil.
Cordova, H. S. (1990). "Estimacion de parametros de estabilidad para determinar la respuesta de hibridos de maiz (*Zea mays* L.) a ambientes contrastantes de Centro America." Paper presented at the XXXVI Reunion de PCCMCA, 26–30 March, 1990, San Salvador, El Salvador.
Cordova, H. S., Poey, F. R., Soto, J .G., Castellanos, J. S., and Dardon, M. A. (1979). "Heterosis del rendimiento y la altura de la planta en variedades mejoradas de maiz." Paper presented at the XXV Reunion PCCMCA, 19–23 March, 1979, Tegucigalpa, Honduras.
Cordova, H., Urbina, R., Aguiluz, A., and Celado, R. (1989). "CIMMYT 1989 Annual Report." El Batan, Mexico.
Cordova, H. S., Velasques, R., Poey, F. R., and Perez, C. (1980). "Efecto de diferentes fuentes de planta baja sobre el rendimiento y altura de planta de maiz (*Zea mays.* L.)." Paper presented at the XXVI Reunion PCCMCA, 24–28 March 1980, Guatemala City, Guatemala.
Cramer, H. H. (1967). *In* "Defensa vegetal y cosecha mundial." Bayerischer Pflanzenschutz, Leverkusen, Germany.
Crossa, J., Gardner, C. O., and Mumm, R. F. (1987). *Theor. Appl. Genet.* **13,** 445–450.
Crossa, J., Taba, S., and Wellhausen, E. J. (1990a). *Crop Sci.* **30,** 1182–1190.
Crossa, J., Vasal, S. K., and Beck, D. L. (1990b). *Maydica* **35,** 273–278.
Dabrowsky, Z. T. (1989). *In* "Toward Insect Resistant Maize for the Third World" (CIMMYT, ed.), pp. 84–93. CIMMYT, El Batan, Mexico.
Darrah, L. L. (1986). *In* "To Feed Ourselves" (B. Gelaw, ed.), pp. 160–178. Proceedings, First Eastern, Central, and Southern Africa Regional Maize Workshop, Lusaka, Zambia, 10–17 March 1985. CIMMYT, El Batan, Mexico.
Darrah, L. L., Eberhart, S. A., and Penny, L. H. (1978). *Euphytica* **27,** 191–204.
De Leon, C. (1982). *In* "Proceedings of the X Reunion de Especialistas en Maiz de la Zona Andina, 28 March–2 April, 1982," pp. 187–198, Santa Cruz, Bolivia.

De Leon, C., and Granados, R. (1991). "Simultaneous Improvement for Agronomic Characters and Resistance to Downy Mildew (*Pernosclerospora* spp) in Four Tropical Maize Populations." Paper presented at the XII International Plant Protection Congress of Brazil, 11–16 August, Rio de Janiero.

De Leon, C., and Pandey, S. (1989). *Crop Sci.* **29,** 12–17.

De Leon, D. G., Hoisington, D. A., Jewell, D., Deutsch, J. A., and Mihm, J. A. (1990). *Agron. Abstr.*, 196.

Delgado, H. L. V., and Marquez, F. S. (1984). *Agrociencia* **58,** 29–43.

Diener, U. L., and Davis, N. D. (1987). *In* "Aflatoxin in Maize" (M. S. Zuber, E. B. Lillehoj, and B. L. Renfro, eds.), pp 33–40. CIMMYT, Mexico.

Diwakar, M. C., and Payak, M. M. (1975). *Indian Phytopathol.* **28,** 548–549.

Eberhart, S. A. (1970). *African Soils* **15,** 669–680.

Eberhart, S. A., and Harrison, M. N. (1973). *East African Agric. For. Res. J.* **39,** 12–16.

Eberhart, S. A., Harrison, M. N., and Ogada, F. (1967). *Zuchter* **37,** 169–174.

Eberhart, S. A., and Russell, W. A. (1966). *Crop Sci.* **6,** 36–40.

Edmeades, G. O., Bolanos, J., Laffitte, H. R., Rajaram, S., Pfeiffer, W., and Fischer, R. A. (1989). *In* "Drought Resistance in Cereals—Theory and Practice" (F. W. G. Baker, ed.), pp. 27–52. ICSU Press, Paris.

Edmeades, G. O., and Lafitte, H. R. (1987). "CIMMYT Research Highlights 1986." CIMMYT, El Batan, Mexico.

Edmeades, G. O., and Tollenaar, M. (1990). *In* "Proceedings, International Congress of Plant Physiology" (S. K. Sinha, P. V. Sane, S. C. Bhargava, and P. K. Agrawal, eds.), February 15–20, 1988, pp. 164–180. New Delhi.

Empig, L. T., Gardner, C. O., and Compton, W. A. (1981). "Nebraska Agricultural Experimental Station MP26" (revised).

Faluyi, J. O., and Olorode, O. (1984). *Theor. Appl. Genet.* **67,** 341–344.

Fasoulas, A. (1973). "Pub. No. 7, Department of Genetics and Plant Breeding," Aristotelian University of Thessaloniki, Thessaloniki, Greece.

Fasoulas, A. (1977). "Pub. No. 8, Department of Genetics and Plant Breeding," Aristotelian University of Thessaloniki, Thessaloniki, Greece.

Fasoulas, A., and Tsaftaris, A. (1975). "Pub. No. 5, Department of Genetics and Plant Breeding," Aristotelian University of Thessaloniki, Thessaloniki, Greece.

Fischer, K. S., Edmeades, G. O., and Johnson, E. C. (1987). *Crop Sci.* **27,** 1150–1156.

Fischer, K. S., Edmeades, G. O., and Johnson, E. C. (1989). *Field Crops Res.* **22,** 227–243.

Fischer, K. S., Johnson, E. C., and Edmeades, G. O. (1983). Breeding and Selection for Drought Resistance in Tropical Maize." CIMMYT, El Batan, Mexico.

Frederiksen, R. A., and Renfro, B. L. (1977). *Annu. Rev. Phytopathol.* **15,** 249–275.

Gardner, C. O. (1961). *Crop Sci.* **1,** 241–245.

Gardner, C. O. (1978). *In* "Maize Breeding and Genetics" (D. B. Walden, ed.). Wiley, New York.

Gardner, C. O. (1986). *In* "Proceedings, XII Reunion de Maiceros de la Zona Andina, 29 September–3 October, 1986," pp. 8–18. CIMMYT, El Batan, Mexico.

Gardner, C. O., and Paterniani, E. (1967). *Ciencia v Cultura* **19,** 95–101.

Gardner, C. O., Thomas-Compton, M. A., Compton, W. A., and Johnson, B. E. (1990). "Final Report to USAID, UNL/IANR/ARD Project NEB 12–159," University of Nebraska, Lincoln.

Genter, C. F. (1976). *Crop Sci.* **16,** 556–558.

Gevers, H. O. (1974). "Reciprocal Recurrent Selection in Maize under Two Systems of Parent Selection." Paper presented at the V Genetics Congress of Republic of South Africa.

Gomez, L. G. (1990). *In* "Proceedings, XIV Reunion de Maiceros de la Zona Andina, 17–21 September 1990," Maracay, Venezuela.
Goodman, M. M. (1985). *Iowa State J. Res.* **59,** 497–527.
Goodman, M. M., (1988). *CRC Crit. Rev. Plant Sci.* **7,** 197–220.
Goodman, M. M., and Brown, W. L. (1988). *In* "Corn and Corn Improvement" (G. F. Sprague and J. W. Dudley, eds.), pp. 33–79. Agronomy monograph, ASA, Madison, Wisconsin.
Greenblat, I. M. (1968). *Maize Genet. Coop. Newslett.* **42,** 144.
Gutierrez, G. M. A., Cortez, H., Wathika, E. N., Gardner, C. O., Oyervides, G. M., Hallauer, A. R., and Darrah, L. L. (1986). *Crop Sci.* **26,** 99–104.
Hakim, R., and Dhalan, M. (1973). *In* "Proceedings, IX Inter-Asian Corn Imp. Workshop," Vol. 9, pp. 54–58.
Hallauer, A. R. (1985). *CRC Crit. Rev. Plant Sci.* **3,** 1–33.
Hallauer, A. R., and Eberhart, S. A. (1970). *Crop Sci.* **10,** 315–316.
Hallauer, A. R., and Miranda, J. B. (1988). "Quantitative Genetics in Maize Breeding," 2nd ed. Iowa State Univ. Press, Ames, Iowa.
Hallauer, A. R., Russell, W. A., and Lamkey, K. R. (1988). *In* "Corn and Corn Improvement" (G. F. Sprague and J. W. Dudley, eds.), pp. 463–565. Agronomy Monograph 18, ASA, Madison, Wisconsin.
Hallauer, A. R., and Sears, J. H. (1972). *Crop Sci.* **12,** 203–206.
Hamilton, P. B. (1987). *In* "Aflatoxin in Maize" (M. S. Zuber, E. B. Lillehoz, and B. L. Renfro, eds.), pp. 51–57. CIMMYT, El Batan, Mexico.
Han, G. C., Vasal, S. K., Beck, D. L., and Elias, E. (1991). *Maydica* **36,** 57–64.
Handoo, M. I., Renfro, B. L., and Payak, M. M. (1970). *Indian Phytopathol.* **23,** 231–249.
Hardacre, A. K., and Eagles, H. A. (1980). *Crop Sci.* **20,** 780–784.
Hardacre, A. K., Eagles, H. A., and Gardner, C. O. (1990). *Maydica* **35,** 215–219.
Hooker, A. L. (1978). *In* "Proceedings, III International Congress of Plant Pathology," Munchen, Vol. 16, p. 23.
Hooker, A. L., and Le Roux, P. M. (1967). *Phytopathology* **47,** 187–191.
Hopkins, C. G. (1899). *Illinois Agric. Exp. Sta. Bull.* **55,** 205–240.
Horner, E. S., Lundy, H. W., Lutrick, M.C., and Chapman, W. H. (1973). *Crop Sci.* **13,** 485–489.
Horner, E. S., Lundy, H. W., Lutrick, M. C., and Wallace, R. W. (1963). *Crop Sci.* **3,** 63–66.
Hull, F. H. (1945). *Agron. J.* **37,** 134–145.
Hull, F. H. (1952). *In* "Heterosis" (J. W. Gowen, ed.). Iowa State Univ. Press, Ames, Iowa.
International Maize and Wheat Improvement Center (CIMMYT) (1974). "Symp. Proc. Worldwide Maize Improvement in the 70's and the Role for CIMMYT". El Batan, Mexico.
International Maize and Wheat Improvement Center (CIMMYT) (1979). "CIMMYT Report on Maize Improvement 1976–77. "CIMMYT, El Batan, Mexico.
International Maize and Wheat Improvement Center (CIMMYT) (1981). "CIMMYT Report on Maize Improvement 1978–79." CIMMYT, El Batan, Mexico. .
International Maize and Wheat Improvement Center (CIMMYT) (1982). "CIMMYT's Maize Program: An Ovewview." CIMMYT, El Batan, Mexico.
International Maize and Wheat Improvement Center (CIMMYT) (1984). "CIMMYT Maize Facts and Trends Report 2: An Analysis of Changes in the World Food and Feed Uses of Maize." CIMMYT, El Batan, Mexico.
International Maize and Wheat Improvement Center (CIMMYT) (1987). "1986 CIMMYT World Maize Facts and Trends: The Economics of Commerical Maize Seed Production in Developing Countries." El Batan, Mexico.

International Maize and Wheat Improvement Center (CIMMYT) (1988). "CIMMYT Resena de la Investigacion." El Batan, Mexico.
International Maize and Wheat Improvement Center (CIMMYT) (1989). "Towards the 21st Century: CIMMYT's Strategy." Mexico.
International Maize and Wheat Improvement Center (CIMMYT) (1990). "1989/90 CIMMYT World Maize Facts and Trends: Realizing the Potential of Maize in Sub-Saharan Africa." Mexico.
Ivbijaro, M. F. (1981). *J. Agric. Sci. Camb.* **96,** 479–481.
Jenkins, M. T. (1940). *Agron. J.* **32,** 55–63.
Jinahyon, S. (1973). *In* "Proceedings, IX Inter-Asian Corn Impr. Workshop," Vol. 9, pp. 30–39.
Jinahyon, S., and Moore, C. L. (1973). *Agron. Abstr.*, 7.
Johnson, E. C. (1963). *Am. Soc. Agron. Abstr.*, 82.
Johnson, E. C., Fischer, K. S., Edmeades, G. O., and Palmer, A. F. E. (1986). *Crop Sci.* **26,** 253–260.
Jones, L. P., Compton, W. A., and Gardner, C. O. (1971). *Theor. Appl. Genet.* **41,** 36–39.
Kaneko, K., and Aday, B. A. (1980). *Crop Sci.* **20,** 590–594.
Kasim, F. 1989. "Field Studies for Evaluating Performance of Corn Population in Acid Soils." Ph.D. thesis, Kansas State University, Manhattan, Kansas.
Khan, A. Q., and Paliwal, R. L. (1980). *Indian J. Genet. Plant Breeding* **40,** 427–431.
Khera, A. S., Dhillon, B. S., Malhotra, V. V., Brar, H. S., Saxena, V. K., Kapoor, W. R., Sharma, R. K., Pal, S. S., and Malhi, N. S. (1988). *In* "Proceedings of the III Asian Regional Maize Workshop, 8–15 June 1988," pp. 58–71. Kunming and Nanning, Peoples' Republic of China.
Khera, A. S., Dhillon, B. S., Saxena, V. K., and Malhi, N. S. (1981). *Crop. Impr.* **8,** 60–62.
Kim, S. K., Efron, Y., Fajemisin, J., and Buddenhagen, I. W. (1982). *Am. Soc. Agron. Abstr.*, 72.
Kim, S. K., Efron, Y., Fajemisin, J. M., and Buddenhagen, I. W. (1989). *Crop Sci.* **29,** 890–894.
Kim, S. K., Efron, Y., Khadr, F., Fajemisin, J., and Lee, M. H. (1987). *Crop Sci.* **27,** 824–825.
Kim, S. K., Lee, M. H., Khadr, F., and Efron, Y. (1990). "Combining Ability Estimates for Tropical Maize Germplasm in West Africa," in press.
Lima, M., and Paterniani, E. (1977). "*Rel. Cient. Dept. Genet. ESALQ, Piracicaba,*" **11,** 82–83.
Lima, M., Paterniani, E., and Miranda, J. B. (1974). *Rel. Cient. Inst. Genet.* (ESALQ-USP) **8,** 78–85.
Londono, N. R., and Andersen, P. P. (1975). "Descripcion de factores asociados con bajos rendimientos de maiz en fincas pequenas de tres departamentos de Colombia." Serie ES-No. 18, CIAT, Cali, Colombia.
Lonnquist, J. H. (1964). *Crop Sci.* **4,** 227–228.
Lonnquist, J. H., and Williams, N. E. (1967). *Crop Sci.* **7,** 369–370.
Machado, A. T., and Paterniani, E. (1988). "Avaliacao de germoplasma de milho (*Zea mays* L.) com relacao de eficiencia e/ou fixacao biologica do nitrogenio II." Paper presented at the XVII Congreso Nacional de Milho e Sorgo of Brazil, Piracicaba, Brazil.
Machado, A. T., Purcino, A. A. C., Magalhaes, P. C., Magnavaca, R., and Magalhaes, J. R. (1990). *In* "Proceedings, XIV Reunion de Maiceros de la Zona Andina and I Reunion Suramericana de Maiceros," 17–21 September 1990, Maracay, Venezuela.
Magnavaca, R. (1983). *Diss. Abstr.* **43,** 2073B.
Magnavaca, R., Gardner, C. O., and Clark, R. B. (1987a). *In* "Genetic Aspects of Plant

Mineral Nutrition" (W. H. Gabelman and B. C. Loughman, eds.), pp. 201–212. Martinus Nijhoff, The Hague, The Netherlands.
Magnavaca, R., Gardner, C. O., and Clark, R. B. (1987b). *In* "Genetic Aspects of Plant Mineral Nutrition" (W. H. Gabelman and B. C. Loughman, eds.), pp. 189–199. Martinus Nijhoff, The Hague, The Netherlands.
Magnavaca, R., Gardner, C. O., and Clark, R. B. (1987c). *In* "Genetic Aspects of Plant Mineral Nutrition" (W. H. Gabelman and B. C. Loughman, eds.), pp. 255–265. Martinus Nijhoff, The Hague, The Netherlands.
Mangelsdorf, P. C., and Reeves, R. G. (1939). *Texas Agric. Exp. Sta. Bull.*, 574.
Marquez-Sanchez, F. (1982). *Crop Sci.* **22,** 314–318.
Mashingaidze, K. (1984). *Zimbabwe Agric. J.* **81,** 147–152.
McMillian, W. W. (1987). *In* "Aflatoxin in Maize" (M. S. Zuber, E. B. Lillehoj, and B. L. Renfro, eds.), pp. 250–253. CIMMYT, El Batan, Mexico.
Miederna, P. (1979). *Euphytica* **28,** 661–664.
Mihm, J. A. (1989). *In* "Toward Insect Resistant Maize for the Third World" (CIMMYT, ed.), pp. 109–121. CIMMYT, El Batan, Mexico.
Miranda, J. B. (1985). *In* "Breeding Strategies for Maize Production Improvement in the Tropics" (A. Brandolini and F. Salamini, eds.), pp. 177–206. FAO and Instituto Agronomico per L'Oltremare, Firenze.
Miranda, L. T. (1982). *Maize Genet. Coop. Newslett.* **56,** 28.
Miranda, L. T., Miranda, L. E. C., Pommer, C. V., and Sawazaki, E. (1977). *Bragantia* **36,** 187–196.
Molina G. (1980). *Agrociencia* **42,** 67–76.
Moll, R. H., and Robinson, H. F. (1966). *Crop Sci.* **6,** 319–324.
Moll, R. H., and Stuber, C. W. (1971). *Crop Sci.* **11,** 706–711.
Montgomery, E. G. (1909). *Univ. Nebraska Agric. Exp. Sta. Bull.*, 112.
Munoz, O. A., Stevenson, K. R., Ortiz, J. C., Thurtell, G. W., and Carballo, A. C. (1983). *Agrociencia* **51,** 115–153.
Mwenda, E. W. (1986). *In* "To Feed Ourselves" (B. Gelaw, ed.), pp. 32–42. Proceedings, First Eastern, Central and Southern Africa Regional Maize Workshop. CIMMYT, Mexico.
Narro, L. N. (1988). "Evaluation of Half-Sib Family Selection in a Maize Population." M.S. thesis, Iowa State University, Ames, Iowa.
Naspolini, V., Bahia, A. F. C., Viana, R. T., Gama, E. E. G., Vasconcellos, C. A., and Magnavaca, R. (1981). *Ciencia y Cultura* **33,** 722.
National Research Council (1988). "Quality-Protein Maize." National Academy Press, Washington, D.C.
Nault, L. R. (1980). *Phytopathology* **70,** 656–662.
Nelson, H. G. (1972). *Annu. Corn Sorghum Ind. Res. Conf. Proc.* **27,** 115–118.
Ochieng, J. A. W. (1986). *In* "To Feed Ourselves" (B. Gelaw, ed.), pp. 26–31. Proceedings, First Eastern, Central and Southern Africa Regional Maize Workshop. CIMMYT, Mexico.
Ochieng, J. A. W., and Kamidi, R. E. (1992). In press.
Odhiambo, M. O., and Compton, W. A. (1989). *Crop Sci.* **29,** 314–319.
Oliver, R. C. (1986). *In* "To Feed Ourselves" (B. Gelaw, ed.), pp. 138–142. Proceedings, First Eastern, Central and Southern Africa Regional Maize Workshop. CIMMYT, Mexico.
Omolo, E. O. (1983). *Insect Sci. Appl.* **4,** 105–108.
Overdal, A. C., and Ackerson, K. T. (1972). "Agricultural Soil Resources of the World." U.S.D.A. Soil Conservation Service, Washington, D. C.
Oyervides, G. M., Hallauer, A. R., and Cortez, H. M. (1985). *Crop Sci.* **25,** 115–120.

Paliwal, R. L., and Sprague, E. W. (1981). "Improving Adaptation and Yield Dependability in Maize in the Developing World." CIMMYT, El Batan, Mexico.
Pandey, S., Diallo, A. O., Islam, T. M. T., and Deutsch, J. (1986). *Crop Sci.* **26,** 879–884.
Pandey, S., Diallo, A. O., Islam, T. M. T., and Deutsch, J. (1987). *Crop Sci.* **27,** 617–622.
Pandey, S., Vasal, S. K., De Leon, C., Ortega, A. C., Granados, G., and Villegas, E. (1984). *Genetika* (Yugoslavia) **16,** 23–42.
Pandey, S., Vasal, S. K., and Deutsch, J. A. (1991). *Crop Sci.* **31,** 285–290.
Pant, J. C., Kapoor, S., and Pant, N. C. (1964). *Indian J. Entomol.* **26,** 434–437.
Paterniani, E. (1964). *Fitotec. Latinoam.* **1,** 15.
Paterniani, E. (1967). *Crop Sci.* **7,** 212–216.
Paterniani, E. (1968). "Catedra Escola Superior de Agricultura Luis de Queiroz/USP, Piracicaba."
Paterniani, E. (1970). *Rel. Cient. Inst. Genet.* ESALQ/USP **4,** 83.
Paterniani, E. (1978). *Maydica* **23,** 29–34.
Paterniani, E. (1981). *Maydica* **26,** 85–91.
Paterniani, E. (1985). *In* "Breeding Strategies for Maize Production Improvement in the Tropics" (A. Brandolini and F. Salamini, eds.), pp. 329–339. FAO and Istituto Agronomico per L'Oltremare, Firenze.
Paterniani, E. (1990). *CRC Crit. Rev. Plant Sci.* **9,** 125–154.
Paterniani, E., and Goodman, M. M. (1977). "Races of Maize in Brazil and Adjacent Areas." CIMMYT, El Batan, Mexico.
Paterniani, E., and Lonnquist, J. H. (1963). *Crop Sci.* **3,** 504–507.
Paterniani, E., and Vencovsky, R. (1977). *Maydica* **22,** 141–152.
Paterniani, E., and Vencovsky, R. (1978). *Maydica* **23,** 209–219.
Payak, M. M., and Renfro, B. L. (1966). *Indian Phytopathol.* **19,** 121–132.
Payak, M. M., and Sharma, R. C. (1985). *Trop. Pest Management* **31,** 302–310.
Perez, C., Alvarado, A., Soto, N., Aguiluz, A., Brizuela,., Celado, R., and Cordova, H. (1988). "Efectos de aptitud combinatoria general e identificacion de hibridos trilineales de maiz de grano amarillo. Centro America, Panama, y el Caribe." Paper presented at the XXXV Reunion PCCMCA, 2–7 April 1988, San Pedro Sula, Honduras.
Pier, A. C. (1987). *In* "Aflatoxin in Maize" (M. S. Zuber, E. B. Lillehoj, and B. L. Renfro, eds.), pp. 58–65. Proceedings, Workshop, CIMMYT, Mexico, D. F.
Poey, F. R., Grajeda, J. E., Fernandez, O. J., and Soto, F. (1977). *Agron. Abstr.*, 44.
Renfro, B. L. (1985). *In* "Breeding Strategies of Maize Production Improvement in the Tropics" (A. Brandolini and F. Salamini, eds.), pp. 341–365. FAO and Instituto Agronomico per L'Oltremare, Firenze.
Renfro, B. L., and Ullstrup, A. J. (1976). *PNAS* **22,** 491–498.
Russell, W. A., and Eberhart, S. A. (1975). *Crop Sci.* **15,** 1–4.
Russell, W. A., Eberthart, S. A., and Vega, U. A. (1973). *Crop Sci.* **13,** 257–261.
Safeeulla, K. M. (1975). *Japan. Agric. Res. O. Rept.* **8,** 93–102.
Sanchez, H. C. (1990). *In* "Proceedings, XIV Reunion de Maiceros de la Zona Andina y I Reunion Suramericana de Maiceros," 17–21 September 1990, Maracay, Venezuela.
Sanchez, P. A. (1977). *In* "Int. Seminar on Soil, Environment, and Fertility Management in Intensive Agriculture," pp. 535–566. Tokyo.
Santos, M. X., Geraldi, I. O., and de Souza, C. L. (1988). *Pesq. Agropec. Bras.* **23,** 519–523.
Santos, M. X., and Naspolini-Filho, V. (1986a). *Braz. J. Genet.* **9,** 307–319.
Santos, M. X., and Naspolini-Filho, V. (1986b). *Pesq. Agropec. Bras.* **21,** 739–746.
Sarup, P. (1980). *In* "Breeding Production, and Protection Methodologies of Maize in India" (J. Singh, ed.), pp. 193–197. All India Coordinated Maize Improvement Project, IARI, New Delhi.
Sawazaki, E. (1979). "Tres ciclos de selecao entre e dentro de familias meios irmaos para

producao de graos no milho IAC-Maya." Mestrado Escol. Superior de Agricultura Luiz de Queiroz/USP, Piracicaba.
Segovia, R. T. (1976). "Sies ciclos de selecao entre e dentro de familias de meios irmaos no milho (*Zea mays* L.) Centralmex." Tese de Doutoramento, ESALQ-USP, Piracicaba, Brazil.
Seth, M. L., and Singh, S. (1975). *Indian Phytopathol.* **28,** 144–145.
Sevilla, R. (1975). *Inf. Maiz* **5,** 11.
Short, K., Wedderburn, R., and Pham, H. (1990). "CIMMYT's maize research program in Africa." Paper presented at the IX South African Maize Breeding Symposium, 20–22 March 1990, Pietermaritzburg.
Singburaudom, N., and Renfro, B. L. (1982). *Crop Protection* **1,** 323–332.
Singh, J. (1986). *In* "Proceedings, II Asian Regional Maize Workshop, 27 April–3 May 1986," pp. 48–69. Jakarta and East Java, Indonesia.
Singh, M., Khera, A. S., and Dhillon, B. S. (1986). *Crop Sci.* **26,** 275–278.
Singh, P. (1979). "Inheritance Studies on Stalk Rot Caused by *Erwinia chrysanthemi* Pathotype *zeae* in Maize (*Zea mays* L.)." Ph. D. thesis submitted to G. B. Pant Univ. of Agric. and Technol., Pantnagar, India.
Smith, M. E., Mihm, J. A., and Jewell, D. C. (1989). *In* "Toward Insect Resistant Maize for the Third World" (CIMMYT, ed.), pp. 222–234. CIMMYT, El Batan, Mexico.
Sosa, G., Eaton, D., and Byrne, P. (1990). *Agron. Abstr.*, 111.
Sprague, E. W. (1974). *In* "Proceedings, Worldwide Maize Improvement in the 70's and the Role for CIMMYT, 22–26 April 1974," pp. 2.1–2.22. CIMMYT, El Batan, Mexico.
Sprague, G. F., and Eberhart, S. A. (1977). *In* "Corn and Corn Improvement" (G. F. Sprague, ed.), pp. 305–363. ASA, Madison, Wisconsin.
Storey, H. H., and Howland, A. K. (1967). *Ann. Appl. Biol.* **59,** 429.
Straub, R. W., and Fairchild, M. L. (1970). *J. Econ. Entomol.* **63,** 1901–1903.
Thome, C. R., Smith, M. E., and Mihm, J. A. (1990). *Agron. Abstr.*, 113.
Torregroza, M. (1973). *Agron. Abstr.*, 16.
Torregroza, M., Arias, E., Diaz, C., and Arboleda, F. (1972). *Agron. Abstr.*, 20.
Torregroza, M., Diaz, C., Arias, E., Rivera, J. A., and Ramirez, C. (1976). *Inf. Maiz* **16,** 12–13.
Tragesser, S. L. (1991). "Generation Means Estimation of Unbiased Genetic Effects from Five Cycles of Replicated S_1 and Reciprocal Full-Sib Recurrent Selection. Ph.D. thesis, University of Nebraska, Lincoln.
Troyer, A. F., and Brown, W. L. (1972). *Crop Sci.* **12,** 301–304.
Troyer, A. F., and Brown, W. L. (1976). *Crop Sci.* **16,** 767–772.
Usua, E. J. (1986). *J. Econ. Entomol.* **61,** 375–376.
Vargas-Sanchez, J. E. (1990). "Response to Selection by Yield and Prolificacy at Different Environments in a Tropical Maize Population." Ph.D. thesis, Iowa State University, Ames, Iowa.
Vasal, S. K., Beck, D. L., and Crossa, J. (1987). "CIMMYT Research Highlights 1986." CIMMYT, El Batan, Mexico.
Vasal, S. K., Moore, C. L., and Pupipat, U. (1970). *SABRAO Newslett.*, 81–89, Mishima 2.
Vasal, S. K., Ortega, A. C., and Pandey, S. (1982). "CIMMYT's Maize Germplasm Management, Improvement, and Utilization Program." CIMMYT, El Batan, Mexico.
Vasal, S. K., and Taba, S. (1988). *In* "Plant Genet. Resources Indian Perspective" (R. S. Paroda, R. K. Arora, and K. P. S. Chandel, eds.), pp. 91–107. NBPGR, New Delhi.
Vasal, S. K., Villegas, E., and Tang, C. Y. (1984). *In* "Panel Proceedings Series: Cereal Grain Protein Improvement," pp. 167–189. 6–10 December 1982, IAEA, Vienna, Austria.

Vencovsky, R., Zinsly, J. R., and Vello, N. A. (1970). "VIII Reunion Brasilera de Milho e Sorgo Brazil, (Abstr.).

Villena, W. (1965). *In* "Proceedings of the XI Reunion PCCMCA, 16–19 March 1965," Panama City, Panama.

Villena, W. (1974). "Resultados preliminares de respuestas a dos ciclos de seleccion en tres poblaciones tropicales de maiz. "Paper presented at the XX Reunion PCCMCA, 11–15 February 1974, San Pedro Sula, Honduras.

Waiss, A. C., Jr. Chan, B. J., Elliger, C. A., Wiseman, B. R., McMillian, W. W. Widstrom, N. W., Zuber, M. S., and Keaster, A. J. (1979). *J. Econ. Entomol.* **72,** 256–258.

Weatherwax, P. (1955). *In* "Corn and Corn Improvement" (G.F. Sprague, ed.), pp. 1–16. Academic Press, New York.

Wellhausen, E. J. (1965). *In* "Proceedings of the Annual Hybrid Corn Res. Conf." Vol. 20, pp. 31–45.

Widstrom, N. W. (1972). *Crop Sci.* **12,** 245–247.

Widstrom, N. W. (1987). *In* "Aflatoxin in Maize" (M. S. Zuber, E. B. Lillehoj, and B. L. Renfro, eds.), pp. 212–220. CIMMYT, El Batan, Mexico.

Widstrom, N. W. (1989). *In* "Toward Insect Resistance Maize for the Third World" (CIMMYT, ed.), pp. 211–221. CIMMYT, EL Batan, Mexico.

Widstrom, N. W., and Hamm, J. J. (1969). *Crop Sci.* **9,** 216–219.

Widstrom, N. W., Wiseman, B. R., and McMillian, W. W. (1982). *Crop Sci.* **22,** 843–846

Widstrom, N. W., Wiser, W. J., and Bauman, L. F. (1970). *Crop Sci.* **10,** 674–676.

Williams, W. P., and Davis, F. M. (1985). *Crop Sci.* **25,** 317–319.

Williams, W. P., and Davis, F. M. (1989). *In* "Toward Insect Resistant Maize for the Third World" (CIMMYT, ed.), pp. 207–210. CIMMYT, El Batan, Mexico.

Williams, W. P., and Davis, F. M. (1990). *South. Entomol.* **15,** 163–166.

Williams, W. P., Davis, F. M., and Scott, G. E. (1978). *Crop Sci.* **18**, 861–863.

Younis, S. E. A., Dahab, A., and Mallah, G. S. (1969). *Indian J. Genet. Plant Breeding* **29,** 418–425.

Pearl Millet for Food, Feed, and Forage

D. J. Andrews[1] and K. A. Kumar[2]

[1]Department of Agronomy,
University of Nebraska,
Lincoln, Nebraska 68583-0915

[2]International Crops Research Institute for the Semi-arid
Tropics (ICRISAT), Sahelien Centre, BP 12.404, Niamey, Niger

Advances in Agronomy, Volume 48

I. INTRODUCTION

Pearl millet [*Pennisetum glaucum* (L.) R. Br.] is widely grown as a food crop in subsistence agriculture in Africa and on the Indian subcontinent on a total of about 26 million ha (Rachie and Majudar, 1980) where grain yields average 500–600 kg/ha. Relatively little is grown (mostly as a forage crop) in intensive agriculture in other continents. Pearl millet has a number of advantages that have made it the traditional staple cereal crop in subsistence or low-resource agriculture in hot semiarid regions like the West African Sahel and Rajasthan in northwestern India. These advantages include tolerance to drought, heat, and leached acid sandy soils with very low clay and organic matter content. However, it has the ability to grow rapidly in response to brief periods of favorable conditions—a feature of such semiarid tropical regions. In ideal conditions, it has one of the highest growth rates of all cereals (Kassam and Kowal, 1975; Craufurd and Bidinger, 1989) (Fig. 1). Its grain is generally superior to sorghum as human food and at least equals maize in value as a feed grain. Whereas grain is the main purpose of cultivation in Africa and Asia, the forage, or stover, at harvest is an important secondary product in subsistence agriculture for animal feed, fuel, or construction. Thus vigorous tall or semitall relatively late varieties with a high biomass production are preferred. High-yielding, early semidwarf hybrids are grown in India on about 30% of the cultivated area; however, lower biomass production and unstable disease resistance have limited their spread.

Two principal types of food are traditionally made from pearl millet—porridges and flat unleavened breads. Both of these are made from flour; however, because pearl millet flour deteriorates after a few days and acquires a "mousy" odor, fresh flour must be ground frequently. Other products include rice-like foods made from pearled grain, couscous, foods from blends with legume flour, and beer.

Pearl millet gives a productive pasture for grazing, especially with dwarf varieties, and silage is easily made. Several cuts can be taken. Although pearl millet does not produce a cyanogenic glucoside like dhurrin in sorghum, it is a strong nitrogen accumulator and can produce potentially toxic levels of nitrates if not well managed. Pearl millet is used as a forage crop in the United States, Australia, and southern Africa, but the hybrid with elephant grass (napier grass), *P. purpureum* Schum., is widely used as a perennial forage crop in east and southern Africa, Brazil, and India where it is principally propagated by cuttings.

Several of the attributes that made pearl millet the best adapted food cereal for the stressful production conditions of West African Sahel are valuable for developing a combine feed grain crop for intensive cultivation

Figure 1. Crop of pearl millet grain variety Ex Bornu at Samaru, Nigeria, producing 22 t/ha above-ground dry matter in 90 days, of which 3.2 t (14.5%) was grain.

in warm-temperate regions. Principal of these is a high growth rate, which confers excellent response to increases in soil fertility. Under ideal conditions, grain yields of 3.5–8 t/ha have been obtained in India from early hybrids maturing in 85 days (Burton *et al.*, 1972). Other factors include an enormous range of genetic variability already collected but largely unused in the primary germplasm pool of the species, from which traits of major importance are still being identified. Existing characteristics, which are already being used in developing combine phenotypes, are major dwarfing genes, maturity control, through both photoperiodicity and independent maturity genes, and high levels of tolerance to heat and moisture stress. Several systems of cytoplasmic-genic male sterility are available to exploit well-manifested hybrid vigor. Though good yields are possible from varieties, typically 20-30% more grain yield can be expected from a hybrid of the same maturity class. Pearl millet grain crops can, when properly dry, be harvested with sorghum equipment. Grain test weights are about 10% higher than those for sorghum, and feeding tests show that pearl millet grain is generally slightly superior to sorghum in feed value, particularly for poultry. The better feed value is due to lower levels of polyphenols without tannins, higher protein levels with a slightly better amino acid profile, and 3–6% oil providing an increased energy content. Experimental dwarf combine hybrids have been recently developed in the United States. These hybrids can produce a grain crop in a temperate summer season as short as that in Carrington, North Dakota (Andrews and Rajewski, 1991).

II. THE PLANT

A. Description and Origin of Pearl Millet

Pearl millet has had a varied taxonomic history. It is currently again known as *Pennisetum glaucum* (L.) R. Br. (USDA, 1986; IBPGR, 1987) but was variously classified as *P. americanum* (L.) Leeke, *P. typhoides* Stapf. and Hubb., and *P. glaucum* (Brunken, 1977; de Wet, 1987). The position of the cross-fertile wild and weedy relatives that were previously given different specific names remains unclear. Since *glaucum* has been adopted at the species level, and if the attributions of Brunken (1977) are followed, then the weedy subspecies become *P. glaucum* ssp. *stenostachyum* and the wild subspecies *P. glaucum* ssp. *monodii*. Common names include bulrush or cattail millet and mil aux chandelles. In Africa, gero, maiwa, souna, dukhn, and sanio and in India bajra and cumbu are a few of many names.

Pearl millet is an annual tillering diploid ($2n = 14$), highly cross-pollinating cereal. Three gene pools have been recognized in respect of pearl millet (Harlan and de Wet, 1971). The primary pool contains cultivated, wild, and weedy pearl millets (above) which interbreed freely. The secondary pool contains only elephant grass, *P. purpureum* ($2n = 28$). This species can be crossed with pearl millet but although there is some homology between the pearl millet genome and the A′ genome of elephant grass (Hanna, 1987, 1990), the progeny are sterile unless the chromosome number is artificially doubled. This interspecific cross can also be achieved by first doubling pearl millet, which will reproduce at the tetraploid level. The tertiary gene pool contains numerous distantly related *Pennisetum* species with various ploidy levels that do not naturally interbreed with the primary pool. Each pool has potential for the improvement of cultivated pearl millet. The primary pool has an enormous range of variability (Kumar and Appa Rao, 1987), but this has been inadequately evaluated and even less used. Important genes for forage quality, disease resistance, and male sterility systems have been recently recognized. The interspecific cross between pearl millet and elephant grass is widely used in the tropics as a vegetatively propagated multicut perennial forage. Elephant grass contributes perenniality, drought and disease resistance, and high biomass production to this cross while pearl millet improves forage quality and palatability. This interspecific cross has been used to derive genes for fertility restoration, stiff stalk, maturity, and height from the A′ *purpureum* genome (Hanna, 1990) and is also of potential importance in accessing traits from the tertiary gene pool. Dujardin and Hanna (1989, 1990) have shown that the *P. glaucum* × *purpureum* hybrid can be used as a genetic bridge to make crosses with species such as *P. squamulatum*, a source of apomixis, which cannot be crossed directly with pearl millet.

Pearl millet was probably domesticated in Africa in the savannah south of the Sahara and west of the Nile possibly 5000 years ago (Brunken *et al.*, 1977; Porteres, 1976). Domestication involved relatively few gene changes (Bilquez and LeComte, 1969; Marchais and Tostain, 1985). The crop subsequently spread to east and southern Africa, and about 3000 years ago to the Indian subcontinent. It is generally agreed that the ancestral type for pearl millet resembled *P. violaceum* (a race of ssp. *monodii*, according to Brunken, 1977), which is still currently distributed on the southern fringes of the Sahara (Fig. 2). This concept of domestication, developed from

Figure 2. Wild pearl millet *Pennisetum glaucum* ssp. *monodii* (syn. *P. violaceum*), Northern Niger. (Courtesy S. Tostain.)

archeological, ecological, morphological, cross-compatability and genetic studies, is supported by relationships based on isozyme and restriction fragment length polymorphism (RFLP) analyses (Tostain *et al.*, 1987; Gepts and Clegg, 1989; Tostain and Marchais, 1989) and protein fractions (Chanda and Matta, 1990). Possibly there were several domestication events (Porteres, 1976; Tostain and Marchais, 1987).

A great range of diversity has developed, particularly in west and central Africa, which has been collected, maintained, and partly classified by ICRISAT in conjunction with IBPGR. The World Collection of pearl millet and some of its wild and weedy relatives now stands at 22,000 accessions. It is generally believed that wild relatives of pearl millet continue to intercross with cultivated varieties in west and central Africa to form hybrid swarms, part of which, called "shibras," mimic and survive in the host cultivar (Fig. 3). However, recent research indicates the presence of barriers that restrict but do not entirely prevent gene flow between the wild and the cultivated species (Robert *et al.*, 1991). Although the shibras are a nuisance to the farmer because their grain shatters, the ongoing genetic

Figure 3. Variability in a farmer's pearl millet crop, Bankass, Mali, including a weedy segregant (a shibra—taller plant with many thin heads, back right.)

transfer from the wild and weedy types since domestication has probably been of much evolutionary value in terms of adaptation and stability of production of the cultivated crop in the long term. A similar situation has been described in sorghum (Doggett and Majisu, 1968). Bramel-Cox *et al.* (1986) showed that progeny derived from crosses of pearl millet cultivars × wild or weedy species had higher growth rates than those from cultivated × cultivated crosses.

Since time of maturation is an important factor in the adaptation of tropical cereals, particularly in respect to yield and quality, flowering in almost all pearl millet landrace varieties is retarded by long days and induced by short days (Burton, 1965a; Ong and Everard, 1979). This photoperoid sensitivity, which differs minutely between cultivars, permits flowering and, hence, grain maturation to coincide with the time when the season usually ends each year, largely irrespective of the date of planting. Bilquez (1963) classed pearl millet varieties as either facultative (flowering occurs but is delayed by long days) or obligate (only flowers when short days occur).

Photoperoid response is one of many environmental factors that are of critical importance in the utilization of pearl millet germplasm and in the characterization of many traits. Whereas a few important traits, such as grain color, are relatively independent of environmental effects many others, such as grain and forage yield and quality, are strongly affected. Variation in the period of time between seedling emergence and floral initiation obviously has a large influence on performance in both grain and forage varieties (Craufurd and Bidinger, 1988, 1989). It is essential, therefore, to recognize the effect of environment both in breeding and in the assessment of quality values.

The relationship between pearl millet and sorghum [*Sorghum bicolor* (L.) Moench] is relevant when discussing the existing adaptation and potential use of pearl millet for food, feed, and forage in both tropical and warm-temperate agriculture. Sorghum was also domesticated in Africa and is widely grown as a food cereal there and in other semiarid regions that are similar to those where pearl millet is dominant but with more ensured rainfall and better soils. The interface between the adaptation zones of the two crops in Africa is, however, not all that distinct. There are a few specialized sorghums adapted to the drought stress and heat peaks at seedling establishment and floral development that characterize the zone where pearl millet is the dominant cereal. In contrast, there are many millet cultivars of both long and short duration found where sorghum is the predominant food cereal. They are used together to stabilize food supply over varied soil types and unpredictable rainfall regimes. Indeed short-season millet and long-season sorghum are common traditional intercrops in Nigeria, which more efficiently utilizes season-long resources (Andrews, 1972; Andrews and Kassam, 1976).

Despite the similarity between pearl millet and sorghum in terms of adaptation and use as food, and though both crops were introduced into the United States in the last century and used as forage crops, pearl millet has not been developed into a feed grain crop, as sorghum was in a process that commenced around the 1930s (Duncan *et al.*, 1991). The reason for this is not clear, but it may initially have been because mutants, especially for reduced plant stature, are easier to extract and multiply in sorghum, a largely self-pollinated species, than in pearl millet. Following the success of hybrid development in India, and the realization that pearl millet generally has a more nutritious grain than sorghum, breeding pearl millet for feed grain production has commenced in the United States.

B. Breeding

Pearl millet is a naturally cross-pollinating species in which traditional cultivars are random-mating populations with considerable internal variability. As much as 30% inbreeding depression may occur in these after one generation of selfing (Khadr and El-Rouby, 1978; Rai *et al.*, 1984).

The floral biology of pearl millet permits many breeding techniques to be used, ranging from various types of population improvement to strict pedigree selection. Pearl millet is protogynous and the interval between the emergence of all stigmas and anthesis on one head may extend from 1 to 6 days (Fig. 4). This facilitates natural cross-pollination, but bagging emerging heads allows easily controlled crossing or selfing. Each head produces 500 to 1500 seeds and, depending on density, one plant may produce many heads.

Good levels of heterosis are expressed in pearl millet. Typically a single-cross hybrid between two inbred parent lines yields 20–30% higher than an adapted variety of comparable maturity, though much higher levels have been reported in the literature (Rachie and Majmudar, 1980; Kumar, 1987). Good line × variety hybrids can also be made. Though inbreeding depression is significant, productive inbred lines can be selected with sufficiently high seed yields so that three-way crosses are not economically necessary for hybrid seed production.

Several cytoplasmic–genic male sterility (cms) systems are available in pearl millet (Kumar and Andrews, 1984; Hanna, 1989). The release of the first and currently the most widely used source, Tift23A_1 from Tifton, Georgia, in 1965 (Burton, 1958, 1965b) permitted forage hybrids to be developed in the United States and grain hybrids to be widely grown in India.

The possibility of using protogyny to make grain hybrids in pearl millet is being investigated (Andrews, 1990). This method allows quicker hybrid

Figure 4. Complete protogyny in a pearl millet head.

development, is less restrictive of the range of parental combinations possible, and avoids diseases that are associated with the use of cms seed parents, particularly in Africa, where these hybrids, especially of the topcross type, would be of most utility. Some "seed parent" selfing may occur when making hybrid seed by protogyny. Tests with mechanical mixtures of seed parent and hybrid seed showed that up to 20% of seed from an inbred female parent had no significant effect on hybrid performance, provided the hybrid has a dominant phenotype (Andrews, 1990). The use of a variety as a male parent reduces female parent selfing through a profuse and prolonged pollen supply and confers some of the stability of performance characteristic of varieties into the resulting topcross hybrid.

The development of forage cultivars in the United States in the past 50 years (see Table II) has progressed from open-pollinated varieties, through synthetic varieties and polycross F_1's, to single-cross hybrids. Advances have been made in both biomass productivity and digestibility, the latter largely through the use of dwarfing genes to increase leaf/stem ratio. Tolerance to nematodes and diseases has been incorporated.

Most of the breeding for grain production in pearl millet has so far been done in India, although the cytoplasm used to produce all Indian hybrids derives from Tift23A_1.The longevity of individual hybrids in India until recently has been short, 3–5 years, because of the instability of their resistance to pearl millet downy mildew, a disease that is not known in the New World. The Indian hybrids, though their yield potential is high, are semidwarf, 1.3–1.8 m, and do not possess the persistent stem strength needed for mechanical harvesting. Many are also partly photosensitive, and thus mature too late when planted more than about 30° latitude from the equator (Bidinger and Rai, 1989). New phenotypes are, therefore, required for use in the U.S. Midwest. These phenotypes should be non-photoperiod-sensitive and early to very early with sufficient stalk and peduncle strength and with an upright tiller habit to confer lodging resistance that will persist after frost. Experimental hybrids approaching the required phenotypes have been produced in Nebraska (Andrews, 1990) (Fig. 5) and Kansas (Christensen *et al.*, 1984; Stegmeier, 1990) and have been jointly tested, with hybrids from Tifton, Georgia, in 1988–1990 regional tests. Test locations (years) have been in Mississippi State, Mississippi (1); Tifton, Georgia (2); Hays, Kansas (3); Lincoln and Sidney, Nebraska (3); Lafayette, Indiana (3); and Carrington, North Dakota (1). Mean location yields ranged from 2300 to 3800 kg/ha. Across tests, the best millet hybrids averaged 85% of sorghum check yields; however, in locations where the season was short as in North Dakota and in double-cropping after wheat in Indiana, millet yields exceeded those of sorghum. Similar results were obtained earlier in western Kansas (Christensen *et al.*, 1984) here the highest experimental millet grain yield was 5300 kg/ha.

C. Biotechnology

The term "biotechnology" encompasses several applications to manipulate genetic material—ranging from incorporation of desired genes into plants and use of DNA markers such as RFLPs to select desirable genes to the determination of genetic relationships and the production of superior individuals. Tanksley *et al.* (1989) suggested that integration of RFLP techniques into plant breeding would (*a*) hasten the movement of desirable

Figure 5. Dwarf pearl millet grain hybrid in regional test, Sidney, Nebraska.

genes among varieties; (*b*) permit transfer of genes from related wild species; (*c*) allow analysis of complex polygenic characters as groups of single Medelian factors; and (*d*) establish genetic relationships between sexually incompatible crops plants.

Smith *et al.* (1989) identified 64 RFLP markers linked to genes of 26 plant traits in elephant grass, most of which are quantitative in nature. The RFLP markers were linked to genes affecting *in vitro* organic matter digestibility (IVODM), neutral detergent fiber, and nitrogen and phosphorous concentration. In addition to RFLP analysis, a newer technique, random amplified polymorphic DNA (RAPD), that relies on polymerase chain reaction (PCR) appears to be more promising in applied breeding programs, as it is quicker and eliminates several tedious steps that are involved in the RFLP technique (Williams *et al.*, 1990; Welsh and McClelland, 1990).

In a practical breeding program, DNA markers would allow the definitive identification of plants carrying a recessive gene in segregating populations independent of evironment. In addition, establishment of correlations between quantitative trait loci (QTL) of interest and specific RFLP markers allows selection of specific chromosome segments affecting a quantitative trait from a population of plants and incorporation into single plants with high efficiency (Helentjaris and Burr, 1989). In pearl millet, specific areas where these techniques could be used would include selection for complex traits such as drought and disease resistance and forage quality attributes. A random genomic probe library in pearl millet is now available (R. L. Smith, 1991, personal communication) and could be used to establish linkages between QTL and RFLP markers. To take advantage of the new technologies available and achieve desired goals, close cooperation among biotechnologists, plant breeders, and agronomists is essential.

D. Consumer Quality and Yield

Pearl millet is principally grown for human food in the drier tropical regions of the world where agricultural production is at most risk from pest, disease, and highly variable weather conditions. Farmers have, therefore, historically selected their varieties for consistency of production in the face of the occurrence of stress caused by drought and low soil fertility and for resistance to pests, birds, and storage insects, as well as for characteristics associated with food preparation and organoleptic qualities. For characteristics that are associated with these objectives, compromises have resulted. For instance, harder grain (more vitreous endosperm) is associated with resistance to damage from storage insects, whereas for many food products a softer endosperm would be better. In blind preparation or tasting tests the most widely grown local variety is usually rated as acceptable, but it may not be the most preferred (this has consequences both for the breeder and for the interpretation of consumer test results without reference to cultivar performance). Additionally, nutritional quality values as such have not been selected for traditional agriculture, only in as much as they are associated with some other evidently desirable trait. However, research into the genetic variability present in germplasm resources of pearl millet has revealed a wide range of values for traits affecting nutritional quality. In consequence, where variability is demonstrated, the probability of further improving nutritional quality traits in pearl millet through breeding exists. A relevant example is that of protein and lysine levels. Variability for protein content was demonstrated in pearl millet genotypes by Kumar *et al.* (1983). Selection resulted in inbreds with higher protein levels. Lysine in protein percentage declined, but there was still a net increase in lysine per sample. This contributed to a higher protein efficiency ratio, and thus an increase in total nutritional value as determined by rat-feeding tests (Singh *et al.*, 1987). Hybrids made between these inbreds and normal parents showed some elevation in protein level (ICRISAT, 1984), indicating partial dominance for the expression of protein content.

A review of information on the nutritional status of humans in the West African Sahel (IDRC, 1981) concluded that there are seasonal energy deficiencies, accentuated in women by the additional labor needed to find fuel and prepare food, protein-energy deficiencies in 20–30% of the children, and nutritional anemia in 40% of the children and up to 60% of the women. The latter had many causes—including deficiencies in iron, folic acid, vitamin B_{12}, and protein. Annegers (1973) had earlier noted that the nutritional status of the population in West Africa primarily dependent on pearl millet was better than that of people mainly using sorghum, maize, or rice. Many socioeconomic factors contribute to malnutrition, but higher and more stable cereal production levels, more protein (vegetable and animal), and dietary education are needed.

In a nutritional survey in south Indian villages, in a sorghum-, pearl-, millet-, and rice-growing area, Ryan *et* al. (1984) noted calorie and vitamin deficiences especially among rice eaters. They concluded that the prime need was more energy supply and that breeding for increased protein content or quality of cereals should not be undertaken if it would hinder progress for yield. Protein and vitamins could more easily be obtained by other means.

E. Grain Characteristics

1. Physical

Pearl millet grain is about one-third the size of sorghum (Fig. 6) and 100 grain weights range from 0.5 g to over 2.0 g. While average grain weights of common cultivars vary greatly in different regions, over about 1.0 g/100 is normally acceptable; however, it appears feasible to breed for slightly larger grain without sacrificing yield potential. Grain shape can vary considerably from globular to lanceolate (IBPGR, 1981) where the length-to-width ratio ranges from equality to nearly 4:1. Long thin grains are the result of a high grain number relative to the surface area of the panicle, and thus are usually angular in cross section. Grain of such shapes are harder to

Figure 6. Pearl millet grain. (Left) Cream/white cultivar; (right) large gray-grained cultivar; (bottom) red-grained sorghum.

decorticate and give lower flour yields. Grain color in pearl millet is a combination of the color and thickness of the pericarp, particularly the mesocarp, and the color and vitreosity of the endosperm. Colors are creamy-white to yellow, light to dark brown, various shades of light to dark blue/gray, and purple. Sunlight causes the lighter colors to fade and humid conditions during grain ripening dull the color of the grain through superficial mold and bacterial infection in the pericarp.

The slate gray color in pearl millet is due to the presence of flavonoids (Reichert, 1979), which can be present in both the pericarp and the peripheral endosperm. While no condensed tannins have been found in pearl millet (Hulse *et al.*, 1980; Reichert *et al.*, 1979), such as those that interfere with protein utilization in sorghum, differences in total phenol content have been detected (McDonough and Rooney, 1985) with bronze (brown) seeds having higher levels than yellow or blue gray. The slate gray color in millet product is pH dependent (Reichert and Youngs, 1979) and can be reduced by using acidic additives such as tamarind extract, or sour milk during food preparation.

Pearl millet grain averages 75% endosperm, 17% germ, and 8% bran (Abdelrahman and Hoseney, 1984) (Fig. 7). The proportion of germ in pearl millet is thus about twice that of sorghum, which is a factor contributing to the higher nutritive value of pearl millet grain. The germ is firmly embedded in the endosperm and may not be completely removed by milling.

The pericarp is composed of three layers of different cell structures (Figs. 7 and 8). The epicarp has one or two layers of thick cubic cells with a thin layer of cutin on the outer surface. The epicarp is most important in resisting "weather" damage (Sullins and Rooney, 1977). The mesocarp may vary in thickness, being composed of one to several layers of cells that collapse at maturity. Pericarps with thin mesocarps are more translucent and allow the endosperm color and texture to show through. The term "pearl" in pearl millet is derived from the glistening appearance of unblemished grains with a translucent pericarp and vitreous endosperm. The innermost component of the pericarp is the endocarp composed of both cross and tube cells below which is the outer layer of endosperm aleurone cells, which may be pigmented. On decortication the bran, which is composed of all three layers of the pericarp, is reported to separate from the endosperm either above or below the layer of the aleurone cells (McDonough and Rooney, 1989). This difference is attributed to the method of decortication and is important because the aleurone layer is relatively rich in protein and vitamins. Thick pericarp varieties better resist superficial

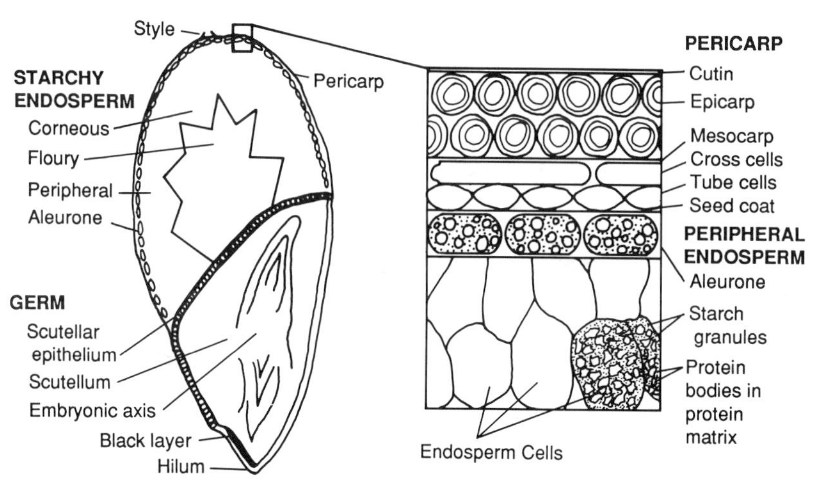

Figure 7. Diagram of longitudinal section of pearl millet grain. Modified from Rooney and McDonough (1987); reproduced with permission from the publisher.

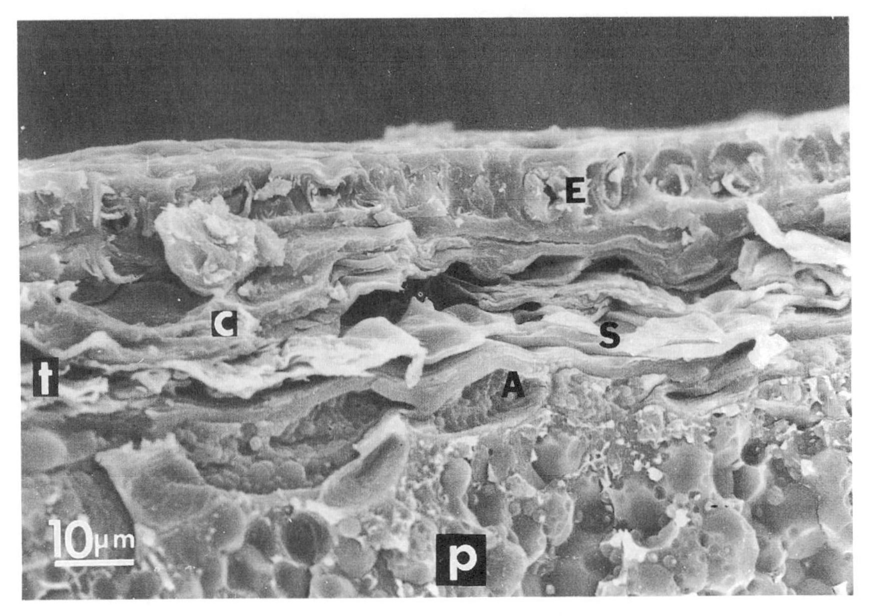

Figure 8. SEM micrograph of pearl millet pericarp and peripheral endosperm. e, epicarp cells; c, cross cells; t, tube cells; s, seed coat; a, aleurone; p, peripheral endopserm. Reproduced with permission from ICRISAT (see Rooney and McDonough, 1987).

mold damage in moist conditions and may be easier to decorticate (Kante *et al.*, 1984).

Three parts to the starchy endosperm, which may vary in proportion in different varieties and which all contribute to the flour, are recognized by Rooney and McDonough (1987). The peripheral region contains many protein bodies, in a matrix surrounding small starch granules. Below that is a corneous layer in which large, uniform-size starch granules are closely packed in a protein matrix with a few protein bodies. The innermost part is the floury endosperm with loose large round starch granules in a thin protein matrix with a few protein bodies. The flour consists of free starch granules from the floury endosperm and pieces of the other two parts. The fineness of these pieces determines quality in many food products where there is not adequate time given in preparation to soften these pieces. Indeed this may be a contributory reason why products such as the porridges are left to ferment or stand overnight before consumption. Bread made from wheat flour extended with pearl millet flour may be noticeably gritty unless the millet is milled very finely (Hoseney, 1988).

2. Composition

The averages of proximate analyses conducted on pearl millet indicate a protein content of about 12%, carbohydrates 69%, lipids 5%, fiber and ash around 2.5% each, and the remainder being moisture (Rooney ad McDonough, 1987; Hoseney *et al.*, 1987; Hulse *et al.*, 1980). However, considerable ranges—especially in protein content, from 8 to 24% (Hulse *et al.*, 1980; Rooney and McDonough, 1987), and lipids, from 3.0 to 7.4% (Hoseney, *et al.*, 1987)—have been reported (Table I). It is likely, however, that in normal germplasm protein levels much over 18% are caused by partial grain development or derived from unusually low yielding sources—weak inbred lines or where seed-set was incomplete. Kumar *et al.* (1983) indicate that grain protein level is subject to strong environmental effects. Several authors conclude (Burton *et al.*, 1972; Rooney and McDonough, 1987; Hoseney *et al.*, 1987) that pearl millet tends to have a higher mean protein level than sorghum grown under similar conditions. In pearl millet variety trials grown at three locations in the U.S. Midwest, grain protein levels in pearl millet average 10.6%, whereas the sorghum checks recorded 9.5% (Andrews, 1990). However, protein levels of 12 to 14% are commonly reported for grain pearl millet grown in the Midwest (Walker, 1987).

The fatty acid composition of the free and bound lipids in pearl millet and sorghum are quite similar (Rooney, 1978) but the total level is considerably higher in pearl millet, 3 to 7%, a factor that contributes to the higher calorific values of pearl millet.

The amino acid profile of pearl millet is better than that of normal sorghum and normal maize and is comparable to the small grains such as wheat, barley, and rice (Ejeta *et al.*, 1987) with less disparate leucine/isoleucine ratio (Hoseney *et al.*, 1987; Rooney ad McDonough, 1987). The percentage of lysine in protein as reported in pearl millet ranges from 1.9 to 3.9 g/100 g protein (Ejeta *et al.*, 1987; Hoseney *et al.*, 1987).

Despite analytical surveys of the World Collection by ICRISAT, mutants have not been discovered in pearl millet where, as in sorghum, maize, and barley, the lysine content of grain protein is substantially increased by the action of major genes. Although it appears possible through selection to increase grain protein content, percentage lysine in protein decreases at higher grain protein levels, though percentage lysine in the sample still increases (Kumar *et al.*, 1983). Additionally, as grain yields are increased generally grain protein content declines, though, as with lysine content, net protein yield per hectare increases. Pearl millet contains a lower proportion of cross-linked prolamins which are slightly

Table I

Protein Content and Essential Amino Acid Composition of Pearl Millet and Sorghum Grains[a,b]

Amino acids	Pearl Millet					Sorghum			
	No. of samples	Range (g 16 g^{-1} N)	Mean (g 16 g^{-1} N)	Amino acid score[c]	Pattern[d] (g 16 g^{-1} N)	No. of samples	Range (g 16 g^{-1} N)	Mean (g 16 g^{-1} N)	Amino acid score
Lysine	280	1.59–3.80	2.84	52	5.5	412	1.06–3.64	2.09	38
Threonine	29	3.17–5.66	4.07	102	4.0	29	2.12–3.94	3.21	80
Valine	29	4.38–7.67	6.01	120	5.0	29	3.84–6.93	5.40	108
Methionine + cystine	29	1.43–3.96	2.71	77	3.5	24	1.80–2.69	2.36	67
Isoleucine	29	3.70–6.34	4.56	114	4.0	29	2.85–5.05	4.17	104
Leucine	29	8.62–14.80	12.42	177	7.0	29	10.12–17.60	14.67	210
Phenylalanine + tyrosine	29	6.54–10.81	8.49	142	6.0	29	6.11–10.72	8.87	148
Protein (%)	280	6.40–24.25	12.30			412	4.60–20.25	11.98	

[a] Determined by ion exchange chromotography.
[b] Adapted from Jambunathan *et al.* (1984) *in* Rooney and McDonough (1987). Reprinted with permission from the publisher (see ICRISAT, 1987).
[c] Percentage of recommended pattern.
[d] From FAO/WHO (1973), p. 67.

higher in lysine and tryptophan content than sorghum (Jambunathan and Subramanian, 1988), which may be an additional factor contributing to the higher digestability of pearl millet proteins (Hoseney *et al.*, 1987). Yellow endosperm has been reported in pearl millet (Curtis *et al.*, 1966), but although its carotene content was about the same (1.0–0.2 ppm) as that for yellow sorghum, it is low compared to that for yellow maize.

Pearl millet appears to be generally free of any major antinutritional factors, such as the condensed tannins in sorghum which reduce protein availability. As with other cereals, pearl millet contains phytic and nicotinic acids mainly in the germ (Simwemba *et al.*, 1984; McDonough, 1986). Two trypsin inhibitors have been reported in pearl millet (Chandrasekhar and Pattabiraman, 1981). Osman and Fatah (1981) reported that in western Sudan, where iodine intake is low, goiter in humans was associated with high millet as against wheat-based diets. Klopfenstein *et al.* (1983, 1985) showed that in rats thyroid hormones were affected, but that symptoms could be alleviated by dietary adjustments.

III. FOOD PRODUCTS

A range of food products with numerous local names (Appa Rao, 1987; Sautier and O'Deyes, 1989) are made from pearl millet, which are well described by Hulse *et al.* (1980), Hoseney *et al.* (1987), Rooney and McDonough (1987), and Serna-Saldivar *et al.* (1990).

The major types of food are (*a*) porridges, either thick or thin, which are common in Africa and (*b*) flat bread, either unfermented (mostly Asian) or fermented (Ethiopia and Sudan). Other products are couscous, boiled rice-like preparations, snacks from blends with legume flours, and non-fermented or fermented beverages in Africa.

All these products are made from either coarsely or finely ground millet flour (the degree of fineness is often important) usually with separation and removal of the bran. A major constraint with the utilization of pearl millet is the propensity of the flour (or damaged grain) to acquire a mousy rancid odor within a few days of milling, which is accentuated when water is used to temper the grain before milling. This odor was previously attributed to enzymatic degradation of grain lipids (Chaudhary and Kapoor, 1984) but subsequently Reddy *et al* (1986) have shown that it is associated with apigenin, a flavanoid compound in the usual blue/gray pigmentation in pearl millet. However, varieties with white or cream-colored grain also give the mousy odor, but to a lesser extent. It is probably that both oxidation of lipids and apigenin contribute to the "off"-odors in pearl millet. Dry milling or heat treatment after milling produces a flour with a longer shelf life.

In general there are less-pronounced cultivar preferences in pearl millet than there are in sorghum. However, consumers prefer lighter colored rather than darker pearl millet foods. This is achieved through a combination of grain type, degree of decortication, and change in pH during preparation. Reichert *et al.* (1979) showed that the pigments of gray or yellow-gray grain varieties are easily bleached by traditional methods of lowering pH, whereas pigments of yellow, brown, or purple are not. Since some pigments are located in the external layers of the endosperm, decortication is often excessive in traditional methods to remove these pigmented layers, which reduces both the extraction rate and the nutritive value of the flour.

A. Porridges

Porridges are made in both India and Africa, and a variety of methods are used. Rooney and McDonough (1987) describe the preparation steps in detail for the more widely used types.

Thick porridges or pastes known as "tô" or "tuwo" in West Africa are generally solid enough to be broken into convenient pieces for eating and dipped in a stew or sauce. Since thick porridges are normally eaten with the fingers, it is important that there be sufficient gelatinization of the starch to prevent crumbling, but not enough to make it too sticky and therefore difficult to handle. Although genotype has some effect, the most important factors are the fineness to which the flour is ground and the preparation technique, which involves cooking part of the flour first to get a thin well-gelatinized liquid and then adding the remaining flour and cooking further to get the correct consistency. Both alkaline and acid porridges may be made. In some preparations, the flour may be soaked overnight with sour milk or fermented briefly, before cooking.

Thin porridges, or gruels, are made to be drunk or eaten with utensils. In Nigeria, the whole grain may be steeped for 2 or 3 days in water before it is crushed and its bran removed. This allows the germination process to begin (but not to the stage of malting) with resultant changes in the endosperm and germ. The process, which involves further fermentation and decanting before cooking, produes a smooth-textured porridge with a preferred sour taste. Germination and fermentation have been shown to improve nutritional value through improved protein quality and digestibility and increased vitamin content (Serna-Saldivar *et al.*, 1990).

B. Flat Breads

Roti, a flat unleavened bread, is the usual way in which pearl millet and other grains such as sorghum and maize are eaten on the Indian subcontinent. This type of unleavened bread is uncommon in Africa, but fermented breads are used on both continents. Roti can be made from whole millet flour, without separation of the bran, simply with the addition of warm water. In India, flour is traditionally ground between rotating or reciprocating stones, which can grind more finely than the wooden mortar and pestle used in Africa (though in Africa saddlestones may also be used). In both regions, of course, motor-driven plate or hammer mills are now commonplace at the village level, but preferences for characteristics in food products change slowly. In roti, particularly, fineness of the flour is most important. Pearl millet, as is the case with other tropical cereals, has no gluten, and cohesiveness of the dough is dependent on the surface tension between particles, which, when the correct amount of water is used, is greatest as particle size decreases (Olewik *et al.*, 1984). The dough for roti should be capable of being flattened by hand into a disk 1–3 mm thick and 12–25 cm in diameter, and after cooking briefly (about 2 min, including turning) at a high temperature, it should, if kept in a covered container, retain flexibility for several hours.

To make fermented breads—galettes of West Africa, kisra of Sudan and Ethiopia, and dosai of India—the flour is first mixed with water and a starter and left to ferment for 12 to 24 hr and then cooked quickly at a high temperature on a metal sheet or in a clay oven. For kisra, just before cooking, sufficient water is added to make a very runny paste that can be spread very thinly. For dosai bean flour up to one-third of the total may be added before fermenting.

C. Other Traditional Foods

Couscous or "arraw" is a steamed product made by gelatinizing and agglomerating pearl millet flour with additives that may include sugar and mucilaginous gums from okra or baobab. Steaming prevents off-odors from forming and couscous is one of the few products that stores well and can be conveniently prepared by rehydrating with milk or steam. The drawback with couscous is that it is complicated and energy-expensive to prepare.

Idli is a steamed product made in India, usually for breakfast, from a fermented mixture of pearl millet and legume flour.

D. Traditional Processing

The processes of fermenting, malting, and brewing (below) each increase nutritive value relative to the direct use of unprocessed flour.

1. Fermenting and Malting

Fermenting without prior malting, hydrolyzes starch, softens flour particles, and lowers the pH, which helps bleach the flour, and slightly increases protein digestibility (Hemanalini *et al.*, 1980).

Malting (germination) begins the process of mobilizing seed reserves, both starch and protein, and initiating shoot and root growth. The vitamin content of the grain is improved and the levels of lipids, phytates, and oxalates are lowered (Opoku *et al.*, 1981). Intrinsic grain enzymes improve protein quality and digestibility and increase the availability of free sugars, B-vitamins, and ascorbic acid (Hamad and Fields, 1979; Aliya and Geervani, 1981). Brewing continues the biochemical processes commenced in malting, on both the malt and any other starch base which may be added, assisted by enzymatic activity of lactobacilli and yeasts which may be present or added in a starter kept from previous brewings. The microorganisms continue to improve protein and vitamin availability (Hulse *et al.*, 1980; Novellie, 1982). Pearl millet malt is made in the usual way, germinating the grain by keeping it moist, warm, and aerated. At the correct time the germination process is halted and the grain sun-dried and ground. The diastatic activity was found to be the highest 32 hr after germination (Jain and Date, 1975). Apart from brewing, the malt may be used, together with legume flour in other preparations of value as weaning food and for pregnant women and for convalescent people.

2. Brewing

Beers are important nutritional adjuncts to the diets of people who are principally cereal dependent, as the malting and brewing processes increase bioavailability and vitamin contents. Beer production is common throughout Africa and any cereal that is locally available may be used, though sorghum is perhaps most widely used. Sorghum varieties preferred for brewing have moderate but not high tannin levels and produce a malt with high diastatic power. Pearl millet does not have tannins and in eastern and southern Africa malt from finger millet (*Eleusine coracana* Gaertn.) may be used as an additive to provide the bitter taste and increase diastatic levels. Both lactobacilli, which contribute the acid sour taste through the formation of lactic acid, and wild yeasts are involved in fermentation

(Doggett, 1988). Temperatures used during the stages of brewing are important in controlling the activity of the enzymes and organisms (Rooney and McDonough, 1987). Two types of beer are made: a sour opaque alcoholic beverage that is still fermenting when consumed and a clear sweet or slightly sour beer.

E. New Processes and New Foods

Ease of food preparation is becoming a more important consideration in communities dependent on subsistence agriculture. Compared with the alternatives of preparing food from sorghum, wheat flour, or rice, which store longer, pearl millet flour, which needs to be prepared from grain (which itself has to be threshed, as pearl millet is usually stored at on the head in household granaries) every few days, is a less attractive prospect. This may be one of the reasons why the area cultivated with pearl millet has declined by about 1% a year in India over the last 20 years and has remained static in Africa despite increases in population. More knowledge about the causes of rancidity is obviously needed, to determine what genetic variation may exist and whether economically viable processing control methods can be developed.

A major source of nutrient loss is in decortication. Excessive decortication reduces extraction rates and lowers nutritive value of the flour, since protein and vitamin levels are higher at the periphery of the endosperm. Manual decortication methods are highly variable, depending on the operator and utensils, and in general give extraction rates of 70 to 80%, which are lower than those obtained with mechanical means. The main drawback to manual methods, however, is the time and effort involved (Eastman, 1980). Varietal differences have been linked to ease of decortication (Kante *et al.*, 1984) and grain shape also affects extraction rates, with nearly sperical grains being best (Rooney and McDonough, 1987). Plate and hammer mills are now commonly being used at the village level, and, for the same extraction rates, little difference in flour nutrient levels between manual and mechanical methods was observed (Reichert and Youngs, 1977).

The most effective and easily controlled method of decortication is an abrasion process, as in the Tangential Abrasion Decortication Device (TADD) developed by IDRC (Reichert *et al.*, 1986) for village level use. Either batch or continuous flow versions are available. Akingbala (1991) noted while studying pigments distributed mainly in the pericarp of pearl millet grain, that 8% decortication by dry milling with the TADD dehuller was equal to 20% by traditional methods in terms of pigment removal.

The traditional method of making couscous from coursely or finely milled flour is a protracted and skilled process and involves a relatively high amount of fuel. A standardized batch process involving a V-blender of making "arraw," a form of Senegalese couscous with sugar, has been developed by Walker (1987). This should reduce product cost and make couscous an available food of a more uniform quality.

Blending up to 20% pearl millet flour with wheat flour has been shown to be practical (Dendy *et al.*, 1970; Sautier and O'Deyes, 1989) and acceptable to consumers in Senegal and the Sudan (Perten, 1983). Fine milling of pearl millet flour is necessary to avoid a gritty texture in the bread. Blending has not been widely adopted, however, as imported subsidized wheat is cheaper and of a more consistent quality than pearl millet at the flour mill gate.

After decortication, whole or broken pearl millet grain can be used in rice-like preparations in Africa and India. Rooney (1989), while researching ways that this might be done on a large scale, found that decortication was easier and resulted in less loss if the grain was first parboiled and slowly dried. The pearled parboiled grain can be easily cooked and produces a rice-like product called "milri." Cream or yellow varieties of pearl millet make the most attractive milri. Parboiling gives good control of the development of off-odors and thus prolongs shelf life. Recent work (S. D. Serna-Saldivar, C. Clegg, and L. W. Rooney, 1991, personal communications) show that only slight differences exist in chemical composition and nutritional value of parboiled and raw grains decorticated to the same extent. In either case 17.5% decortication increased protein and dry matter digestibilities. Milri, therefore, appears to meet several of the requirements of a new food from pearl millet—it can be produced in bulk, it stores better, and food is easily prepared from it.

Germination and malting increase the food value of cereal grains, and additions of legume flour and vitamins can produce a balanced weaning food (Malleshi and Desikachar, 1986). Badi *et al.* (1990) compared pearl millet- and sorghum-based baby foods each made from 70% flour, 13% malt, and 17% milk powder. The use of part of the cereal as malt lowered the viscosity of the foods as well as slightly improving digestibility. There were no differences in energy values but the pearl millet food contained 20% more protein and was considered to provide adequate protein and energy levels for 9-month-old children, whereas the sorghum food would provide sufficient levels for 1-year-old children. Almedia-Dominguez *et al.* (1990a) described the production by extrustion or puffing of baby food from 70% pearl millet and 30% cowpea flour, which supplied 17, 72, and 110%, respectively, of the daily needs of protein, lysine, and threonine of a

2-year-old child. Extrusion of moist pearl millet flour at 200–240° C greatly increased water solubility and would be suitable for preparing snack foods (Almeida-Dominguez *et al.*, 1990b).

IV. PEARL MILLET GRAIN AS FEED

Though pearl millet is grown only for its forage in the United States, it has the potential to be used like maize and sorghum in rations of poultry, swine, and beef cattle (Hanna *et al.*, 1991).

A. Poultry Feeds

Research on formulation of poultry feeds using pearl millet has been carried out to ease competition for energy sources between humans and monogastrics and reduce feed costs. Studies conducted by several workers (French, 1948; Singh and Barsaul, 1976; Sharma *et al.*, 1979) have shown that millets compared favorably with maize in poultry diets. Fancher *et al.* (1987) reported that the metabolizable energy (ME_n) content of ground pearl millet varied from 2.891 to 3.204 kcal g^{-1} dry matter, depending on the cultivar, and suggested that previously reported ME_n values for pearl millet were underestimated by up to 21%.

French (1948) included up to 60% pearl millet or white seeded finger millet in layer's diets without any reduction in egg production. Sanford *et al.* (1973) found that at equal levels, the amino acid profile, rate of chick gain, and efficiency of utilization of feed were favorable for pearl millet compared with sorghum grain as a source of energy and protein for broiler-strain chicks.

Lloyd (1964) observed that broilers fed on millet rations were heavier and had better feed conversion than those fed on maize rations. No significant differences were found between the diets in slaughter weights and yields and both were equivalent in the production of good quality carcasses. However, it was noted that millet-fed carcasses had a slightly higher pigmentation than those fed on maize. Abate and Gomez (1983/1984) partly substituted maize with pearl millet and finger millet in both broiler starter and finisher feeds. The chicks fed on pearl millet diets had highest overall body weight gain and finger millet was comparable to maize. They also showed that in broiler diets pearl millet could effectively replace part of the vegetable protein supplement provided the diet was supplemented with up to 0.3% lysine.

Smith *et al.* (1989) conducted trials with pearl millet, grain sorghum, and triticale substituting each grain for 50 or 100% of maize in the control diet. Pearl millet and grain sorghum replaced maize in the diet of chicks without adversely affecting gain or feed efficiency. They concluded that an economic assessment of dietary treatments would be premature because no fair market price or established production practices exist for pearl millet in the United States.

Sullivan *et al.* (1990) reported three chick-feeding tests conducted in Nebraska and Kansas. In each test, chick growth rates were similar on diets containing pearl millet, low-tannin sorghum, and maize. High-tannin sorghum gave lower growth rates. Pearl millet had higher ME_n values. In 1989, when pearl millet was produced in the same field as the sorghum, the crude protein content of the pearl millet grain was 0.9 to 1.7% points higher than that of sorghum varieties.

The studies reviewed indicate that pearl millet, if properly supplemented with protein sources, could be used in poultry feeds as an energy source. The protein in pearl millet grain provides all essential amino acids except lysine. The decision to use pearl millet in poultry feeds is primarily one of economics, as feeds would still require, as for maize, some added protein or essential amino acids from more expensive ingredients.

The potential of dehydrated and pelleted pearl millet forage has been explored in the United States (Wilkinson *et al.*, 1968). Wilkinson and Barbee (1968) reported that dehydrated products of coastal Bermuda grass and pearl millet compared favorably in metabolizable energy values to dehydrated alfalfa and can be used as adjuncts to dehydrated alfalfa in supplying supplementary xanthophyll in poultry feeds. Butler *et al.* (1969) observed that millet can be dehydrated and pelleted at ages up to 45 days, despite the high proportion of large-diameter stems, and that it can be processed in to a high-quality feed additive. The potential of dehydrated forage for use as an ancillary to dehydrated alfalfa in supplying supplementary xanthophyll in poultry feeds needs further investigations.

B. Beef Cattle

Studies at Fort Hays, Kansas, have shown that finishing steers on pearl millet diet gained as well as those fed sorghum (Christensen *et al.*, 1984). Compared to sorghum grain, pearl millet grain had higher levels of both fat and protein and the protein had a better balance of essential amino acids. Estimated net energy of millet was 4% higher than that of finely rolled sorghum. Steers gained 1.32 kg/day on pearl millet compared to 1.26 kg/day on sorghum. Pearl millet grain used with Rumensin in growing rations for calves gave significantly higher gains that sorghum. The authors concluded that pearl millet grain is an excellent source of protein for beef cattle rations. Preliminary results from Zimbabwe (S. C. Gupta, personal communication) indicate that sorghum and pearl millet could be used as high-energy substitutes for maize in pen-fattening diets of steers.

In a metabolism trial with six steers, Hill and Hanna (1990) found that apparent digestibility of dry matter (DM), organic matter, and dietary total digestible nutrients were higher for the control, 73% maize + 6% soybean meal (C), than for 76.2% grain sorghum + 2.8% soybean meal (GS) or 79% pearl millet (PM) diets. Ether extract and crude protein digestibilities were higher for C and PM than GS and retained nitrogen level was similar for all three diets. In a growth trial with yearling heifers, they observed higher average daily gain on C compared with PM; however, feed : gain ratios were similar for all three diets (8.2, 9.1, and 8.5 kg feed/kg gain, respectively).

C. Pigs

Calder (1955, 1961) reported that for pig feeding the millet grains should be finely ground so that the risk of internal irritation caused by the hard hull of the grain is reduced to a minimum. Pigs fed ad libitum on diets containing 75 and 50% pearl millet reached the average slaughter weight of 90.8 kg 10 days earlier than the maize control group. His experiments established that pearl millet has high value for pig feeding. Pearl millet promoted the formation of firm white fat comparable to that resulting from barley feed.

Pearl millet could also be used for grazing by pigs to save on concentrates. Burton (1980) reported that 45.5-kg pigs on full feed of a balanced concentrate grazed young pearl millet (var. Tifleaf 1) for 35 days, gained as well as those in dry lot, and required less concentrate per kilogram of gain. The concentrate saved by grazing made the pearl millet crop worth $250 per hectare.

D. Sheep

There are not many reports on the use of pearl millet grain and forage in the feeding of sheep. French (1948) observed that whole grains of pearl millet fed to sheep were less digested and a fair percentage passed whole into the feces. He suggested that grinding the grain can bring its feeding value nearly to that of crushed maize.

Martinez (1988) allowed lambs 4, 6, 9, or 12 kg herbage/100 kg live weight. Average live weight gains per lamb were 65, 68, 90, and 100 g/day respectively, corresponding to live weight gains of 663, 478, 434, and 374 kg/ha. A 6- to 7-kg allowance gave a reasonable balance between live weight gain per lamb and carrying capacity per hectare.

V. PEARL MILLET FORAGE

A. Varieties and Hybrids for Forage

Pearl millet varieties and hybrids with improved forage production potential have been developed and released for use in the United States and Australia. A list of the varieties and hybrids released for forage production is found in Table II. Recent varieties and hybrids furnish good summer grazing for milk cows and all classes of livestock; give a desirable seasonal distribution of forage; and are suitable for green chop, dehydration, pelleting, and production of quality silage.

In the development of new varieties and hybrids, two major genes, *d*, and *tr*, have been used in the improvement of forage quality (Burton, 1983). Increases in the proportion of leaf, and thus the nutritive value of the forage, are obtained by the use of the d_2, dwarfing gene which reduces internode length and therefore plant height by 50%. It is inherited as a simple recessive gene (Burton and Fortson, 1966; Burton *et al.*, 1969; Johnson *et al.*, 1968). Because of higher leaf proportion, forage from dwarf plants is higher in IVDMD than forage from tall plants. Johnson *et al.* (1968) observed that though the dwarf millets produce less dry matter per hectare they have 50% more leaves, 15% more protein, and 17% less lignin, and animal gains were more than those from tall millet. The trichomeless (*tr*) gene suppresses trichomes on all plant parts and is inherited as a single recessive gene (Powell and Burton, 1971). Though the *tr* gene increases forage palatability for cattle, because of a lower number of cracks on the adaxial and abaxial leaf surfaces, penetration of rumen microbes is reduced, resulting in slower digestion and thus reduced intake (Hanna and Akin, 1978). The loss in IVDMD associated with the *tr* gene tends to be compensated for by associated drought tolerance and better palatability (Burton *et al.*, 1977, 1988).

The brown-midrib (*bmr*) trait in pearl millet has potential to contribute to increased digestibility, as it is associated with reduced lignin concentration and increased IVDMD (Cherney *et al.*, 1988). Using wether sheep,

Table II

Pearl Millet Varieties and Hybrids for Forage Production

Designation	Type	Developed from	Main attributes (reference)
		Australia	
Katherine Pearl	Variety	An introduction from Ghana	Tall, late maturing, provides "wet season" grazing during growth and a standover forage for dry season (CSIRO, 1972)
Tamworth	Variety	Selection made following of a F_3 plant of Gahi 1	Mid-season to late in maturity, tillers well, suitable for summer/autumn grazing (CSIRO, 1972)
Ingrid Pearl	Variety	Introduction from Bambey, Senegal	Earlier than Katherine Pearl
		United States	
Starr	Synthetic	Crossing a leafy short plant discovered in Russian introductions with broadleafed and palatable "common" millet	Percentage and yield of leaves higher, good for fattening cattle (Burton and DeVane, 1951)
Gahi 1	Hybrid	Mixture of about 75% of 6 possible hybrids from 4 inbreds and 25% selfed and sibbed seed of these inbreds	Leafier, more forage, better seasonal distribution (Burton, 1962)
Gahi 2	Hybrid	Same as Gahi 1, except 4 different dwarf inbreds used; hybrid is tall	Produced less forage than Gahi 1 (Burton and Powell, 1968)
Tiflate	Synthetic	54 short-day photoperiod-sensitive introductions from West Africa	Remains vegetative much longer, better distribution of forage than Gahi 1, higher leaf and percentage IVDMD than Gahi 1 (Burton, 1972)
Gahi 3	Hybrid	cms Tift23A (or Tift23DA) × inbred Tift186	Matures later than Gahi 1, provides grazing for longer period, immune to *Pyricularia*, resistant to nematodes (Burton, 1977)
Tifleaf 1	Hybrid	cms Tift23DA × inbred Tift383	Easier to manage, matures later than Gahi 1, provides grazing for longer period, immune to *Pyricularia*, resistant to nematodes (Burton, 1980)
Tifleaf 2	Hybrid	cm Tift85D_2A × inbred Tift383	Resistant to rust and *Pyricularia*, better performance than Tifleaf 1 under rust infection (Hanna *et al.*, 1988)

Cherney *et al.* (1990) reported that the digestibility of DM, neutral detergent fiber, and acid detergent fiber were uniformly higher in the *bmr* genotype than in the normal, and lambs spent an average of 2.6 min on *bmr* genotype for every minute spent on the normal genotype, indicating its good palatability. The orange node trait (*on*) controlled by a single recessive gene resembles the *bmr* trait. The earhead, stem and leaf sheath were more digestible in the *onon* plants than normal plants and leaf blade digestibility did not differ. Results from crosses between *on* × *bmr* indicated that they were affected by the same gene (Degenhart *et al.*, 1991).

Burton (1962) demonstrated the potential of heterosis for forage production with the four-parent hybrid Gahi 1. Gahi 1 is a mixture of approximately 75% of the six possible hybrids from four inbred lines and 25% of selfed and sibbed seed of these lines. The hybrid seedlings, being more vigorous than their selfed parents, crowd out the inbreds and usually give yields comparable to 100% hybrid seed (Burton 1948, 1989). Gahi 3, a hybrid between the cms Tift23DA (or Tift23A) and inbred Tift186, eliminates this problem of mixtures of hybrid and selfed seed (Burton, 1977). In the production of cms forage hybrids, male parents (pollinators) that fail to restore fertility of the F_1, hybrid are preferred because they reduce the weed potential of the hybrid and improve forage quality, provide longer grazing, and yield under stress (Burton, 1981).

In Australia, improved varieties have potential for wet-season grazing during growth, as stand-over mature forage for the dry season, and in the production of palatable silage. In South Africa, pearl millet is called "Babala" and is used for summer grazing by cows and is also valued as a silage crop because it usually yields better than other silage crops (Penzhorn and Lesch, 1965; Hammes, 1972). In Korea, recent research has indicated that pearl millet has excellent potential as a forage crop (Choi *et al.*, 1990a). A pearl millet hybrid "Chungaecho" (Tift23DA × Tift186 = Gahi 3) was recently recommended to livestock farmers following extensive evaluations (Choi *et al.*, 1990b). Pearl millet is also being advocated for use as pasture in Brazil (Coser and Maraschin, 1983; Moraes and Maraschin, 1988). Breeding for forage yields per se has not received any attention in India and West Africa. Local varieties are dual purpose—provide both grain and dry fodder (Rachie and Majmudar, 1980). Forage breeding in India has concentrated on the interspecific hybrid between pearl millet and elephant grass (Gupta and Sidhu, 1972).

B. Diseases and Forage Production

Diseases are usually of minor importance where pearl millet is grown for forage. The initial growth is generally free from diseases but forage production from the second growth late in the season is affected by them (Burton, 1951). Diseases when severe cause substantial reductions in forage quantity and quality by lowering the protein content, acceptability, and digestibility (Burton, 1954; Burton and Wells, 1981).

Rust caused by *Puccinia substriata* var. *indica* was recognized as a serious disease on late-planted pearl millet following outbreaks in the southeastern Great Plains in the United States in 1972 (Wells *et al.*, 1973). The effects of rust are very severe and range from death of young plants from early infection to premature desiccation and or death of leaves with later infection.

Comparisons of rust-resistant and -susceptible plants showed significant reductions in DM concentration, DM yield, and IVDMD in diseased plants (Monson *et al.*, 1986). The combined effect of lower yields and lower IVDMD led to a mean 51% reduction in IVDMD from the infected plants. The yield of leaves was reduced less by rust than was the yield of stems, but the opposite was true for IVDMD concentrations. In contrast to the results of Monson *et al.* (1986), Wilson *et al.* (1991) did not observe an effect on DM concentration and this was explained to have resulted from differences in stages of harvest. They have observed a decrease in DM yield and digestibility with increased rust infection. Their results suggested that rapid loss of digestible DM yield at low rust severities represents a loss from the more highly digestible leaves.

The rust situation has been admirably managed by breeding varieties and male-sterile lines that are resistant to rust—an important step in maintaining forage quality. Andrews *et al.* (1985) identified a single dominant gene (Rpp_1) for rust resistance in an accession from Chad. Hanna *et al.* (1985) reported that a single dominant gene (Rr_1) controlled rust resistance in *P. americanum* subsp. *monodii* from Senegal, which was transferred to pearl millet by backcrossing.

Leaf spots caused by *Bipolaris* (*Cochliobolus*) *setariae, Cercospora penniseti, Helminthosporium stenospilum, Phyllosticta penicillariae*, and *Pyricularia grisea* (*Magnaporthe grisea*) generally appear after pearl millet flowers (Luttrell, 1954; Hanna and Wells, 1989; Wells and Hanna, 1988; Wilson *et al.*, 1989). Burton and Wells (1981) estimated that *Cercospora* leaf spot when severe reduced forage yields by 20–25% and brown mottle (unknown etiology) had no effect on the first forage yield, but reduced the second harvest by 23% and the third by 30%. Resistance to *B. setariae* leaf spot is controlled by a four-independent gene system (Wells and Hanna, 1988) and to *P. grisea* by three independent and dominant genes (Hanna and Wells, 1989).

Insect pests that infest pearl millet include the fall armyworm, *Spodoptera frugiperda*, larvae of the corn earworm *Heliothis zea*, the lesser cornstalk borer (*Elasmopalpus lignosellus*), and chinch bugs (*Blissus leucopterus leucopterus*). Lines resistant to fall armyworm and chinch bug have been identified (Leuck *et al.*, 1968; Merkle *et al.*, 1983). Pearl millet suffers rare infestations by the lesion nematode (*Pratylenchus* spp.) and sting nematode (*Belonolaimus* sp.). New varieties carry resistance to these nematodes (Burton, 1977).

C. Management for Forage

Since Voorhees (1907) first described management practices for obtaining palatable forage from pearl millet in the United States, several studies have been carried out on the management of this crop for grazing and forage. These essentially include experiments on intensity of grazing and frequency, number and height of clipping on quality, and interaction between stubble height and cutting frequency and regrowth.

Burton (1965a, 1966) and Burton *et al.* (1986) have demonstrated that later-maturing millet varieties produce leafier forage for a longer duration, have a better seasonal distribution of forage, are higher in protein content, and are easier to manage and more digestible than earlier varieties. Burton (1951) suggested that when about 45 cm tall, the crop could be grazed rotationally and mowing to prevent heading could extend the productive season and improve forage quality.

The stage of development at harvest greatly influences yield and regrowth habit. Beaty *et al.* (1965), Begg (1965), Fribourg (1966), and Clapp and Chamblee (1970) reported that as stubble height was raised regrowth from terminal buds increased while axillary and basal tillering remained constant. Stephenson and Posler (1984) observed that more tillers are initiated at the vegetative stage and at taller stubble heights. Basal tillering increased as the stubble height was lowered and regrowth at the boot stage was more dependent on reserve carbohydrates than at the vegetative stage. The root system has to be well developed to maintain new growth during the early phases of tillering (Clapp and Chamblee, 1970).

Total nonstructural carbohydrate has been shown to be used in part for regrowth following defoliation by Mays and Washko (1962). They reported substantial dependence on carbohydrate reserves when pearl millet was harvested to a 5-cm stubble height, and dependence decreased as the stubble height was raised to 15 cm. Plants cut at 15 or 20 cm had sufficient remaining photosynthetic tissue to supply the plant requirements for regrowth, whereas those cut at 5 or 10 cm utilized more reserve material.

Higher yields of palatable and digestible feed are obtained if pearl millet is harvested just as it comes to head. The yields reported in the literature vary widely, ranging from 18 to 45 t/ha, the latter when the season is favorable and the crop is allowed to reach maturity.

Hoveland and McCloud (1957) found that rows spaced 45.7 to 50.8 cm apart produced highest yields. Among several treatments investigated the best combination for production and quality was obtained when 76-cm plants were clipped to 45-cm stubble. Broyles and Fribourg (1959) suggested that pearl millet to be used for pasture or silage should be allowed to reach a height of 76 cm before it is grazed or cut down to 15.2 to 25.4 cm. Regrowth was more rapid from 15.2- to 20.3-cm stubble than from 7.6- to 10-cm stubble.

Beaty *et al.* (1965) reported that Gahi 1 pearl millet responded to a wider range of harvest conditions that Tift Sudan grass or Sudax 11, a hybrid between Sudan grass and sorghum. Harvesting at 5-week intervals increased forage production by 46% over harvesting every 2 weeks. Harvesting seven-eights of the plant increased production by 18% over harvesting one-third of available height. Mays *et al.* (1966) cut Sudan grass–sorghum hybrids, Sudan grass, and pearl millet to stubble heights of either 10 or 20 cm on reaching heights of 50, 86, and 120 cm. Yield averages for all crops indicated that those cut when 50 and 86 cm high, respectively, yielded 42 and 71% as much as those cut when 120 cm high. Cutting to a stubble height of 20 cm resulted in about 7% lower yields than cutting to a height of 10 cm.

Burger and Hittle (1967), using sorghum × Sudan grass hybrids, Sudan grass hybrids, pearl millet hybrids, and Sudan grass found that all produced superior yields at three harvests per year compared to four harvests. Better yields were obtained with a 7.6-cm stubble than with a 15.2-cm stubble. Hoveland *et al.* (1967) reported the protein content of sorghum–Sudan grass to be higher than that of pearl millet at several rates of nitrogen when harvested in the preboot stage with similar DM digestibility. They found a response to high rates of nitrogen from both species, but did not observe this to affect either DM digestibility or leaf percentage. In New South Wales, Australia, Ferraris and Norman (1973) suggested that for obtaining high, well-distributed yields coupled with quality, frequent harvests leaving a tall stubble of 30 cm was desirable. Although total productivity was favored by less intensive management, quality was improved by intensive cutting.

Management influences not only forage yields but also forage quality. Rusoff *et al.* (1961) found that lignin content progressively increased with plant maturity. Burton *et al.* (1964) reported that young leaf blades contained higher levels of CP, true protein, and lower levels of lignin than older leaves. Improvements in protein content were reported with frequent and severe cutting (Burger and Hittle, 1967) and with delay in sowing, and decreased stubble height was also reported (Hoveland and McCloud, 1957; Westphalen and Jacques, 1978). Decrease in protein content in forage with increasing age has also been reported (Hill, 1969; Patel *et al.*, 1958; Sehagal and Goswami, 1969).

Because forages are grown to feed animals, their digestibility, composition, and intake are of prime importance. Generally a good quality forage has a high leaf–stem ratio, is high in protein and digestible nutrients, and is low in fiber and lignin. Hart (1967) observed positive correlations between leafiness and DM digestibility. Lignin content was found to be a good predictor of digestibility. DM digestibility of leaves was significantly correlated with CP and crude fiber content, but less closely with lignin content. Achacoso *et al.* (1960) observed that as DM increased there were increases in both crude fiber and lignin content. They recorded highly significant and negative correlations for CP content with lignin, crude fiber, and DM.

D. Grazing by Livestock

Pastures provide the least expensive source of nutrients, as grazing of crops with cattle eliminates costs and losses of nutrients associated with harvesting, processing, storing, and feeding.

The palatability of pearl millet forage for dairy cows is high (Ball, 1903). Burton et al. (1964) reported that young leaves are much more palatable than older leaves. Norman and Phillips (1968) observed that cattle grazing on a mature and near-mature crop consumed first the earheads, followed by stems and leaves, and suggested that the order of preference was probably associated with digestible carbohydrate content.

Pearl millet pasture grazed rotationally by dairy cows provides total digestible nutrients (TDN) in the range 1400–2300 kg/ha, a quantity generally superior to that for Sudan grass and sorghum (Faires *et al.*, 1941; Roark *et al.*, 1952; Marshall *et al.*, 1953). Marshall *et al.* (1953) found that at an annual average of 2360 kg/ha of TDN, lactating cows derived 60% of their TDN intake while on millet pasture, which was adequate to support the requirement for body maintenance and 4.5 kg out of a daily production of 13.8 kg of 4% fat-corrected milk.

Miles *et al.* (1956) have shown that Tift sudan has consistently produced more DM, milk, and TDN than pearl millet but pearl millet consistently provided higher quality pasture than permanent pastures. Rollins *et al.* (1963) reported that intensively managed Gahi 1 pearl millet was the best forage for maintaining lactation, though a combination of pearl millet and Bermuda grass gave an equivalent production; both were superior to Bermuda grass alone. Experiments with lactating Jersey cows showed that pearl millet did not differ in the amount of grazing provided and was superior to Sudan grass in yield of dry forage (Baxter *et al.*, 1959). Average milk production was between 41.7 to 43.4 kg/day.

In South Africa, Penzhorn and Lesch (1965) observed that dairy cows found sweet Sudan grass more palatable than pearl millet. However, they grazed pearl millet readily and it gave a higher carrying capacity than sweet Sudan grass with similar average milk production.

McCartor and Rouquette (1977) studied livestock gains of weanling cattle and the factors affecting profitability of grazing pearl millet. The most important factor was the differential between the buying and selling price of the cattle, which the farmer seldom controls. Besides this factor, grazing pressure was an important determinant of profit. They concluded that greatest profit or least loss occurs at medium grazing pressures (approximately 2 kg of available forage per kilogram of animal live weight). They observed that pearl millet was difficult to manage as a grazing crop because of fluctuations in forage production over a relatively short period of time. Maximum live weight gains were achieved at low stocking rates, resulting in low forage utilization.

1. Weight gains of beef cattle on forage

Pearl millet proved to be the best temporary grazing pasture tested in Tifton, Georgia. Pearl millet grown on good soil, fertilized liberally, and grazed rotationally required only 0.12 ha to provide all the forage a dairy cow would consume. A succession of plantings of pearl millet is desirable for continuous grazing through the summer. Pearl millet planted in rows and cultivated once yielded 229 kg/ha of live weight gain and when grazed for about 80 days compared to 212 kg/ha for a broadcast crop (Georgia Coastal Plain Experiment Station, 1947).

Dunavin (1980) reported that total gains of 643 kg/ha when yearling cattle grazed on two successive plantings of pearl millet in the same season; animal gains produced by grazing the first plantings were significantly greater than those for the second planting. Norman and Stewart (1964) found that cattle grazing mature standing pearl millet in the dry season made an average live weight gain of 296 kg/ha in 16 weeks. Wet-season grazing by beef cattle in northern Australia gave gains of 102 kg/head in 20–24 weeks at 2.5 animals/ha (Norman, 1963).

In a 3-year study, Dunavin (1970) compared Gahi 1 pearl millet with two sorghum × Sudan grass hybrids as pasture for yearling beef cattle. Gahi 1 millet produced superior gains per hectare per day on both early (3.70 kg) and late (2.60 kg) planted pastures. Hoveland *et al.* (1967) reported only 0.45 to 0.57 kg/day on pearl millet and attributed these low gains to high moisture content in the forage. Johnson *et al.* (1978) evaluated the performance of dairy heifers grazing on Gahi 1, Gahi 3, and Tifleaf 1. Daily gain per animal averaged 0.63, 0.76 and 0.84 kg from Gahi 1, Gahi 3, and Tifleaf 1, respectively.

2. Effect of pearl millet grazing on milk fat

Grazing lactating cows on pearl millet is associated with depression in butter fat content of milk, a problem accentuated by high grain supplement levels (Clark *et al.*, 1965). Miller *et al.* (1963) found that the milk produced on millet pasture had only 2.85% butterfat compared with 3.64% from Sudan grass grazing. However, total production, nonfat solids, and protein content were similar. Hemken *et al.* (1968) indicated that fat depression was influenced by cation fertilization levels. Bucholtz *et al.* (1969) observed that cows grazing on pearl millet produced milk that is significantly lower in fat content and the fat contained a higher degree of unsaturation than when the cows were grazed on Sudan grass. The molar percentage of rumen butyrate was significantly reduced in cows grazed on pearl millet. The concentration of oxalic acid was significantly higher in pearl millet herbage than in Sudan grass. There was a trend toward higher concentrations of all minerals (Mn, K, Na) in pearl millet than in Sudan grass. Schneider and Clark (1970) suggested limiting K fertilization, correcting Ca and Mg deficiencies, and monitoring nitrate levels during periods of moisture stress to restrict nitrate toxicity. The same conditions also appeared to reduce oxalate and succinate levels with consequent increases in butterfat content.

A method to reduce oxalic acid content in pearl millet forage was suggested by Parveen *et al.* (1988). When pearl millet fodder cut at preflowering stage with 2.12% oxalic acid was soaked in water (1:10) for 30 min twice in succession, the oxalic acid content was reduced to 0.69%.

E. Nitrate Toxicity

Poisoning is caused when cattle graze millet that is abnormally high in nitrates (Green, 1973). High amounts of nitrates are likely to occur in crops grown under stress situations such as drought, low temperatures, and diseases. Rouquette *et al.* (1980) reported that pearl millet grown under apparent drought stress conditions was unpalatable to grazing cattle and contained potentially toxic levels of nitrate and high levels of total alkaloids. In a study of 11 pearl millet lines by Krejsa *et al.* (1984), alkaloid levels ranged from 17 to 101 mg/kg and nitrate levels from 2.4 to 9.8 g/kg. Leaf blades contained more total alkaloids than stem plus sheaths, and stem plus sheaths contained more nitrate than leaf blades.

Stage of growth markedly changes the nitrate content of forages. Nitrate concentrations are higher in young plants and decrease as the plant matures. Leaves contain less nitrate than stems and harvest near maturity normally leads to lower levels of nitrate (Burger and Hittle, 1967). Nitrogen fertilizer also affects nitrate levels. Fribourg (1974) reported that nitrogen accumulation was highest in Sudan grass and pearl millet, especially in drought periods with high available soil nitrogen and potassium levels and with molybdenum deficiency. Oxalate was found to increase with increases in soil potassium level and drought stress. Lemon and McMurphy (1984) observed that nitrate levels increased with higher nitrogen fertilization, but decreased with later maturity stages. Lower portions of the plant contained 2.7 to 3.5 times more nitrate than the upper portions, suggesting that raising the cutting height would reduce the forage nitrate content.

Mefluidide, a quality-enhancing plant growth regulator in pearl millet, was shown to inhibit vegetative growth and increase nitrogen content and nitrate nitrogen. Concentrations of total nitrogen and nitrate nitrogen were inversely related to DM yield, suggesting that accumulations were a result of prolonged nitrate uptake in the absence of growth (Fales and Wilkinson, 1984). They advised caution in the use of mefluidide, as levels of accumulated nitrate nitrogen could be potentially hazardous to livestock.

F. Silage

Pearl millet produces good silage, as it is high in carbohydrates and DM content is sufficiently high to result in little or no excess moisture (Boyle and Johnson, 1968). The stage of maturity at harvest is an important factor influencing the composition and nutritive value of silage. Generally, addition of a carbohydrate preservative such as ground snapped corn or citrus pulp is required for making good silage.

Johnson and Southwell (1960) reported on the performance of animals fed on millet and maize silages. On DM basis the intake of millet silage by lactating Jersey cows was greater than that of corn silage. Though milk production from millet silage was significantly higher, differences in body weight gains were not significant. It appeared that cows fed millet silage obtained more net energy per unit DM intake than those fed corn silage. Similar results were reported by Lansbury (1959) and Sisk *et al.* (1960). Working with yearling steers, Baker (1970/1971) reported that pearl millet and sorghum × Sudan grass hybrid silage was more satisfactory than small-grain and perennial grass silage for yield and quality. However, summer annual grass silage often did not give as good results as comparable pasture.

Bertrand and Dunavin (1973) observed that steers grazing Gahi 1 millet pasture gained weight quicker than steers receiving Gahi 1 millet silage

(0.75 and 0.62 kg/head/day, respectively) and beef yield was slightly higher for steers receiving millet silage (495 versus 455 kg/ha). They concluded that the small increase in amount of beef produced would not compensate for the additional expenditures required for harvesting, storing, and feeding millet silage. Jaster *et al.* (1985) concluded that heifers consuming pearl millet and sorghum silages showed higher DM intake and DM digestibility than those consuming cool-season silages following evaluations under a forage double-cropping system.

Gupta *et al.* (1981) using cross-bred (Haryana × Jersey) cattle studied the improvements brought about in the nutritional quality of fodder when pearl millet and cowpea (*Vigna ungiculata*) were ensiled together. The digestibility coefficients for DM and CP content, IVDMD, digestible CP, and TDN were more for pearl millet–cowpea silage than for pearl millet alone. Similar results were reported by Freitas (1988).

For sheep, Silveira *et al.* (1981) reported that silage of pearl millet harvested at boot leaf stage in mixture with cowpea had a higher organic matter intake and digestibility than other silages. Crude protein increased from 7.2 in the fresh material to 8.6% in the silage. As the cell wall content was lower and digestibility higher, energy intake was higher and the sheep were able to retain consumed nitrogen. On the other hand, working with sheep, Singh and Mudgal (1980) observed that although cowpea and pearl millet can be successfully conserved as silage, protein and energy supplementation will be required for a practical feeding of animals. Andrade and Andrade (1982a) reported that a maximum dry matter yield of 21.9 t/ha was obtained at 134 days vegetative growth to produce acceptable silage. Sugarcane and molasses improved the quality of silage by reducing butyric acid and increasing lactic acid contents. However, in tests with sheep, CP, crude fiber, and DM digestibility were not significantly increased when sugarcane or molasses were added (Andrade and Andrade, 1982b).

VI. CONCLUSIONS

The potential benefits from the application of existing knowledge and from further research in pearl millet are substantial both for the food crop in low-resource agriculture and for the forage or feed grain crop in warm-temperate agriculture.

In low-resource agriculture the main benefits will come from research into grain processing and food product research as well as from plant breeding. Pearl millet foods must be as easy to make as those from rice and wheat, and the flour, or partially prepared grain, must have a longer shelf life. In India approximately 25% of the pearl millet crop is marketed, which is two to five times as much as in Africa (FAO, 1990). This has been a crucial factor in stimulating continued research and supporting a hybrid seed industry in India. In Africa, general economic factors and poor cereal markets have been major constraints to the adoption of yield-increasing technologies for dryland cereal production (OTA, 1988). Although the main constraints to production for pearl millet in Africa are recognized as low soil nutrient levels as well as low rainfall (Fussell *et al.*, 1987), the highest cost/benefit return comes from the adoption of new cultivars. The increase in pearl millet productivity of 2.3% per annum over the past 20 years in India following the adoption of new cultivars (Harinarayana, 1987) points to what can be achieved in Africa. However, producing cereal grain surplus to family needs for marketing must be equally or more attractive to the African farmer than alternative cash crop options. The increase in cultivar yield potential in either hybrids or varieties in India shows no indication of plateauing (Harinarayana, 1987; ICRISAT, 1991). Research in Africa has shown that both single-cross and topcross hybrids are substantially higher yielding than varieties (Kumar, 1987).

The steady development in the quality and yield of pearl millet as a forage crop in the United States over the last 50 years is a remarkable testimony both to the species and to plant breeding. The incorporation of the low lignin factor, currently under way in several breeding programs, will result in a significant improvement in forage digestibility. New sources of disease resistance, the identification of improved heterotic patterns, and the potential use of genes from related *Pennisetum* species may further improve productivity and use of pearl millet.

Possibly the greatest advances in the next decade will come from the development of pearl millet as feed grain crop, adapted to warm-temperate regions. In some respects, the development of pearl millet for the U.S. Midwest is following a course similar to that of soybeans 40 years ago where there was very little germplasm in which yield potential was not strongly associated with photoperiod sensitivity. Existing cultivars were then too late, tall, and weak-stemmed for use in the Midwest. Similarly, a market had not been developed. However, rapid progress has been made in reorganizing the pearl millet plant to a combine type and in building up yield levels in the dwarf early-maturing background. Though the good nutritional status of the grain is already established, opportunities for further genetic improvement in feed value are known to exist.

The benefits from basic research on pearl millet and related species and the application of biotechnology offer less certain but potentially far-reaching impacts. The possibility of transferring apomixis from *P. squamulatum* into pearl millet (Dujardin and Hanna, 1989) would essentially mean that hybrids could be cloned by seed increase. Gene mapping, both nuclear and cytoplasmic, by various methods is currently under way in several laboratories and will assist in locating important genes or quantitative trait loci, enhancing breeding efficiency. Anther culture and haploid development have been reported in pearl millet (Bui-Dang-Ha and Pernes, 1985; Bui-Dang-Ha *et al.*, 1986), as has protoplast formation and embryo regeneration (Vasil and Vasil, 1980; Lorz *et al.*, 1981), but there are no reports of further progress.

Increases in production and product quality are dependent in the long-term on the discoveries made from basic research. Research interest in pearl millet has been steadily growing as evidenced by the number and scope of recent publications, and Hanna (1987) comments that it is the "drosophila" of cereals for research. From discoveries already made and used in the crops' improvement, it is apparent that pearl millet has the potential to have a larger role in world agriculture, both as food and forage in the developing world and as feed and forage in temperate agriculture.

Acknowledgments

This work was partially supported by USAID Grant No. DAN 1254-G-00-0021-00 through INTSORMIL, the International Sorghum and Millet CRSP, by Program Support Grant No. AID-DSAN-XII-G-0124.

References

Abdelrahman, A. A., and Hoseney, R. C. (1984). Basis for hardness in pearl millet, grain sorghum and corn. *Cereal Chem.* **61,** 232–235.

Abate, A. N., and Gomez, M. (1983/1984). Substitution of finger millet (*Eleusine coracana*) and bulrush millet (*Pennisetum typhoides*) for maize in broiler feeds. *Anim. Feed Sci. Technol.* **10,** 291–299.

Achacoso, A. S., Mondart, C. L., Jr., Bonner, F. L., and Rusoff, L. L. (1960). Relationship of lignin to other chemical constituents in sudan and millet forages. *J. Dairy Sci.* **43,** 443. (abstract).

Akingbala, J. O. (1991). Effect of processing on flavenoids in millet (*Pennisetum americanum*) flour. *Cereal Chem.* **68,** 180–183.

Aliya, S., and Geervani, P. (1981). An assessment of the protein quality and vitamin B content of commonly used fermented products of legumes and millets. *J. Sci. Food Agric.* **32,** 837.

Almeida-Dominguez, H. D., Gomez, M. H., Serna-Saldivar, S. O., and Rooney, L. (1990a). Weaning food from composite of extruded or press dried pearl millet and cowpea. *Cereal Food World* **35,** 831. (abstract)

Almeida-Dominguez, H. D., Gomez, M. H., Serna-Saldivar, S. O., Rooney, L., and Lusas, E. (1990b). Extrusion of pearl millet. *Inst. Food Technol.* p. 114 (abstract).

Andrade, J. B. De, and Andrade, P. De (1982a). Silage production from pearl millet (*Pennisetum americanum* (L.) K. Schum.) [Pt]. *Bol. Ind. Anim.* **39,** 155–165.

Andrade, J. B. De, and Andrade, P. De (1982b). In vivo digestibility of pearl millet (*Pennisetum americanum* (L.) K. Schum.) silage [Pt]. *Bol. Ind. Anim.* **39,** 67–73.

Andrews, D. J. (1972). Intercropping with sorghum in Nigeria. *Expl. Agric.* **8,** 139–150.

Andrews, D. J. (1990). Breeding pearl millet for developing countries. *In* "INTSORMIL Annual Report," pp. 114–118. University of Nebraska–Lincoln.

Andrews, D. J., and Kassam, A. H. (1976). The importance of multiple cropping in increasing world food supplies. *In* "Multiple Cropping," ASA Special Publ. No. 27, pp. 1–10.

Andrews, D. J., Rai, K. N., and Singh, S. D. (1985). A single dominant gene for rust resistance in pearl millet. *Crop Sci.* **25,** 565–566.

Andrews, D. J., and Rajewski, J. F. (1991). "1990 Pearl Millet Regional Grain Yield Trials: Preliminary Report," pp. 9. Univ. of Nebraska-Lincoln. (mimeo)

Annegers, J. F. (1973). The protein–calorie ratio of West African diets and their relationship to protein-calorie malnutrition. *Ecol. Food Nutr.* **2,** 225–235.

Appa Rao, S. (1987). Traditional food preparations of pearl millet in Asia and Africa. *In* "Proceedings International Pearl Millet Workshop" (J. R. Witcombe and S. R. Beckerman, eds.), pp. 289. ICRISAT, Patancheru, India.

Badi, S., Pedersen, B., Manowar, L., and Eggum, B. O. (1990). The nutritive value of new and traditional sorghum and millet foods from Sudan. *Plant Foods Hum. Nutri.* **40,** 5–19.

Baker, F. S., Jr. (1970/1971). Mimeographed data. Florida Agricultural Experiment Station, AREC, Quincy, Florida (quoted by Bertrand and Dunavin, 1973).

Ball, C. R. (1903). "Pearl Millet." Farmers' Bulletin No. 168, USDA, Washington, D.C.

Baxter, H. D., Owen, J. R., and Ratcliff, L. (1959). Comparison of two varieties of pearl millet for dairy cattle. *In* "Tennessee Farm Home Science Progress Report No. 32," pp. 4, 5.

Beaty, E. R., Smith, Y. C., McCreery, R. A., Ethredge, W. J., and Beasley, K. (1965). Effect of cutting height and frequency on forage production of summer annuals. *Agron. J.* **57,** 277–279.

Begg, J. E. (1965). The growth and development of a crop of bulrush millet (*Pennisetum typhoides* S. & H.). *J. Agric. Sci.* **65,** 341–349.

Bertrand, J. E., and Dunavin, L. S. (1973). Millet pasture and millet silage supplemented and unsupplemented for beef steers. *Proc. Soil Crop Sci. Soc. Florida* **32,** 23–25.

Bidinger, F. R., and Rai, K. N. (1989). Photoperiodic response of parental lines and F1 hybrids in pearl millet. *Indian J. Genet.* **49,** 257–264.

Bilquez, A. F. (1963). Etude du mode d'heredite de la precocite chez la mil penicillaire (*Pennisetum typhoides*). I. Determinisme genetique des differences a la longeur du jour existant entre les mils due group Sanio et ceux du groupe Souna. *Agron. Trop.* **18,** 1249–1253.

Bilquez, A. F., and LeComte, J. (1969). Relations entre mils sauvages et mils cultives: Etude de l'hybride. *Agron. Trop.* **24,** 249–257.

Boyle, J. W., and Johnson, R. I. (1968). New summer forage crop in New South Wales. *Agric. Gazette N.S.W.* **79,** 513–515.

Bramel-Cox, P. J., Andrews, D. J., and Frey, K. J. (1986). Exotic germplasm for improving grain yield and growth rate in pearl millet. *Crop Sci.* **26,** 687–690.

Broyles, K. R., and Fribourg, H. A. (1959). Nitrogen fertilization and cutting management of sudangrass and millets. *Agron. J.* **51,** 277–279.

Brunken, J. N. (1977). A systematic study of *Pennisetum* sect. *Pennisetum* (Gramineae). *Am. J. Bot.* **64,** 161–176.

Brunken, J. N., de Wet, J. M. J., and Harlan, J. R. (1977). The morphology and domestication of pearl millet. *Econ. Bot.* **31,** 163–174.

Bucholtz, H. F., Davis, C. L., Palinquist, D. L., and Kendall, K. A. (1969). Study of the low-fat milk phenomenon in cows grazing pearl millet pastures. *J. Dairy Sci.* **52,** 1385–1394.

Bui-Dang-Ha, M., Mequinon, M. J., and Pernes, J. (1986). Genetic changes after androgenesis in pearl millet, *Pennisetum americanum*: Studies from pollen culture of an F1 hybrid (Massue × Ligui). "Proceeding Nuclear Techniques and in Vitro Culture for Plant Improvement," August 1985, IAEA, Vienna.

Bui-Dang-Ha, D., and Pernes, J. (1985). Variability observed in millet (*Pennisetum typhoides*) after anther culture. *Bull. Soc. Bot. France* **132,** 150.

Burger, A. W., and Hittle, C. N. (1967). Yield, protein, nitrate and prussic acid content of sudangrass, sudangrass hybrids and pearl millets harvested at two cutting frequencies and two stubble heights. *Agron. J.* **59,** 259–262.

Burton, G. W. (1948). The performance of various mixtures of hybrid and parent inbred pearl millet, *Pennisetum glaucum* (L.) R. Br. *J. Am. Soc. Agron.* **40,** 908–915.

Burton, G. W. (1951). The adaptability and breeding of suitable grasses for the southeastern states. *In* "Advances in Agronomy" (A. Norman, ed.), Vol. 3, pp. 197–241. Academic Press, San Diego.

Burton, G. W. (1954). Does disease resistance affect forage quality? *Agron. J.* **46,** 99.

Burton, G. W. (1958). Cytoplasmic male-sterility in pearl millet *Pennisetum glaucum* (L.) R. Br. *Agron. J.* **50,** 230.

Burton, G. W. (1962). Registration of varieties of other grasses: Gahi-1 pearl millet. *Crop Sci.* **2,** 355–356.

Burton, G. W. (1965a). Photoperiodism in pearl millet, *Pennisetum typhoides. Crop Sci.* **5,** 333–335.

Burton, G. W. (1965b). Pearl millet Tift 23A released. *Crops Soils* **17,** 19.

Burton, G. W. (1966). Photoperiodism in pearl millet, *Pennisetum typhoides*, its inheritance and use in forage improvement. *In* "Proceedings, X International Grassland Congress," Helsinki, Finland, pp. 720–723.

Burton, G. W. (1972). Registration of Tiflate pearl millet. *Crop Sci.* **12,** 128.

Burton, G. W. (1977). Registration of Gahi 3 pearl millet. *Crop Sci.* **17,** 345–346.

Burton, G. W. (1980). Registration of pearl millet inbred Tift 383 and Tifleaf 1 pearl millet. *Crop Sci.* **20,** 292.

Burton, G. W. (1981). Improving the efficiency of forage-crop breeding. *In* "Proceedings, XIV International Grassland Congress" (J. A. Smith and V. W. Hays, eds.), pp. 138–140. Westview Press, Boulder, Colorado.

Burton, G. W. (1983). Breeding pearl millet. *Plant Breeding Rev.* **1,** 162–182.

Burton, G. W. (1989). Composition and forage yield of hybrid-inbred mixtures of pearl millet. *Crop Sci.* **29,** 252–255.

Burton, G. W., and DeVane, E. H. (1951). Starr millet—synthetic cattail lasts longer and produces more beef per acre. *South. Seedsman* **14,** 17, 68, 69.

Burton, G. W., and Fortson, J. C. (1966). Inheritance and utilization of five dwarfs in pearl millet (*Pennisetum typhoides*) breeding. *Crop Sci.* **6,** 69–72.

Burton, G. W., Hanna, W. W., Johnson, J. C., Jr., Leuck, D. B., Monson, W. G., Powell, J. B., Wells, H. D., and Widstrom, N. W. (1977). Pleiotropic effects of the *tr* trichomeless gene in pearl millet on transpiration, forage quality, and pest resistance. *Crop Sci.* **17,** 613–616.

Burton, G. W., Knox, F. E., and Beardsley, D. W. (1964). Effect of age on the chemical composition, palatability, and digestibility of grass leaves. *Agron. J.* **56,** 160–161.

Burton, G. W., Kvien, C. S., and Maw, B. W. (1988). Effect of drought stress on productivity of trichomeless pearl millet. *Crop Sci.* **28,** 809–811.

Burton, G. W., Monson, W. G., Johnson, J. C., Jr., Lowery, R. S., Chapman, H. D., and Marchant, W. H. (1969). Effect of the d_2 dwarfing gene on the forage yield and quality of pearl millet. *Agron. J.* **61,** 607–612.

Burton, G. W., and Powell, J. B. (1968). Pearl millet breeding and cytogenetics. *In* Advances in Agronomy," (N. Brady, ed.), Vol. 20, pp. 49–89. Academic Press, San Diego.

Burton, G. W., Primo, A. T., and Lowrey, R. S. (1986). Effect of clipping frequency and maturity on the yield and quality of four pearl millets. *Crop Sci.* **26,** 79–81.

Burton, G. W., Wallace, A. T., and Rachie, K. O. (1972). Chemical composition and nutritive value of pearl millet. *Crop Sci.* **12,** 187, 188.

Burton, G. W., and Wells, H. D. (1981). Use of near-isogenic host populations to estimate the effect of three foliage diseases on pearl millet forage yield. *Phytopathology* **71,** 331–333.

Butler, J. L., Wilkinson, W. S., Hellwig, R. E., Barbee, C., and Knox, F. E. (1969). Factors involved in production and storage of high-quality dehydrated coastal bermudagrass and pearl millet. *Trans. Am. Soc. Agric. Eng.* **12,** 552–555.

Calder, A. (1955). Value of munga (millet) for pig feeding. *Rhodesia Agric. J.* **52,** 161–170.

Calder, A. (1960). The value of ropoko (millet) for pig feeding. *Rhodesia Agric. J.* **57,** 116–119.

Calder, A. (1961). The production of pork pigs comparing maize, munga (millet) and pollards. *Rhodesia Agric. J.* **56,** 363–364.

Chanda, M., and Matta, N. K. (1990). Characterization of pearl millet protein fractions. *Phytochemistry* **29,** 3395–3399.

Chandrasekhar, G., and Pattabiraman, T. N. (1981). Natural plant enzyme inhibitors: Isolation and characterization of two trypsin inhibitors from bajra (*Pennisetum typhoideum*). *Indian J. Biochem. Biophys.* **19,** 1–7.

Chaudhary, P., and Kapoor, A. C. (1984). Changes in the nutritional value of pearl millet flour during storage. *J. Sci. Food Agric.* **35,** 1219–1224.

Cherney, D. J. R., Patterson, J. A., and Johnson, K. D. (1990). Digestibility and feeding value of pearl millet as influenced by the brown-midrib, low-lignin trait. *J. Anim. Sci.* **68,** 4345–4351.

Cherney, J. H., Axtell, J. D., Hassen, M. M., and Anliker, K. S. (1988). Forage quality characterization of a chemically induced brown-midrib mutant in pearl millet. *Crop Sci.* **28,** 783–787.

Choi, H. B., Park, K. Y., and Park, R. K. (1990a). Productivity and feed value of pearl millet (*Pennisetum americanum* (L.) Leeke) grown as a newly introduced forage crop. *J. Korean Soc. Int. Agric.* **1,** 71–84.

Choi, H. B., Park, K. Y., and Park, R. K. (1990b). A new pearl millet hybrid "Chungaecho" of high quality and high forage yield. *Res. Rep. Rural Dev. Admin.* (South Korea) **32,** 45–52.

Christensen, N. B., Palmer, J. C., Praeger, H. A., Jr., Stegmeier, W. D., and Vanderlip, R. L. (1984). Pearl millet: A potential crop for Kansas. *In* "Keeping up with Research," No. 77, Kansas Agric. Exp. Sta., Manhattan.

Clapp, J. G., and Chamblee, D. S. (1970). Influence of different defoliation systems on the regrowth of pearl millet, hybrid sudangrass, and two sorghum-sudangrass hybrids from terminal, axillary and basal buds. *Crop Sci.* **10,** 345–349.

Clark, N. A., Hemken, R. W., and Vandersall, J. H. (1965). A comparison of pearl millet, sudangrass and sorghum–sudangrass hybrid as pasture for lactating dairy cows. *Agron. J.* **57,** 266–269.

Coser, A. C., and Maraschin, E. G. (1983). Desempenho animal em pastagens de milheto comum e sorgho. *Pesquisa Agropecuaria Brasileira* **18,** 421–426.

Craufurd, P. Q., and Bidinger, F. R. (1988). Effect of the duration of the vegetative phase on shoot growth, development and yield in pearl millet (*Pennisetum americanum* (L.) Leeke). *J. Exp. Bot.* **39,** 124–139.

Craufurd, P. Q., and Bidinger, F. R. (1989). Potential and realized yield in pearl millet (*Pennisetum americanum*) as influenced by plant population density and life-cycle duration. *Field Crops Res.* **22,** 211–225.

CSIRO (1972). "Register of Australian Herbage Plant Cultivars." Australian Herbage Plant Registration Authority, Division of Plant Industry, Commonwealth Scientific and Industrial Research Organization, Canberra, Australia.

Curtis, D. L., Burton, G. W., and Webster, O. J. (1966). Carotinoids in pearl millet seed. *Crop Sci.* **6,** 300, 301.

de Wet, J. M. J. (1987). Pearl millet (*Pennisetum glaucum*) in Africa and India. *In* Proceedings, International Pearl Millet Workshop (J. R. Witcombe and S. R. Beckerman, eds.), pp. 3, 4. ICRISAT, Patancheru, India.

Degenhart, N. R., Werner, B. K., and Burton, G. W. (1991). An orange node trait in pearl millet: Its inheritance and effect on digestibility and herbage yield. *In* "American Society of Agronomy Abstracts, Southern Branch," Fort Worth, Texas.

Dendy, D. A. V., Clarke, P. A., and James, A. W. (1970). The use of wheat and non-wheat flours in breadmaking. *Trop. Sci.* **12,** 131–141.

Doggett, H. 1988. "Sorghum." pp. 411–427. Longman, Essex, England.

Doggett, H., and Majisu, B. N. (1968). Disruptive selection in crop improvement. *Heredity* **23,** 1–23.

Dujardin, M., and Hanna, W. W. (1989). Developing apomictic pearl millet—Characterization of a BC_3 plant. *J. Genet. Breeding* **43,** 145–151.

Dujardin, M., and Hanna, W. W. (1990). Cytogenetics and reproductive behavior of 48 chromosome pearl millet × *Pennisetum squamulatum* derivatives. *Crop Sci.* **30,** 1015–1016.

Dunavin, L. S. (1970). Gahi-1 pearl millet and two sorghum × sudangrass hybrids as pasture for yearling beef cattle. *Agron. J.* **62,** 375–377.

Dunavin, L. S. (1980). Forage production from pearl millet following rye and ryegrass fertilized with sulfur-coated urea. *Proc. Soil Crop Sci. Soc. Florida* **39,** 92–95.

Duncan, R. R., Bramel-Cox, P. J., and Miller, F. R. (1991). Contributions of introduced sorghum germplasm to hybrid development in the U.S.A. *In* "Use of Plant Introductions in Cultivar Development, Part I," CSSA Special Publ. No. 17, pp. 69–102. Madison, Wisconsin.

Eastman, P. (1980). "An End to Pounding. " IDRC-152e, p. 63. International Development Research Centre, Ottawa.

Ejeta, G., Hansen, M. M., and Mertz, E. T. (1987). *In vitro* digestibility and amino acid composition of pearl millet (*Pennisetum typhoides*) and other cereals. *Proc. Nat. Acad. Sci.* U.S.A. **84,** 6016–6019.

Faires, E. W., Dawson, J. R., LaMaster, J. P., Wise, G. H. (1941). Experiments with annual crops and permanent pastures to provide grazing for dairy cows in the Sandhill region of the southeast. *In* "USDA Technical Bulletin No. 805," Washington, D.C.

Fales, S. L., and Wilkinson, R. E. (1984). Mefluidide-induced nitrate accumulation in pearl millet forage. *Agron. J.* **76,** 857–860.

Fancher, B. I., Jensen, L. S., Smith, R. L., and Hanna, W. W. (1987). Metabolizable energy content of pearl millet [*Pennisetum americanum* (L.) Leeke]. *Poult. Sci.* **66,** 1693–1696.

FAO (1990). Structure and characteristics of the world millet economy. *In* "Report of the 24th Session of the Committee on Commodity Problems, GR 90/4." FAO, Rome. (mimeo)

FAO/WHO (1973) Energy and protein requirements: Report of a joint FAO/WHO ad hoc Expert Committee. *In* "WHO Technical Report Series No. 522;" "FAO Nutrition Meetings Report Series No. 52." Geneva, Switzerland.

Ferraris, R., and Norman, M. J. T. (1973). Adaptation of pearl millet (*Pennisetum typhoides*) to coastal New South Wales. 2. Productivity under defoliation. *Austr. J. Exp. Agric. Anim. Husbandry* **13,** 692–199.

Freitas, E. A. G. De (1988). Millet for milk production [Pt]. *Agropecuariària Catarinense* **1,** 20–22.

French, M. H. (1948). Local millets as substitutes for maize in feeding of domestic animals. *East Afr. Agric. J.* **13,** 217–220.

Fribourg, H. A. (1966). The effect of morphology and defoliation intensity on the tillering, regrowth and leafiness of pearl millet, *Pennisetum typhoides* (Burm.) Stapf & C. E.

Hubb. *In* "Proceedings IX International Grassland Congress, Sao Paulo, Brazil, 1965," pp. 489–491.
Fribourg, H. A. (1974). Fertilization of summer annual grasses and silage crops. *In* "Forage Fertilization" (D. A. Mays, ed.). American Society of Agronomy, Madison, Wisconsin.
Fussell, L. K., Serafini, P. G., Bationo, A., and Klaij, M. C. (1987). Management practices to increase yield and yield stability of pearl millet in Africa. *In* "Proceedings of International Pearl Millet Workshop" (J. R. Witcombe and S. R. Beckerman, eds.), pp. 255–268. ICRISAT, Patancheru, India.
Georgia Coastal Plain Experiment Station (1947). Temporary grazing crops recommended for economical milk production. *In* "27th Annual Report (Bulletin 44)," pp. 29, 36.
Gepts, P., and Clegg, M. T. (1989). Genetic diversity in pearl millet [*Pennistum glaucum* (L.) R. Br.] at the DNA sequence level. *J. Hered.* **80,** 203–208.
Green, V. E., Jr., (1973). Pearl millet, *Pennisetum typhoides* (Burm.) Stapf and Hubb.: Research and observations in Florida—1888–1968. *Proc. Soil Crop Sci. Soc. Florida* **32,** 25–29.
Gupta, P. C., Singh, K., and Sharda, D. P. (1981). Note on the in vivo studies on the nutritive value of pearl millet, pearl millet–cowpea forage mixture and its silage. *Indian J. of Anim. Sci.* **51,** 1166–1167.
Gupta, V. P., and Sidhu, P. S. (1972). N.B.-21—The new hybrid napier (*Pennisetum purpureum*). *Prog. Farming* **8,** 7.
Hamad, A. M., and Fields, M. L. (1979). Evaluation of the protein quality and available lysine of germinated and fermented cereals. *J. Food Sci.* **44,** 456.
Hammes, P. S. (1972). Pearl millet: The top notch forage crop. *Farming S. Afr.* **48,** 24–26.
Hanna, W. W. (1987). Utilization of wild relatives of pearl millet. *In* "Proceedings of International Pearl Millet Workshop" (J. R. Witcombe and S. R. Beckerman, eds.), pp. 43–61. ICRISAT, Patancheru, India.
Hanna, W. W. (1989). Characteristics and stability of a new cytoplasmic nuclear male-sterile source in pearl millet. *Crop Sci.* **29,** 1457–1459.
Hanna, W. W. (1990). Transfer of germplasm from the secondary to the primary gene pool in *Pennisetum. Theor. Appl. Genet.* **80,** 200–204.
Hanna, W. W., and Akin, D. E. (1978). Microscopic observations on cuticle from trichomeless, *tr*, and normal, *Tr*, pearl millet. *Crop Sci.* **18,** 904–905.
Hanna, W. W., Dove, R., Hill, G. M., and Smith, R. (1991). Pearl millet as an animal feed in the U.S. *In* "American Society of Agronomy Abstracts, Southern Branch." Fort Worth, Texas.
Hanna, W. W., and Wells, H. D. (1989). Inheritance of Pyricularia leaf spot resistance in pearl millet. *J. Hered.* **80,** 145–147.
Hanna, W. W., Wells, H. D., and Burton, G. W. (1985). Dominant gene for rust resistance in pearl millet. *J. Hered.* **76,** 134.
Hanna, W. W., Wells, H. D., Burton, G. W., Hill, G. M., and Monson, W. G. (1988). Registration of 'Tifleaf 2' pearl millet. *Crop Sci.* **28,** 1023.
Harinarayana, G. (1987). Pearl millet in Indian agriculture. *In* "Proceedings of the International Pearl Millet Workshop" (J. R. Witcombe and S. R. Beckerman, eds.), pp. 5–17. ICRISAT, Patancheru, India.
Harlan, J. R., and de Wet, J. M. J. (1971). Toward a rational classification of cultivated plants. *Taxonomy* **20,** 509–517.
Hart, R. H. (1967). Digestibility, morphology, and chemical composition of pearl millet. *Crop Sci.* **7,** 581–584.
Helentjaris, T., and Burr, B. (eds). (1989). Development and application of molecular markers to problems in plant genetics. *In* "Current Communications in Molecular Biology," pp. 165. Long Island, New York.

Hemanalini, G., Padma Umapathy, K., Rao, J. R., and Sarawathi, G. (1980). Nutritional evaluation of sprouted ragi. *Nutr. Rep. Int.* **22,** 271.

Hemken, R. W., Harner, J. P., Vandersall, J. H., Clark, N. A., and Schneider, B. A. (1968). Fertilization of pearl millet forage as related to milk fat test. *J. Dairy Sci.* **51,** 982.

Hill, G. D. (1969). Performance of forage sorghum hybrids and Katherine pearl millet in Bubia. *Papua New Guinea Agric. J.* **21,** 1–6.

Hill, G. M., and Hanna, W. W. (1990). Nutritive characteristics of pearl millet grain in beef cattle diets. *J. Anim. Sci.* **68,** 2061–2066.

Hoseney, R. C. (1988). Overview of sorghum and pearl millet quality utilization and scope for alternative uses. *In* "Proceedings of Workshop on Biotechnology for Tropical Crop Improvement," p. 127. ICRISAT, Patancheru, India.

Hoseney, R. C., Andrews, D. J., and Clark, H. (1987). Sorghum and pearl millet. *In* "Nutritional Quality of Cereal Grains: Genetic and Agronomic Improvement," ASA Monograph 28, pp. 397–456.

Hoveland, C. S., Anthony, W. B., Scarsbrook, C. E. (1967). Effect of management on yield and quality of Sudax sorghum–sudan hybrid and Gahi-1 pearl millet. *In* "Leaflet of the Alabama Agricultural Experimental Station No. 76."

Hoveland, C. S., and McCloud, D. E. (1957). Manage millet for better yields. *In* "Sunshine State Agricultural Research Report" Vol. 2, p. 14.

Hulse, J. H., Laing, E., and Pearson, D. E. (1980). "Sorghum and the Millets: Their Composition and Nutritive Value." Academic Press, New York.

IBPGR (1981). "Descriptors for Pearl Millet." International Board for Plant Genetic Resources (IBPGR) and International Crops Research Institute for the Semi-arid Tropics (ICRISAT), Rome.

IBPGR (1987). Working group on *Pennisetum*. "The Herbarium." Royal Botanic Gardens, Kew, United Kingdom. (mimeo)

ICRISAT. (1984). Pearl millet, high protein hybrids. "ICRISAT Annual Report, 1983," p. 98. Patancheru, India.

ICRISAT (1991). Cereals program. "ICRISAT Annual Report, 1990," p. 137. Patancheru, India.

IDRC (1981). "Nutritional Status of the Rural Population of the Sahel," IDRC 160e. International Research and Development Centre, Ottawa.

Jain, A. K., and Date, W. B. (1975). Relative amylase activity of some malted cereal grains. *J. Food Sci. Technol.* (India) **12,** 131.

Jambunathan, R., Singh, U., and Subramanian, V. (1984). Grain quality of sorghum, pearl millet, pigeon pea and chick pea. *In* "Interfaces between Agriculture, Nutrition, and Food Science" (K. T. Achaya, ed.), Food and Nutritional Bulletin, Supplement No. 9, pp. 47–60. Tokyo, Japan.

Jambunathan, R., and Subramanian, V. (1988). Grain quality and utilization in sorghum and pearl millet. *In* "Proceedings of Workshop on Biotechnology for Tropical Crop Improvement," pp. 133–139. ICRISAT, Patancheru, India.

Jaster, E. H., Fisher, C. M., and Miller, D. A. (1985). Nutritive value of oatlage, barley/pea, pea, oat/pea, pearl millet and sorghum as silage ground under a double cropping forage system for dairy heifers. *J. Dairy Sci.* **68,** 2914–2921.

Johnson, J. C., Jr., Lowrey, R. S., Monson, W. G., and Burton, G. W. (1968). Influence of the dwarf characteristic on composition and feeding value of near-isogenic pearl millets. *J. Dairy Sci.* **51,** 1423–1425.

Johnson, J. C., Jr., McCormick, W. C., Burton, G. W., and Monson, W. G. (1978). Weight gains of heifers grazing Gahi 1, Gahi 3, and Tifleaf 1 hybrid pearl millet. *In* "University of Georgia Research Report No. 274.

Johnson, J. C., Jr., and Southwell, B. L. (1960). Pasture and silage both from Starr millet. *Georgia Agric. Res.* **1,** 6–7.

Kante, A., Coulibaly, S., Scheuring, J. F., and Niangado, O. (1984). The ease of decortication in relation to the grain characteristics of Malian pearl millet. *In* "Proceedings Symposium on the Processing of Sorghum and Millets: Criteria for Quality of Grains and Products for Human Food," pp. 44–47. Int. Assoc. Cereal Chemistry, Vienna, Austria.

Kassam, A. H., and Kowal, J. M. (1975). Water use, energy balance and growth of gero millet at Samaru, Northern Nigeria. *Agric. Met.* **15,** 333–342.

Khadr, F. H., and El-Rouby, M. M. (1978). Inbreeding in quantitative traits of pearl millet (*Pennisetum typhoides*). *Z. Pflanzenzuchtung.* **80,** 149–157.

Klopfenstein, C. F., Hoseney, R. C., and Leipold, H. W. (1983). Goitrogenic effects of pearl millet diets. *Nutr. Rep. Int.* **27,** 1039–1047.

Klopfenstein, C. F., Hoseney, R. C., and Leipold, H. W. (1985). Nutritional quality of pearl millet and sorghum grain diets and pearl millet weanling food. *Nutr. Rep. Int.* **31,** 287–297.

Krejsa, B. B., Rouquette, F. M., Jr., Holt, E. C., Camp, B. J., and Nelson, L. R. (1984). Nitrate and total alkaloid concentration of 11 pearl millet lines. *Argon. J.* **76,** 157–158.

Kumar, K. A. (1987). Brief overview of pearl millet hybrids in Africa. *In* "Proceedings of the International Pearl Millet Workshop" (J. R. Witcombe and S. R. Beckerman, eds.), p. 284. ICRISAT, Patancheru, India.

Kumar, K. A., and Andrews, D. J. (1984). Cytoplasmic male sterility in pearl millet [*Pennisetum americanum* (L.) Leeke]—A review. *Adv. Appl. Biol.* **10,** 113–143.

Kumar, K. A., and Appa Rao, S. (1987). Diversity and utilization of pearl millet germplasm. *In* "Proceedings of the International Pearl Millet Workshop" (J. R. Witcombe and S. R. Beckerman, eds.), pp. 69–82. ICRISAT, Patancheru, India.

Kumar, K. A., Gupta, S. C., and Andrews, D. J. (1983). Relationship between nutritional quality characters and grain yield in pearl millet. *Crop Sci.* **23,** 232–235.

Lansbury, T. J. (1959). The composition and digestibility of some conserved fodder crops for dry season feeding in Ghana. 2. Silage. *Trop. Agric. Trinidad* **36,** 65–69.

Lemon, M. D., and McMurphy, W. E. (1984). Forage NO_3 in sudangrass and pearl millet. *In* "Proceedings. Forage and Grassland Council Conference," pp. 302–306. Houston, Texas.

Leuck, D. B., Taliaferro, C. M., Burton, R. L., Burton, G. W., and Bowman, M. C. (1968). Fall armyworm resistance in pearl millet. *J. Econ. Entomol.* **61,** 693–695.

Lloyd, G. L. (1964). Use of munga to replace maize in a broiler ration. *Rhodesia Agric. J.* **61,** 50–51.

Lorz, H., Brittell, R. S., and Potrykus, I. (1981). Protoplast culture in *Pennisetum americanum. Proc. Int. Bot. Congr. Sidney*, **13,** 309.

Luttrell, E. S. (1954). Diseases of pearl millet in Georgia. *Plant Dis. Rep.* **38,** 507–514.

Malleshi, N. G., and Desikachar, H. S. R. (1986). Nutritive value of malted millet flours. *J. Hum. Nutr.* **36,** 191–196.

Marchais, L., and Tostain, S. (1985). Genetic divergence between wild and cultivated pearl millets (*Pennisetum typhoides*). II. Characters of domestication. *Z. Pflanzenzuchtung* **95,** 245–261.

Marshall, S. P., Sanchez, A. B., Somers, H. L., and Arnold, P. T. D. (1953). Value of pearl millet pasture for dairy cattle. "University of Florida Agricultural Experiment Stations Bulletin No. 527."

Martinez, A. (1988). Performance of lambs grazing pearl millet at four levels of herbage allowance." Dissert. Abstr. Int. *B* (Sci. Eng.) **49,** 578.

Mays, D. A., Peterson, J. R., and Bryant, H. T. (1966). A clipping management study of two sudangrass-sorghum hybrids, sudangrass and Gahi millet for forage production. *In* "Virginia Agricultural Experimental Station Research Report No. 113."

Mays, D. A., and Washko, J. B. (1962). Refractometric determination of "sucrose equivalent" levels in the stubble of sudangrass and pearl millet. *Crop Sci.* **2,** 81–82.

McCartor, M. M., and Rouquette, F. M., Jr. (1977). Grazing pressures and animal performance from pearl millet. *Agron. J.* **69,** 983–987.

McDonough, C. M. (1986). "Structural Characteristics of the Mature Pearl Millet (*Pennisetum americanum*) Caryopsis." M. S. thesis, Texas A&M University, College Station.

McDonough, C. M., and Rooney, L. W. (1985). Structure and phenol content of six species of millets using fluorescence microscopy and HPLC. *Cereal Food World* **30,** 550. (abstract)

McDonough, C. M., and Rooney, L. W. (1989). Structural characteristics of *Pennisetum americanum* using scanning electron and fluorescence microscopies. *Food Microstr.* **8,** 137.

Merkle, O. G., Starks, K. J., and Casady, A. J. (1983). Registration of pearl millet germplasm lines with cinch bug resistance. *Crop Sci.* **23,** 601.

Miles, J. T., Cowsert, W. C., Lusk, J. W., and Owen, J. R. (1956). Most milk from sudan in state college dairy tests. *Mississippi Farm Res.* **19,** 6,7.

Miller, R. W., Waldo, D. R., Okamoto, M., Hemken, R. W., Vandersall, J. H., and Clark, N. A. (1963). Feeding of potassium bicarbonate or magnesium carbonate to cows grazed on sudangrass or pearl millet. *J. Dairy Sci.* **46,** 621.

Monson, W. G., Hanna, W. W., and Gaines, T. P. (1986). Effects of rust on yield and quality of pearl millet forage. *Crop Sci.* **26,** 637–639.

Moraes, A. De, and Maraschin, E. G. (1988). Pressoes de pastejo e producao animal em milheto cv. Comum. *Pesquisa Agropecuaria Brasileira* **23,** 197–205.

Norman, M. J. T. (1963). "Wet-Season Grazing of Sown Pasture and Fodder Crop at Katherine," Technical Memorandum No. 22. Division of Land Research and Regional Survey, CSIRO, Australia.

Norman, M. J. T., and Phillips, L. J. (1968). The effect of time of grazing on bulrush millet (*Pennisetum typhoides*) at Katherine, N. T. *Austr. J. Exp. Agric. Anim. Husbandry* **8,** 288–293.

Norman, M. J. T., and Stewart, G. A. (1964). Investigations on the feeding of beef cattle in the Katherine region, N. T. *J. Austr. Inst. Agric. Sci.* **30,** 39–46.

Novellie, L. (1982). Fermented beverages. *In* "Proceedings, International Grain Sorghum Quality Workshop," (L. W. Rooney and D. S. Murty, eds.), pp. 113–120. ICRISAT, Patancheru, India.

Olewik, M. C., Hoseney, R. C., and Verriano-Marston. (1984). A procedure to produce pearl millet rotis. *Cereal Chem.* **61,** 28–33.

Ong, C. K., and Everard, A. (1979). Short day induction of flowering in pearl millet (*Pennisetum typhoides*) and its effect on plant morphology. *Exp. Agric.* **15,** 401–411.

Opoku, A. R., Ohenehen, S. O., and Ejiofor, N. (1981). Nutrient composition of millet (*Pennisetum typhoides*) in Durfur province western Sudan. *Ann. Nutr. Metab.* **27,** 14.

Osman, A. K., and Fatah, A. A. (1981). Factors other than iodine deficiency contributing to the endemicity of goitre in Durfur province (Sudan). *J. Hum. Nutr.* **35,** 302–309.

OTA, Office of Technology Assessment, U.S. Congress (1988). "Enhancing Agriculture in Africa: A Role for U.S. Development Assistance," OTA-F-356. Govt. Printing Office, Washington, D.C.

Parveen, A., Barsaul, C. S., and Singh, B. B. (1988). Oxalate content of bajra (*Pennisetum typhoides*) fodder and a technique of its removal. *Indian J. Anim. Nutr.* **5,** 41–43.

Patel, B. M., Shah, B. G., and Mistry, V. V. (1958). A study on fodders of Hissar District in the Punjab. *Indian J. Agric. Sci.* **28,** 597–606.

Penzhorn, E. J., and Lesch, S. F. (1965). Babala preferable for summer grazing under irrigation. *Farming S. Afr.* **41,** 57–58.

Perten, H. (1983). Practical experience in processing and use of millet and sorghum in Senegal and Sudan. *Cereal Foods World* **28,** 680–683.

Porteres, R. (1976). African cereals. *In* "Origins of Plant Domestication" (J. R. Harlan, J. M. J. de Wet, and A. B. L. Stemler, eds.), pp. 409–452. Mouton Press, The Hague, The Netherlands.

Powell, J. B., and Burton, G. W. (1971). Genetic suppression of shoot-trichomes in pearl millet, *Pennisetum typhoides. Crop Sci.* **11,** 763–765.

Rachie, K. O., and Majmudar, J. V. M. (1980). "Pearl Millet," Pennsylvania Univ. Press, University Park.

Rai, K. N., Andrews, D. J., and Babu, S. (1984). Inbreeding depression in pearl millet composites. *Z. Pflanzenzuchtung* **94,** 201–207.

Reddy, V., Faubion, J. M., and Hoseney, R. C. (1986). Odor generation in ground, stored pearl millet. *Cereal Chem.* **63,** 403–406.

Reichert, R. D. (1979). The pH-sensitive pigments in pearl millet. *Cereal Chem.* **56,** 291–294.

Reichert, R. D., Tyler, R. T., York, A. E., Schwab, D. J., Tatarynovich, J. E., and Nwasaru, M. A. (1986). Description of a production model of the tangential abrasive dehulling device and its application to breeders' samples. *Cereal Chem.* **63,** 201.

Reichert, R. D., and Youngs, C. G. (1971). Dehulling cereal grains and grain legumes for developing countries. II. Chemical composition of mechanically dehulled and traditionally dehulled sorghum and millet. *Cereal Chem.* **54,** 174–178.

Reichert, R. D., and Youngs, C. G. (1979). Bleaching effect of acid on pearl millet. *Cereal Chem.* **56,** 287–290.

Reichert, R. D., Youngs, C. G., and Christensen, D. A. (1979). Polyphenols in *Pennisetum* millet. *In* "Proceedings, Symposium on Polyphenols in Cereals and Legumes" (J. H. Hulse, ed.), IDRC-145e, pp. 50–60. Ottawa, Canada.

Roark, D. B., Miles, J. T., Lusk, J. W., and Cowsert, W. C. (1952). Milk production from pearl millet, grain sorghum, and tift sudan. *Proc. Assoc. South. Agric. Workers* **49,** 76–77.

Roberts, T., Lespinasse, R., Pernes, J., and Sarr, A. (1991). Gametophytic competition as influencing gene flow between wild and cultivated forms of pearl millet (*Pennisetum typhoides*). *Genome* **34,** 195–200.

Rollins, G. H., Hoveland, C. S., and Autrey, K. M. (1963). "Coastal Bermuda Pastures Compared with Other Forages for Dairy Cows." Auburn University Agricultural Experiment Station Bull. No. 347.

Rooney, L. W. (1978). Sorghum and pearl millet lipids. *Cereal Chem.* **55,** 584–590.

Rooney, L. W. (1989). Food and nutritional quality of sorghum and pearl millet. *In* "INTSORMIL Annual Report," pp. 158–167. University of Nebraska–Lincoln.

Rooney, L. W., and McDonough, C. M. (1987). Food quality and consumer acceptance in pearl millet. *In* "Proceedings, International Pearl Millet Workshop" (J. R. Witcombe and S. R. Beckerman, eds.), pp. 43–61. ICRISAT, Patancheru, India.

Rouquette, F. M., Jr., Keisling, T. C., Camp, B. J., and Smith, K. L. (1980). Characteristics of the occurrence and some factors associated with reduced palatability of pearl millet. *Agron. J.* **72,** 173–174.

Rusoff, L. L., Achacoso, A. S., Mondart, C. L., Jr., and Bonner, F. L. (1961). Relationship of lignin to other chemical constituents in sudan and millet forages. *In* "Louisiana Agricultural Experiment Station Bulletin 542."

Ryan, J. G., Bidinger, P. D., Prahsad Rao, N., and Pushpamma, P. (1984). The determinants of individuals diets and nutritional status of six villages in southern India. *In* "ICRISAT Res. Bull. No. 7," p. 140.

Sanford, P. E., Camacho, F., Knake, R. P., Deyoe, C. W., and Casady, A. J. (1973). Performance of meat-strain chicks fed pearl millet as a source of energy and protein. *Poult. Sci.* **52,** 2081. (abstract)

Sautier, D., and O'Deyes, M. L. (1989). Mil, mais, sorgho. *In* "Techniques et alimentation au Sahel." L'Harmattan, 5–7 rue de l' Ecole Polytechniques, 75005 Paris, France.

Schneider, B. A., and Clark, N. A. (1970). Effect of potassium on the mineral constituents of pearl millet and sudangrass. *Agron. J.* **62,** 474–477.

Sehagal, K. L., and Goswami, A. K. (1969). Composition of pearl millet plants at different stages of growth with special reference to the oxalic acid content. *Indian J. Agric. Sci.* **39,** 72–80.

Serna-Saldivar, S. O., McDonough, C. M., and Rooney, L. W. (1990). The millets. In "Handbook of Cereal Science and Technology" (K. J. Lorenz and K. Kulp, eds.), pp. 271–300. Dekker, New York.

Sharma, B. D., Sadagopan, V. R., and Reddy, V. R. (1979). Utilization of different cereals in broiler rations. *Br. Poult. Sci.* **20,** 371–388.

Silveira, C. A. M., Saibro, J. C. De, Markus, R., and Freitas, E. A. G. De (1981). [Intake, digestibility and nitrogen balance in sheep or pearl millet (*Pennisetum americanum* (L.) Leeke) silage alone or in mixture with cowpea (*Vigna ungiculata* (L.) Walp.)] [Pt]. *Rev. Soc. Brasileira Zootecnia* **10,** 361–380.

Simwemba, C. G., Hoseney, R. C., Varriano-Marston, E., and Zeleznak, K. (1984). Certain B-vitamins and phytic acid content of pearl millet (*Pennisetum americanum* L. Leeke). *J. Agric. Food Chem.* **32,** 31–34.

Singh, N. P., and Mudgal, R. D. (1980). Note on the nutritive value of cowpea (*Vigna ungiculata*) and pearl millet (*Pennisetum typhoides*) silage with or without molasses and sodium chloride for sheep. *Indian J. Anim. Sci.* **50,** 901–903.

Singh, P., Singh, U., Eggum, B. O., Kumar, K. A., and Andrews, D. J. (1987). Nutritional evaluation of high protein genotypes in pearl millet. *J. Food Sci. Agric.* **38,** 41–48.

Singh, S. D., and Barsaul, C. S. (1976). Replacement of maize by coarse grains for growth and production in White Leghorn and Rhode Island Red birds. *Indian J. Anim. Sci.* **46,** 96–99.

Sisk, L. R., McCullough, M. E., and Sell, O. E. (1960). The preservation and feeding value of Starr millet and sudangrass silage for dairy cows. *Proc. Assoc. South. Agric. Workers* **57,** 118. (abstract)

Smith, R. L., Jensen, L. S., Hoveland, C. S., and Hanna, W. W. (1989). Use of pearl millet, sorghum, and triticale grain in broiler diets. *J. Prod. Agric.* **2,** 78–82.

Smith, R. L., Chowdhury, M. K. U., and Shank, S. C. (1989). Use of restriction fragment length polymorphisms (RFLP) markers in genetics and breeding of napiergrass. *Proc. Soil Crop Sci. Soc. Fla.* **48,** 13–18.

Stegmeier, W. M. (1990). Pearl millet breeding. *In* INTSORMIL Annual Report," pp. 48–51. University of Nebraska–Lincoln.

Stephenson, R. J., and Posler, G. L. (1984). Forage yield and regrowth of pearl millet. *Trans. Kansas Acad.* **87,** 91–97.

Sullins, D., and Rooney, L. W. (1977). Pericarp and endosperm structure of of pearl millet (*Pennisetum typhoides*). *In* "Proceedings, Symposium on Sorghum and Millets for Human Foods" (D. A. V. Dendy, ed.), pp. 79–89.

Sullivan, T. W., Douglas, J. H., Andrews, D. J., Bond, P. L., Hancock, J. D., Bramel-Cox, P. J., Stegmeier, W. D., and Brethour, J. R. (1990). Nutritional value of pearl millet for food and feed. *In* "Proceedings International Conference in Sorghum Nutritional Quality," pp. 83–94. Purdue University, Lafayette, Indiana.

Tanksley, S. D., Young, N. D., Paterson, A. H., and Bonierbale, M. W. (1989). RFLP mapping in plant breeding: New tools for an old science. *Bio/Technology* **7,** 257–264.

Tostain, S., and Marchais, L. (1987). General survey of enzymatic diversity in pearl millet. *In* "Proceedings, International Pearl Millet Workshop" (J. R. Witcombe and S. R. Beckerman, eds.), pp. 283, 284. ICRISAT, Patancheru, India.

Tostain, S., and Marchais, L. (1989). Enzyme diversity in pearl millet (*Pennisetum glaucum*). 2. Africa and India. *Theor. Appl. Genet.* **77,** 634–640.

Tostain, S., Riandey, M. F., and Marchais, L. (1987). Enzyme diversity in pearl millet (*Pennisetum glaucum*). 1. West Africa. *Theor. Appl. Genet.* **74,** 188–193.

USDA (1986). A check list of names of 3,000 vascular plants of economic importance. *In* "USDA/ARS Handbook No. 505." Washington, D.C.

Vasil, V., and Vasil, K. (1980). Isolation and culture of cereal protoplasts. II. Embryogenesis and plantlet formation from protoplasts of *Pennisetum americanum. Theor. Appl. Genet.* **56,** 97–100.

Voorhees, E. B. (1907). "Forage Crops for Soiling Silage Hay and Pasture." MacMillan Co., New York.

Walker, C. E. (1987). Evaluating pearl millet for food quality. *In* "INTSORMIL Annual Report," pp. 160–166. University of Nebraska–Lincoln.

Wells, H. D., Burton, G. W., and Hennen, J. W. (1973). *Puccinia substriata* var. *indica* on pearl millet in the Southeast. *Plant Dis. Rep.* **57,** 262.

Wells, H. D., and Hanna, W. W. (1988). Genetics of resistance to *Bipolaris setariae* in pearl millet. *Phytopathology* **78,** 1179–1181.

Welsh, J., and McClelland, M. (1990). Fingerprinting genomes using PCR with arbitrary primers. *Nucleic Acids Res.* **18,** 7213–7218.

Westphalen, S. L., and Jacques, A. V. A. (1978). [Effects of date of sowing, stage of growth and height of cut on the dry matter and protein yields of pearl millet. 1. Late-flowering cultivar.] *Agron. Sulriograndense* **14,** 87–105 [Pt] (from *Sorghum and Millets Abstr.*, 1979, **4**, 131, No. 1002).

Wilkinson, W. S., and Barbee, C. (1968). Metabolizable energy value of dehydrated coastal bermudagrass and pearl millet for the growing chick. *Poult. Sci.* **47,** 1901–1905.

Wilkinson, W. S., Barbee, C., and Knox, F. E. (1968). Nutrient content of dehydrated coastal bermudagrass and pearl millet. *J. Agric. Food Chem.* **16,** 665–668.

Williams, J. G. K., Kubelik, A. R., Livak, K. J., Rafalski, J. A., and Tingery, S. V. (1990). DNA polymorphisms amplified by arbitrary primers are useful as genetic markers. *Nucleic Acids Res.* **18,** 6531–6535.

Wilson, J. P., Gates, R. N., and Hanna, W. W. (1991). Effect of rust on yield and digestibility of pearl millet forage. *Phytopathology* **81,** 233–236.

Wilson, J. P., Wells, H. D., and Burton, G. W. (1989). Inheritance of resistance to *Pyricularia grisea* in pearl millet accessions from Burkina Faso and inbred Tift 85DB. *J. Hered.* **80,** 499–501.

FORMS, REACTIONS, AND AVAILABILITY OF NICKEL IN SOILS

N. C. Uren

School of Agriculture, La Trobe University,
Bundoora, Victoria 3083, Australia

I. INTRODUCTION

The biological significance of Ni is increasing as our awareness of its role in the metabolism of animals, bacteria, and plants increases. The toxicity of Ni to organisms has been known for a long time and most research on Ni has concentrated largely on its possible toxicity to plants growing on

Advances in Agronomy, Volume 48

serpentine soils derived from ultrabasic igneous rocks (Hunter and Vergnano, 1952; Crooke, 1956; Williams, 1967; Halstead, 1968; Anderson *et al.*, 1973); Brooks (1987) has reviewed comprehensively the literature dealing with all aspects of serpentine ecology. The discovery of a biological role and essentiality of Ni in plants, animals, and microorganisms (Thauer *et al.*, 1980; Welch, 1981; Hausinger, 1987) and the concern over the incidental additions of trace metals including Ni to soil in sewage sludge and other wastes (Le Riche, 1968; Purves, 1985; Leeper, 1978; Tiller, 1989) make it important that the behavior of Ni in soils be well understood. Williams (1987) sees the bio-inorganic chemistry of Ni as "a huge challenge."

Because of these more recent developments, that the possibility of deficiency now exists and that the activity of urease in soils may depend on the Ni status, it is timely to review its behavior in soils. Previous reviews (Nielsen *et al.*, 1977; Vanselow, 1966; Hutchinson, 1981) exist, but there are more recent data to be found, particularly in papers that deal collectively with several trace metals. Similarly, more recent reviews on Ni in the natural environment (Boyle and Robinson, 1988) and in plants (Farago and Cole, 1988) do not attempt to reconcile the behavior of Ni in soil with its avalability to plants. The purpose of this review is to synthesize the existing knowledge into a coherent and consistent account of the behavior of Ni in soils and its availability to plants, making where necessary use of comparisons with other trace metals, particularly Cu and Zn.

A. Sources

Igneous rocks are the primary source of the Ni found in soils. Upon weathering, the Ni becomes associated with all the phases of the geochemical cycle: soils and sediments and metamorphic and sedimentary rocks. In soils the total concentration of a trace element such as Ni is related directly to the concentration in the parent material and to weathering processes (Mitchell, 1964), but as weathering and soil formation proceed, the influence of the parent material lessens while the influence of active pedogenic factors gets greater (Tiller, 1963). The losses of Ni from soil occur in solution either by leaching or in run-off, in eroded material and in harvested goods. Gains of Ni to soil occur naturally through the accession of soil eroded from elsewhere and now due to its incidental and intentional (rare) addition in agricultural chemicals and in wastes such as sewage sludge. Atmospheric deposition of airborne particulates (Page and Chang, 1979) and rain (Barrie *et al.*, 1987) add Ni to soils from both natural and anthropogenic sources.

B. Quantitative Aspects

Ni is a trace element in the true sense of the term (Table I), although some unusual and high concentrations are found in polluted soils or in soils formed on ultrabasic igneous rocks. A study of 863 soils and surface materials (0–20 cm) from the United States by Shacklette *et al.* (1971) gave an average Ni concentration of 20 ppm and a range of <5 to 700 ppm, which agrees with the data in Table I and elsewhere (e.g., Mitchell, 1964). The range of Ni concentrations in serpentine soils listed by Brooks (1987) varied from 80 to 7100 ppm.

In plants the concentration of 1 ppm (Table I) is between the limits of the range given by Welch (1981), namely 0.01 to 5 ppm. Brown *et al.* (1987a) estimated a critical concentration of 0.10 ppm of Ni for barley plants deemed to be suffering from Ni deficiency. Toxic concentrations in plants are usually of the order of 25 to 50 ppm, but there is some variation in the reported concentrations associated with toxicity; for example, Slingsby and Brown (1977) found no symptoms were apparent in oats containing between 12 and 44 ppm, and yet Patterson (1971) observed slight symptoms of Ni toxicity in spring wheat at 8.0 ppm. MacNicol and Beckett (1985) concluded that Ni toxicity cannot be the cause of yield reduction if the concentration is less than 10 ppm. In those situations where a multiple toxicity is likely, lower critical concentrations may be more appropriate than those based on studies where only Ni has been added (Smilde, 1981; Wallace, 1982).

The concentrations of Ni in 68 soil solutions from California (Table II) show that the normal concentrations are of the same order as those for other trace metals, but about 10^3 to 10^4 times lower than the concentrations of Ca and Mg. The mean concentrations of Ni in the drainage waters

Table I

Typical Figures and Toxic Limits for Concentrations (ppm) of Trace Metals in Soils and Plants[a]

	Soils		Plants	
Metal	Typical	Range	Common range	Toxic limit
Mn	850	100–4000	15–100	500
Co	8	1–40	0.05–0.5	—
Ni	40	10–1000	1	25
Cu	20	2–100	4–15	30
Zn	50	10–300	8–15	500

[a] From Leeper (1978) by courtesy of Marcel Dekker.

Table II

Concentrations (ppm) of Some Divalent Metals in Saturation Extracts of 68 Californian Soils[a]

Metal	Positive occurrence	Range	Mean	Median
Mg	68	0.4–400	38.0	12.4
Ca	68	1.0–930	128.0	60.0
Mn	26	<0.01–0.95	0.17	<0.01
Co	2	<0.01–0.14	0.06	<0.01
Ni	30	<0.01–0.09	0.02	<0.01
Cu	67	<0.01–0.20	0.04	0.03
Zn	68	0.01–0.40	0.07	0.04

[a] From Bradford *et al.* (1971) by courtesy of Williams and Wilkins.

from a range of soils in Germany varied from 8 to 21 μg/liter while the overall range was from 2 to 51 μg/liter (Böhmer, 1989); Yamasaki *et al.* (1975) measured Ni concentrations of 0.15, 0.56, and 0.061 ppm in the soil solutions of three soils.

In an area suspected of Ni toxicity (Williams, 1967), the concentrations of Ni in the soil solution were found to be between 0.13 and 3.25 ppm (Anderson *et al.*, 1973), whereas in a toxic serpentine soil in Scotland the range was found to be from 0.5 to 0.9 ppm (Johnston and Proctor, 1981). Davis (1979) obtained a range of 0.11 to 1.11 ppm Ni in soil solutions of nine soils with previous histories of sewage sludge deposition, whereas Campbell and Beckett (1988) measured soil solution concentrations between 0.15 and 0.58 μM (0.0088 to 0.034 ppm) in a soil that had been treated with digested sewage sludge. Bradford *et al.* (1975) found the concentration of Ni in saturation extracts of soils exposed to sewage sludge to be up to 100 times that found in untreated soil (0.02 ppm), whereas Leeper (1978) reported 0.06 ppm in drainage water from land irrigated with raw sewage that contained 0.21 ppm Ni. By comparison the soil solution of a sandy loam amended 1 year previously with municipal sludge gave a concentration for Ni of 2.64 ppm, Cu 1.87 ppm, Zn 3.6 ppm, Ca 1667 ppm, and Mg 3380 ppm (Sposito and Bingham, 1981); all these concentrations exceed those given in Table II and the concentration of Ni is about 100 times the concentration for normal soil.

The concentrations of Ni in rain measured by Barrie *et al.* (1987) ranged from 0.17 to 17 μg/liter so that the annual rates of addition of Ni in an area with an annual rainfall of 1000 mm (40 in.) are between 0.002 and 0.2 kg ha^{-1}, enough to increase the total concentration of Ni in the top 10 cm by 0.002 to 0.2 ppm per annum. In comparison, Böhmer (1989) measured annual leaching losses in lysimeters from 0.009 to 0.071 kg ha^{-1} from a range of soils.

Fertilizers usually contain up to 30 ppm Ni although some phosphatic rocks have higher concentrations (Swaine, 1962); an annual rate of addition of 100 kg ha^{-1} of fertilizer containing 30 ppm Ni would add 0.03 kg ha^{-1} yr^{-1}. Crushed serpentine is sometimes blended with superphosphate and used to fertilize Mg-deficient soils in New Zealand (During, 1972; Brooks, 1987); the mixture contains from 500 to 700 ppm Ni and used at a rate of 200 kg ha^{-1} would add about 0.12 kg ha^{-1} Ni.

II. FORMS

Nickel(II) as the nickelous ion, Ni^{2+}, appears to be the valency state of most Ni found in nature (Theis and Richter, 1980), although Ni(III) has been detected in some enzymes that have been reviewed recently by Andrews *et aal.* (1988); Williams (1987) suspects that Ni(I) may be involved in some of these enzymes rather than Ni(III). Only Ni(II) shall be reviewed here and the reader is referred to Sposito (1983) for a more extensive review of the principal forms of trace metals in soils.

A. Soil Solution

The hydrated Ni^{2+} ion, $Ni(H_2O)_6{}^{2+}$, is the most likely form of Ni to be found in the soil solution, but its predominance will tend to decrease as pH increases and as the activities of the various ligands available to form complexes increase (e.g., OH^-, $CO_3{}^=$); each species of Ni has a tendency to be sorbed by the multitude of surfaces (Fig. 1).

Examples of ligands present in the soil solution are anions such as $SO_4{}^{2-}$, which may form an ion-pair, $Ni^{2+} + SO_4{}^{2-} = NiSO_4{}^0$; OH^-, which may form a hydroxy species, $Ni^{2+} + OH^- = NiOH^+$; fulvic acid, which may form an organo-complex, $Ni^{2+} + FA^{n-} = NiFA^{(2-n)+}$; and citric acid, which may chelate the Ni^{2+}, $Ni^{2+} + HCitr^{3-} = NiHCitr^-$. The stability of these complexes depends on the ligand (Table III) and the stability constants for the Ni complexes given probably cover the range of stabilities of Ni complexes found in the soil solution, although most complexes that form are likely to have low values rather than high ones.

If, in the formation of the complex, the ligand displaces one or more water molecules, as in $NiOH^+$ or in a chelate, the association is said to be "inner-sphere" and the stability greater than that when no water molecules

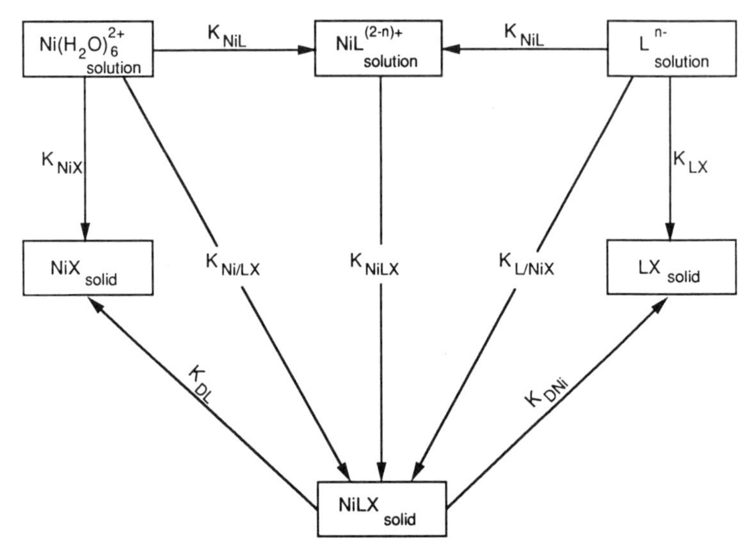

Figure 1. Species of Ni in soil solutions and possible interactions with the solid phase (after Sposito, 1984).

Table III

Some Acidity and Ni(II) Stability Constants for Some Complexing Ligands at 25°C[a]

Ligand	Equilibrium	Log K
OH^-	$K_1 = NiL/Ni \cdot L$	4.1
	$K_{sp} = Ni \cdot L^2$	−15.2
$SO_4^=$	$HL/H \cdot L$	2.2
	$NiL/Ni \cdot L$	2.3
CN^-	$HL/H \cdot L$	9.3
	$NiL_4/Ni \cdot L^4$	31.8
GLY^{-1}	$HL/H \cdot L$	9.9
	$H_2L/H \cdot HL$	2.3
	$NiL/Ni \cdot L$	6.3
	$NiL_2/Ni \cdot L^2$	11.4
CIT^{-4}	$HL/H \cdot L$	16.8
	$H_2L/H \cdot HL$	6.4
	$H_3L/H \cdot H_2L$	4.5
	$H_4L/H \cdot H_3L$	3.0
	$NiHL/Ni \cdot HL$	6.0
	$NiH_2L/Ni \cdot NiH_2L$	3.8
NTA^{-3}	$HL/H \cdot L$	10.5
	$H_2L/H \cdot HL$	3.2
	$H_3L/H \cdot H_2L$	2.2
	$NiL/Ni \cdot L$	13.2

[a] From Theis and Richter (1980) by courtesy of the American Chemical Society.

are displaced "outer-sphere," such as in $NiSO_4^0$. It is believed that all of these complexes of Ni(II) are quite labile (Cotton and Wilkinson, 1988) in that they form instantaneously and that a readily reversible equilibrium is established rapidly (Mattigod *et al.*, 1981), which is in contrast to the hysteresis observed when inner-sphere surface complexes form on solid surfaces (Sposito, 1984). In the soil solution the organic ligands that form associations with Ni vary from simple species in true solution to dispersed colloidal species, which as yet are poorly characterized.

The stability of Ni complexes relative to those of other divalent metal cations can be seen in Fig. 2 where Ni^{2+} is second only to Cu^{2+}. It is believed that S atoms are involved in the coordination of Ni(III) in the several enzymes in which it is found (Cotton and Wilkinson, 1988). In soil solutions the donor groups of the majority of ligands are O and to a lesser extent N and O, and given this and the fact that Ca^{2+} and Mg^{2+} are effective competitors, because of their much greater concentrations, one would expect most Ni to be found in the soil solution as $Ni(H_2O)_6^{2+}$ at low pH and to form complexes with OH^- and organic ligands increasingly as pH increases. Sanders *et al.* (1986a) found that the Ni in displaced soil solutions from a sludge-amended coarse sand was found to exist almost entirely as Ni^{2+} below pH 6, but at pH 7 some 20% was complexed. At all pHs, >90% of the Zn existed as Zn^{2+}, whereas for Cu 40% existed as Cu^{2+} at pH 4.5 but <10% was present as the cation at pH 7. Sposito and Bingham (1981) have determined that the following species were present in the soil solution (pH 6.4) of the polluted sandy loam referred to earlier: Ni, 59% as $Ni(H_2O)_6^{2+}$, 18% as Ni-organic complexes, 7% as $NiCO_3^0$, and

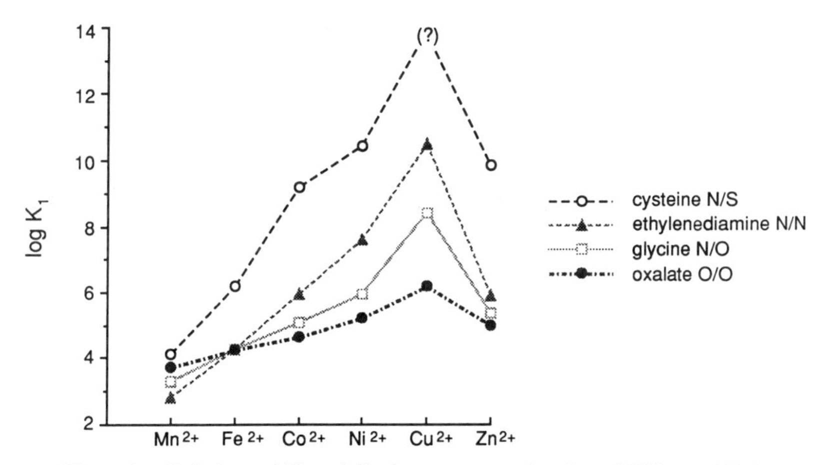

Figure 2. Relative stability of divalent trace metal cations (Williams, 1981).

6% as $NiNO_3^+$; Cu, 98% as Cu-organic complexes; and Zn, 73% as $Zn(H_2O)_6^{2+}$, 12% as $ZnSO_4^0$, 7% as $ZnCO_3^0$, and 6% as $ZnNO_3^+$.

The species present in the soil solution are influenced by the solid phase, since each of the species has an affinity for each of the range of surfaces that are present in soil (Fig. 1), but the weight of evidence available suggests that it is the cationic species that are the most strongly adsorbed.

B. Solid Phases

About 0.001% of the total Ni in a soil is in the soil solution at field capacity, and the remainder is distributed between the insoluble organic and inorganic phases within each of which there may be both occluded and adsorbed forms; precipitates are a special class of inorganic forms of Ni that may exist. Some of the Ni(II) adsorbed onto the surfaces of both phases may exchange readily with other cations in the soil solution or with reagents specifically designed to remove the Ni; these forms represent only a small proportion of the total and shall be discussed in a later section.

1. Inorganic

The inorganic occluded forms of Ni are found in those various silicate minerals and hydrous oxides that provide octahedral sites capable of accommodating Ni^{2+}, whose ionic radius of 70 pm is comparable to that of Mg^{2+} (72 pm), Al^{3+} (53 pm), Ti^{4+} (60.5 pm), and Li^+ (74 pm) and several trace metal cations: Co^{2+} (74 pm), Co^{3+} (61 pm), Cu^{2+} (73), Fe^{2+} (77 pm), Fe^{3+} (65 pm), Mn^{2+} (82 pm), Mn^{3+} (65 pm), and Zn^{2+} (75 pm) (Shannon and Prewitt, 1969). Because Ni^{2+} has the highest crystal field stabilization energy of the common divalent metals, it has the highest octahedral site preference (Burns, 1970). The possibility that Ni may substitute for Si in tetrahedral coordination has been rejected by Manceau and Calas (1987).

In igneous rocks Ni is found in octahedral coordination in early-formed primary silicate minerals, especially olivine (Mason and Moore, 1982)—see Table IV. Mitchell (1964) indicated that Ni, in addition to being found in olivine, hornblende, augite and biotite, is found in ilmenite and magnetite also.

Recent studies have shown that inclusions of nickeliferous Fe–Ni–Cu sulfides may be present in mafic and ultramafic rocks; for example, in picritic basalt from Hawaii the sulfide inclusions that were found in olivine and glass contained as much as 12.0 and 25.3 at.% Ni, respectively (Fleet and Stone, 1990); the concentration of Ni in the olivine was 0.25 wt% and

Table IV

Concentrations (ppm) of Ni, Cu, Co, and Mn in the Constituent Minerals of a Hypersthene Olivine Gabbro[a]

	Constituent minerals			Total from minerals	Found in rock
	Plagioclase	Pyroxene	Olivine		
Ni	<2	150	350	120	135
Co	<2	50	125	40	55
Cu	40	35	20	35	80
Mn	<10	2100	1700	940	700

[a] From Mitchell (1964) by courtesy of Reinhold Publishing.

the inclusions were between 2 and 30 μm in size. The oxidation of these sulfides and the release of the associated metals would be among the first reactions to take place upon weathering.

Because most Ni in igneous rocks is associated with unstable minerals, little is found in the primary minerals that persist in the sand fraction, and most of the Ni in soils is associated with secondary minerals in the clay fraction. Even where magnetite and ilmenite were found in the coarser fractions of serpentine soils, the clay fraction had higher concentrations of Ni in them (Wilson and Berrow, 1978). The secondary minerals in the clay fraction that are most likely to accommodate Ni are the hydrous oxides of Fe and Mn (and possibly Ti and Al) and the trioctahedral species of layer silicates: the serpentines (1:1), the saponites and vermiculites (2:1), and the chlorites (2:1:1). Concretionary forms of Fe and Mn oxides sometimes have high concentrations of Ni.

In the weathering of ultrabasic rocks Ni can form a number of different hydrous magnesian Ni layer silicates (Rankama and Sahama, 1950) with various structures similar to those of kaolinite (Dixon, 1989), talc (Brindley and Hang, 1973), chlorite (Brindley and de Souza, 1975), a regularly stratified chlorite–saponite (Wiewiora and Szpila, 1975), and nontronite (Bosio *et al.*, 1975). The serpentines, where they exist tend to be present as particles coarser than clay particles. They are relatively rare in soils and are unstable in most weathering environments (Dixon, 1989); the Ni varieties of serpentine are called garnierites (Dixon, 1989). Although serpentine minerals commonly weather to smectites (Dixon, 1989), the existence of trioctahedral Ni-rich smectites (i.e., where Ni is the dominant octahedral cation) may be in doubt (Newman and Brown, 1987), yet a Ni-rich smectite was present in the phyllosilicate clay derived from the lateritic weathering of pyroxenes in the Niquelandia area of Brazil (Colin *et al.*, 1985; Decarreau *et al.*, 1987). Some trioctahedral vermiculites studied by Foster (1963) contained from 500 to 2200 ppm Ni, and Ni-bearing layer silicates with concentrations up to 31.4% (pimelite) and 40.4% Ni (nepouite) have been studied by Manceau *et al.* (1985) and Manceau and Calas (1986). In these hydrous Ni layer silicates there is evidence of the tendency of the Ni to "cluster" in Ni-rich domains (Manceau and Calas, 1985, 1986).

In a study of some serpentine soils in Scotland (Wilson and Berrow, 1978), most Ni was associated with vermiculite/chlorite and serpentine minerals and there were only small amounts associated with magnetite and amphiboles. The concentrations of Ni were highest in the clay fraction and in the vermiculite/chlorite; the enrichment process was thought to involve chloritization of vermiculite with a Ni-rich brucite interlayer. Also present in the clay fraction was a ferruginous trioctahedral saponite that increased with depth in the profile and may have acted as a host to Ni in addition to the vermiculite/chlorite.

In less-Ni-rich weathering environments the outcome is not likely to be much different and ultimately the Ni will find its way into octahedral sites of secondary layer silicates as well as those of the hydrous oxides of Fe and Mn. Octahedral sites must exist in the hydrous oxides of Al and Ti, and, although this possibility is usually not considered, the comparatively lower concentrations of Ni in Ti–Fe–(Mn) concretions than in Mn–Fe concretions, studied by Hiller *et al.* (1988), suggest that other sites are more favored. The Ni in ilmenite (Mitchell, 1964) may be a substitute for either Fe or Ti, or both, but only further studies will resolve this question. There is nothing to suggest that Ni may not substitute for Al, but because Ni concentrations are generally higher in trioctahedral layer silicates than dioctahedral ones it is probably safe to assert that substitution for Mg by Ni is more likely than Al by Ni. Lithiophorite, a layer-structured Mn oxide consisting of interlayered MnO_2 and $(Al,Li)(OH)_3$ sheets, provides an example where it is believed that Ni substitutes along with Li for Al (Llorca, 1987) although it is not yet certain whether the Ni substitutes for the Al or the Li or both (Manceau *et al.*, 1990).

Some concretionary forms of Mn from soils have been shown to be enriched with Ni (Taylor and McKenzie, 1966; Childs, 1975; McKenzie, 1975; Sidhu *et al.*, 1977; Dawson *et al.*, 1985; Hiller *et al.*, 1988) by as much as 20 times the concentration in the bulk soil (Dawson *et al.*, 1985). However, in a study of a catenary sequence of soils, the enrichment of Ni in Fe–Mn concretions was not universal and was only apparent in concretions from the conclave footslopes in which the Mn concentrations were highest (Childs and Leslie, 1977); this suggests that the conditions under which the concretions form determines whether enrichment occurs.

The electron microprobe display in Fig. 3 shows the close association of both Ni and Co with Mn in a lateritic soil (Norrish, 1975); in other studies

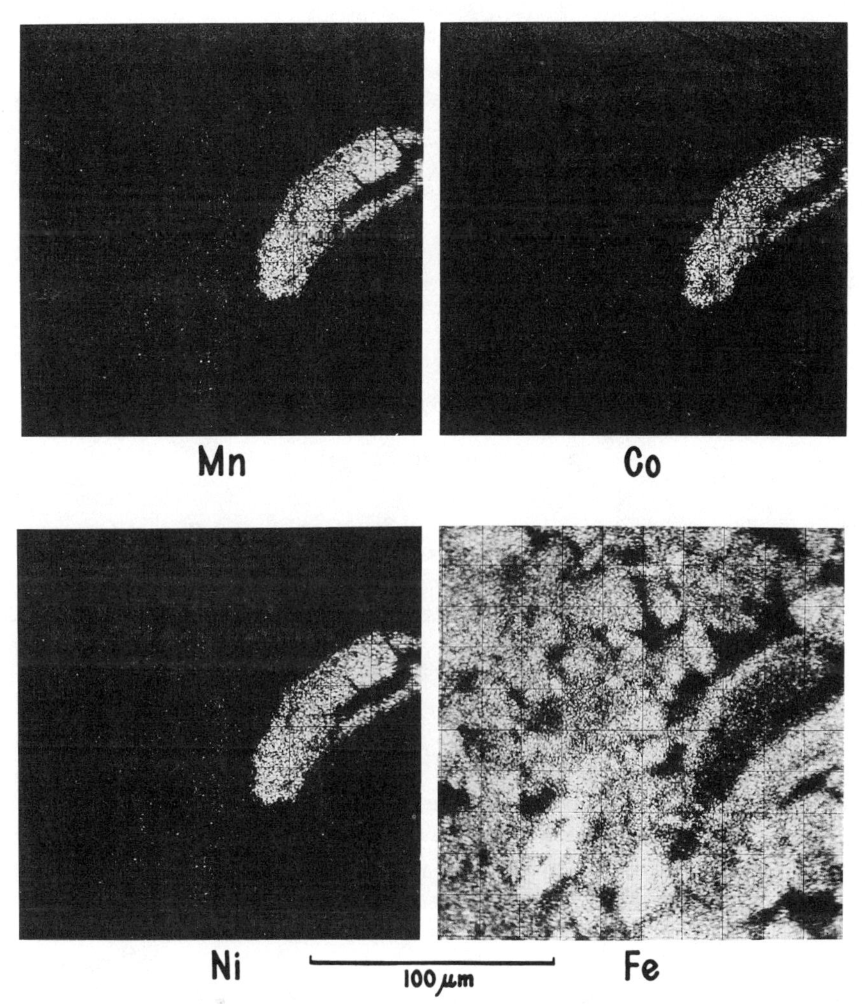

Figure 3. Microprobe displays showing the association of Co and Ni with Mn in laterite from Western Australia (Norrish, 1975).

of nodule-free soils McKenzie (1975) found Ni to be associated with both Mn and Fe, and although the Ni had a greater concentration in the Mn oxides, a greater proportion of the Ni was associated with the Fe oxides (Table V). The proportion of Ni that was not accounted for varied widely between soils and may be partly in Mn or Fe oxides whose particle sizes were too small to be detected by the electron microprobe, or the Ni may be partly in other forms. The extraction of two soils with acidic ammonium

Table V

Ni in Mn and Fe Oxide-Rich Phases of Six Soils Expressed as a Percentage of the Total Concentration of Ni in the Soil[a]

Soil	Mn	Fe	Unaccounted
Urrbrae loam	4	20	76
Rendzina[b]	9	80	11
Terra rossa[b]	5	57	38
Grenfell	<1	18	81
Gray loam	7	23	70
Willunga[b]	19	37	44

[a] From McKenzie (1975) by courtesy of the Australian Journal of Soil Research.
[b] Ni highly correlated with Mn in these soils.

oxalate under UV irradiation extracted ≈70% of the total Fe, ≈90% of the total Mn, ≈30% of the total organic matter, but only ≈25% of the total Ni (Le Riche and Weir, 1963), which means that ≈75% of the total Ni was in forms resistant to the extractant. This result highlights the high proportion of Ni in stable and resistant forms and the unsatisfactory nature of the majority of fractionation schemes, which are discussed later.

Dawson *et al.* (1985) considered adsorption and stabilization through crystal field energy and lattice energy as the factors contributing to the type of enrichment they observed in concretionary forms of Mn. They suggested that the primary enrichment process for Co^{2+}, Ni^{2+}, and Zn^{2+} is sorption to the Fe and Mn oxy–hydroxy phases, presumably as the oxides form rather than by ionic replacement after the oxide has formed; they believe that sorption of Ni^{2+} favors enrichment in Mn oxides over Fe oxides, whereas ionic replacement favors neither one. Solid-state diffusion and subsequent ionic replacement must be considered a strong possibility when the oxide phase is stable and reaction times are periods extending over perhaps hundreds of years or more.

In nature there is an apparent competition between Co and Ni, and their association in soils has been studied (Mitchell, 1945; Makitie, 1962). In the reducing conditions prevailing at the crystallization of igneous rocks from magma, the competition is between Co^{2+} and Ni^{2+}, whereas in the weathering environment Co^{3+} is present as well. As a result Co is more strongly sorbed and enriched in some Mn nodules than Ni^{2+} even though the latter has a greater crystal field stabilization energy and octahedral site preference than Co^{2+}. Murray *et al.* (1968) found that Co^{2+} was specifically sorbed, much more strongly than Cu^{2+} and Ni^{2+}, by 10Å manganite, and they supported Burns' suggestion that this apparently anomalous behavior was due to the surface oxidation of Co^{2+} to Co^{3+} (Burns, 1965); Burns (1976) also suggested that the oxidation was followed by isomorphous substitution of Mn(IV) by the Co^{3+}. Using X-ray photoelectron spectroscopy (XPS), Dillard *et al.* (1982) showed that Co(III) was present in four oceanic ferromanganese nodules and the mechanism was proven by Crowther *et al.* (1983). Dillard *et al.* (1984) found Co(III) in nodules consisting predominantly of δ-MnO_2, which is a common form of Mn oxides identified in soils (Gilkes and McKenzie, 1988); but the same study by Dillard *et al.* (1984) showed that the Ni in an oceanic nodule consisting largely of todorokite, and containing very little Co, was present as Ni(II) in octahedral coordination with O ligands and not OH ligands. Although todorokite, an oxide with tunnel structure, is the main Ni (and Cu)-containing phase in marine nodules (Ostwald and Dubrawski, 1987), it is not commonly found in soils to any extent (McKenzie, 1977), so we must look elsewhere among the Mn oxides for the oxide or oxides in which Ni may be most concentrated. Two oxides that have layer structures, birnessite and lithiophorite, are probably the most common nodular or concretionary forms found in soils (Taylor, 1987; Gilkes and McKenzie, 1988), and in Scotland they were found to contain higher concentrations of trace metals than the two most common tunnel-structured oxides found in soils, hollandite and cryptomelane (Anonymous, 1987).

In New Caledonia the weathering of ultrabasic rocks there gives rise to enrichment of Co and Ni in lithiophorite found at the interface between an uppermost lateritic horizon predominantly of iron oxides and a lower horizon rich in Si and Mg. In lithiophorite the Ni^{2+} is found along with Li^+ in the Al layer while the Co, as Co^{3+}, is found in the Mn layer (Llorca, 1987); although the Ni^{2+} is believed to substitute for Li^+, whose ionic radius is 74 pm (Manceau *et al.*, 1987), it is not yet known whether the Ni substitutes for the Al or the Li or both (Manceau *et al.*, 1990). Manceau *et al.* (1987) drew attention to the contrasting behavior of Co and Ni in the lateritic profiles of New Caledonia, where Co tends to be associated wholly with the Mn minerals asbolane and lithiophorite, whereas Ni is found in garnierites substituting for Mg, in goethite, in lithiophorite substituting for Li (or Al), and in asbolane as layers of $Ni(OH)_2$. It would seem that the oxidation of Co^{2+} to Co^{3+} within the Mn oxides and the persistence of Ni as Ni^{2+} lead to their separation in nature even within the lattices of their host Mn oxides.

In a study of the weathering of orthopyroxenes under lateritic conditions in the West Ivory Coast (Nahon and Colin, 1982), it was found that the end-product was a well-crystallized goethite. Although so-called precursors of the goethite (found below the goethite in the profile) contained about 1% Ni, the goethite itself contained only 0.035% Ni, which suggests that the Ni was excluded from the lattices during aging. It would seem that definite structural evidence of substitution of Fe by Ni in goethite is still lacking (Schwertmann and Latham, 1986), and, although Gerth (1990) found evidence for the isomorphous substitution of Ni^{2+} (and Co^{3+}, Cu^{2+}, Zn^{2+}, Cd^{2+}, and Pb^{4+}) in synthetic goethite prepared under highly alkaline conditions, the nonuniform distribution and presence of Ni in inclusions, probably of $Ni(OH)_2$, suggests that during aging Ni is excluded from goethite as it is when magnetite is oxidized to hematite (Sidhu *et al.*, 1980).

The existence of stoichiometric precipitates such as Ni ferrite ($NiFe_2O_4$) and Ni phosphate $Ni_3(PO_4)_2$ in soils has been proposed and investigated by Sadiq and Enfield (1984a,b); their papers contain discussion and data (equilibrium constants, etc.) on the likely solid-phase transformations of Ni in soils. However, it is considered doubtful that the solubility relationships for such compounds have any relevance except in heavily fertilized and very polluted soils (Tiller, 1983); soils formed on ultrabasic rocks may also be an exception, but even then the pH would need to be alkaline for a nickel phosphate to assume any role in controlling the solubility of Ni (Pratt *et al.*, 1964). For example, the equilibrium concentrations at 50 days in the experiments of Sadiq and Enfield (1984b) ranged from 0.21 to 11.3 ppm for 21 soils; these concentrations are from 10 to 500 times the mean concentration of Ni in saturation extracts of unpolluted soils given in Table II, but comparable to those in a naturally toxic soil (Anderson *et al.*, 1973) and a polluted soil (Sposito and Bingham, 1981). However, even if agreement between experimental results and the predictions based on solubility product did exist one cannot be sure that precipitation is the mechanism responsible for controlling the concentration of the metal ion in the soil solution; adsorption is just as likely but the same difficulty of inferring a mechanism exists (Sposito, 1984).

2. Organic

The associations of trace metals with the organic matter of the solid phase in soils are perceived to be of considerable significance in determining behavior in soils, but despite this perception it has not stimulated a great deal of fruitful research, probably because their investigation is difficult; only when satisfactory methods and techniques exist will the true extent of the associations and their influence be determined. When organic matter is extracted from soil with alkali, the behavior of trace metals is uncertain, not only because of induced changes in the organic matter but also because some metals may change their association with the organic fraction relative to that which existed in the normal soil. For example, adsorbed forms may desorb and precipitate as hydroxides, or they may remain in solution in the extract as an organo-complex or a hydroxy anionic complex such as zincate. Acidification of the extract may cause dissociation of the metal from the humic acid, the fulvic acid, or both, or it may lead to a transfer from one fraction to the other. There are obviously some uncertainties that must include the contrasting claims (see below) concerning the existence of hydroxy species such as nickelate in strongly alkaline and dilute solutions of Ni species.

The associations of trace metals with the humic and fulvic acids extracted with 0.2 *N* NaOH from a soil under pasture (Table VI) may be artifacts or they may represent forms coordinated in very stable complexes. The data indicate that very low proportions (<4%) of the total Co, Mn, and Ni are associated with these extracted forms of organic matter, particularly when compared with Cu, and yet even its degree of association seems low. It is most likely that precipitation of insoluble hydroxides is the cause of the apparent low recovery, although association with humins cannot be discounted entirely. It is unfortunate that Zn was not included in the study because its tendency to form zincate ions would have provided a basis for comparison with Ni, since the nickelate ion $Ni(OH)_3^-$, according to Baes and Mesmer (1976), exists in dilute systems at pH's >10. But since Cotton and Wilkinson state that no nickelate ion exists (Cotton and Wilkinson, 1966) and that $Ni_4(OH)_4^{4+}$ exists at pHs >12 (Cotton and Wilkinson, 1980), uncertainty prevails. These mononuclear and polynuclear hydroxy species may have some significance when strong alkali is used to extract organic matter, but only more systematic study shall determine whether that is the case.

For comparison, some data from work on four highly organic sludge–soil mixtures taken from drying ponds are given in Table VII; the material was extracted with 0.5 *M* NaOH. After acidification and removal of the humic

Table VI

Co, Cu, Mn, and Ni Associated with Humic and Fulvic Acids Extracted with 0.2 *N* NaOH from a Soil under Pasture[a]

Fraction	Organic matter (mg g^{-1} (%))	Co (ppm (%))	Cu (ppm (%))	Mn (ppm (%))	Ni (ppm (%))
Whole soil	123 (100)	47 (100)	28 (100)	2200 (100)	35 (100)
Humic acid	18.2 (15)	0.66 (1.4)	1.9 (6.8)	4.3 (0.2)	0.59 (1.7)
Fulvic acid	54.3 (44)	0.73 (1.6)	2.4 (8.6)	23.8 (1.1)	0.57 (1.6)

[a] From Cheshire *et al.* (1977) by courtesy of Pergamon Press.

Table VII

Cu, Ni, and Zn Extracted from Four Highly Organic Sludge–Soil Mixtures with 0.5 *M* NaOH Expressed as a Percentage of the Total Metal Concentration of Each Mixture

Sludge-soil mixture	Cu			Ni			Zn		
	HA[b]	Ppte	FA	HA	Ppte	FA	HA	Ppte	FA
A	15	4	2	2	23	17	3	8	<1
B	41	<1	<1	8	7	37	9	<1	<1
C	22	<1	<1	10	4	13	1	5	1
D	17	<1	<1	3	20	15	1	2	<1
Mean	24	<2	<1	6	14	21	4	4	<1

[a] Calculated from Holtzclaw *et al.* (1978).
[b] HA, humic acid; Ppte, precipitate; FA, fulvic acid.

acid, the pH was gradually increased stepwise to a pH of 10.5; at each pH the precipitates were collected and analyzed. The material left in solution at pH 10.5 was designated fulvic acid and analyzed also. In the work of Cheshire *et al.* (1977), the precipitated fractions other than the humic acid fraction would have been included in the fulvic acid fraction. The results are surprising in that on average only 26 and 8% of the total Cu and Zn, respectively, were extracted, whereas 41% of the total Ni was extracted. The Cu extracted was associated with the humic acid, whereas the Ni much less so. If zincate were formed during the initial extraction, one would expect more Zn to be extracted, to form Zn^{2+} on acidification and then precipitate as $Zn(OH)_2$ or form zincate or form complexes with the fulvic acid. Why such a high proportion of the total Ni was extracted in the first place is difficult to explain since the low proportion in the humic acid fraction suggests that $Ni(OH)_2$ and nickelate may be the forms of Ni in the precipitate and the fulvic acid fractions, respectively. This discussion only serves to emphasize the uncertain nature of the forms in the original samples, the likelihood of redistribution during extraction, and the uncertain entities in the extract and their chemical behavior as pH is increased. Further, it emphasizes the point that such extractions may have little relevance except perhaps in soils affected by waste disposal.

If the Ni in the fulvic and humic acids resists extraction with both acid and alkali then the Ni may be deemed to be occluded in an experimental sense, but not necessarily in a physical one. Other possible forms of occluded organic Ni may be found in porphyrins (Krauskopf, 1979) or in soil urease (Dalton *et al.*, 1985). Cheshire *et al.* (1977) found the signal of a

Cu porphyrin-type complex in the insoluble acid-hydrolyzed humic acid fraction, a fraction that contained less than 0.5% of the total Cu in the soil; in the same fraction there was less than 0.02% of the total Ni. The geoporphyrins in sedimentary systems are rich in Ni(II) and VO(II) and are thought to be derived from chlorophylls (Filby and Van Berkel, 1987); their existence is attributed to the high stability of the complexes formed between these two metals and the cyclic tetrapyrroles (Quirke, 1987). Highly polar Ni-rich nonporphyrin compounds are also found in heavy crude petroleums and bitumens (Fish *et al.*, 1987).

Another possible source of Ni-containing tetrapyrroles is the cofactor F430 found in all methanogenic bacteria (Hausinger, 1987) such that any sewage sludge derived from an anaerobic process involving methanogenic bacteria may have some of its Ni tied up in Ni-porphyrins. The existence of these very stable forms of Ni indicates that similar forms may exist in soils and may be potential repositries for Ni in soils contaminated with Ni.

Bordons and Jofre (1987a) have investigated the accumulation of Ni from millimolar Ni^{2+} solutions by bacteria isolated from activated sludges and polluted sediments; the highest concentrations of Ni achieved were of the order of 13,000 ppm, much of which was probably due to extracellular adsorption (Bordons and Jofre, 1987b). In a study of the distribution of Cd, Cu, Fe, Ni, and Zn between the principal components of a digested sewage sludge, 97, 90, 80, 84, and 95%, respectively, of the total amount present were in the bacterial detritus or biofloc which made up 84% of the total dry weight (MacNicol and Beckett, 1989). Although much of the Ni is likely to be present in labile forms, some of it is likely to be present in stable forms that resist microbial degradation, particularly once they are adsorbed onto clays.

Urease is an example of a microbial by-product that contains Ni and is present in apparently stable forms in soil. The Ni in urease from jack beans resists isotopic exchange and exhaustive dialysis against chelating agents and is fully active at pH 6 in 0.5 *M* EDTA (Dixon *et al.*, 1976); the Ni concentration in the jack bean urease was found to be 1.21 μg of Ni/mg of urease (Dixon *et al.*, 1980). Since Zantua and Bremner (1977) found that 0.02 mg of urease per gram of soil increased the urease activity of five soils about 10 times (from 27.5 to 266 $\mu g\ hr^{-1}\ g^{-1}$), a crude estimate of the Ni in the native soil urease would be about 2 ng of Ni per gram of soil compared with the total concentration of 40,000 ng g^{-1} given in Table I. Alternatively, since plant and microbial ureases have specific activities from 1000 to 5500 μmol min mg^{-1}, have molecular weights between 200,000 and 250,000, and contain four Ni atoms per molecule (Mobley and Hausinger, 1989), it can be calculated that the native urease in the soils studied by Zantua and Bremner (1977) would have contained much less than 1 ng of

Ni per gram of soil. Thus, if soil urease does contain Ni, then the proportion of the soil's total Ni content is likely to be quantitatively insignificant. Some caution is necessary since Ladd (1985) points out that the hydrolysis of urea in soils may not be due entirely to urease as expected (Bremner and Mulvaney, 1978) but may occur via the allophanate pathway.

Other organic molecules either alone or in association with inorganic matter may provide Ni^{2+} with a stable environment, but these forms, which resist routine extraction with acids, bases, etc., are unlikely to participate in those day-to-day reactions of significance in the availability of Ni to plants. However, their identity and formation are of potential significance in the long-term behavior and fate of Ni added to soil, particularly in wastes such as sewage sludge.

3. Summary of Forms

The main forms and associations of Ni found in soils are summarized in Table VIII. The Ni associated with surfaces of the solid phase have been separated into two operationally defined fractions based on their exchangeability with Ca^{2+} and Cu^{2+}.

Table VIII

Forms and Associations of Ni with Phases of Soil

Phase of soil	Form or association
	Soil solution: $Ni(H_2O)_6{}^{2+}$, $NiOH^+$, $Ni(L^{n-})^{(2-n)+}$
Organic	Exchangeable Ni (Ca^{2+})[a]
	Nonexchangeable Ni (Ca^{2+}); exchangeable Ni (Cu^{2+})[b]
	Nonexchangeable Ni (Cu^{2+}): e.g., porphyrins (?), urease (?)
Inorganic	Exchangeable Ni (Ca^{2+})
	Nonexchangeable Ni (Ca^{2+}); exchangeable Ni (Cu^{2+})
	Nonexchangeable Ni (Cu^{2+}): lattice forms
	Primary minerals
	NiS, olivine, pyroxenes, amphiboles, biotite, ilmenite, magnetite
	Secondary minerals
	Trioctahedral layer silicates
	1:1 serpentines
	2:1 saponites, vermiculties
	2:1:1 chlorite
	Hydrous oxides and oxides of Fe and Mn

[a] Exchangeable with respect to Ca^{2+}.
[b] Exchangeable with respect to Cu^{2+}.

III. REACTIONS

The point of focus here is $Ni(H_2O)_6^{2+}$, which is almost certainly the major form taken up by plant roots (Mishra and Kar, 1974) and microorganisms. The Ni^{2+} released by weathering processes may participate in several processes, including the formation of soluble complexes (Fig. 1), or it may be removed from solution by various sorption or coprecipitation reactions. An extensive review by McBride (1989) gives a thorough account of some of the reactions that are believed to be controlling the solubility of heavy metals in soils. In discussions of reactions in soils, although the scale of things in terms of time and space are of the utmost importance, it is so easy to forget them, and nothing is more extreme than the result of weathering over hundreds and thousands of years and the result of adsorption studies performed in less than a day.

A. Weathering

The ferromagnesian minerals are the least likely of the essential primary minerals to persist in a weathering environment. Thus, Ni is among one of the first metals to be "freed" from its form in the primary mineral, and upon its release from the primary silicate structure it will probably adopt octahedral coordination in an amorphous oxide or silicate, which in turn may be a transient form or the precursor of a secondary mineral. At the time of its release from the ferromagnesian mineral, along with Mg^{2+} and Fe^{2+}, the predominant octahedral sites would most likely be in either trioctahderal layer silicates or Fe oxides. The competition for sites with Mg could be expected to be severe, but presumably it is not overwhelming since Ni is conserved and Mg is lost during serpentinization and the formation of saprolite (Manceau and Calas, 1986; Schellman 1989a,b).

Nickel is a conservative element in the sense that it has the strongest affinity of the common divalent cations for octahedral sites so that it appears to find itself at home in both layer silicates and hydrous oxides of Fe and Mn; irrespective of the weathering conditions there are always octahedral sites available. The highest concentrations in weathered materials are the low-grade Ni ores associated with the formation of laterite. It is curious that, even though Ni in igneous rocks is found in the most easily weathered minerals (ferromagnesian silicates and possibly sulfides), after sequential extraction of the most reactive forms, a high proportion of the Ni in soils is commonly found in the least reactive fraction, the residual fraction. For example, Yamamoto (1984) found in residues from extracted nonvolcanic ash soils that 30 to 53% of the total Ni (and Zn) was in the residual sand plus silt fraction and 15–42% was in the clay fraction. These residual forms have yet to be satisfactorily identified and are discussed further under Fractionation and Speciation.

The process of serpentinization, i.e., the formation of serpentine from olivine or related primary minerals, can be written

$$\underset{\text{olivine}}{6[(Mg_{1.5}Fe_{0.5}SiO_4)]} + 7H_2O \rightarrow \underset{\text{serpentine}}{3[Mg_3Si_2O_5(OH)_4]} + \underset{\text{magnetite}}{Fe_3O_4} + H_2 .$$

If magnetite is a product of serpentinization then some of the Ni released from the olivine may coprecipitate in the magnetite and will be found in the highly magnetic fraction of soils derived from mafic and ultramafic rocks. Although this may happen, it seems that where both are present the Ni prefers the trioctahedral lattices of layer silicates such as saponite and vermiculite/chlorite to those of magnetite (Wilson and Berrow, 1978).

In a study of the weathering of serpentine in a humid warm-temperate climate, Yamamoto (1984) derived the following weathering sequence: serpentine minerals → smectite (saponite) → expanding-lattice clay minerals → chlorite/vermiculite interstratified mineral → chlorite. The soil clays contained from 54 to 74% of the total Ni content and the Ni was associated with chrysotile, chlorite, and the interstratified chlorite/vermiculite mineral. He suggested that the Ni is present in the interlayer spaces of the expanding minerals as $Ni(OH)_2$. In another study of serpentine soils from Japan (Suzuki *et al.*, 1971), a soil, after the destruction of organic matter, was fractionated into sand, slit, and clay. The coarse and fine sand fractions were then separated into magnetic and nonmagnetic fractions with a magnet, and then the nonmagnetic fraction was separated into light and heavy minerals with bromoform (SG 2.9). The concentrations of Ni in the various fractions (Table IX) are all high, but the highest concentration is in the clay fraction, whereas the next highest concentration is in the light nonmagnetic fraction. The data indicate that in this case most of the Ni was in the silt fraction, which is unusual. The main magnetic minerals were ilmenite and magnetite and the heavy nonmagnetic mineral was chromite; the minerals of the light fraction were not identified.

On the basis of slender evidence (Wells, 1960) it would seem that Ni is retained more in soils that have developed on igneous rocks in the tropics than in soils formed in more temperate zones. And, although Ni is among one of the least mobile elements under the conditions of laterite formation (Schellmann, 1986), the laterite is often less enriched with Ni than the zone

Table IX

Concentrations of Ni, Mn, and Zn in Separate Fractions in a Serpentine Soil[a]

			Fine sand			Coarse sand		
				Nonmagnetic			Nonmagnetic	
	Original	Clay	Magnetic	Heavy	Light	Magnetic	Heavy	Light
Element								
Ni	2600	2500	1370	760	1700	1000	817	1960
Mn	1080	500	4030	2730	500	2700	2900	370
Zn	170	280	1610	456	180	318	259	114
			Proportion (%)					
Whole soil	100	33.7	0.20	1.28	11.3	0.60	0.17	6.72
Total Ni	100	32.4	0.1	0.4	7.4	0.2	0.05	5.1

[a] From Suzuki *et al.* (1971) by courtesy of the Japanese Society of Soil Science and Plant Nutrition.

of saprolite below the laterite, which in turn is enriched relative to the unweathered parent material (Schellmann, 1989a, 1989b). It seems that the enrichment of the saprolite is occurring at the expense of Mg and that once weathering has proceeded to the point where most Mg and Si have been lost the Ni is relatively poorly retained by the hydrous Fe(III) oxides, which allows the Ni to be leached into the zone below where the saprolite is forming. Wolfenden (1965) found in the weathering of andesite in Sarawak that Ni was depleted during the formation of both hill and swamp bauxites and appeared to be retained in association with the formation of kaolinitic clay. By comparison, the weathering of andesite and soil formation on Hawaii led to increasing loss of Ni with pedogenic development (increasing rainfall); there was little loss of Ni from developing oxisols, but in ultisols and spodosols the losses were from upper horizons (Kimura and Swindale, 1967). A study of a nickeliferous laterite profile derived from serpentinized lherzolite at Soroako in Indonesia showed that Ni had been enriched by the loss of Si and Mg, which had occurred at a much higher rate than the loss of Fe (Golightly and Arancibia, 1979).

Schreier *et al.* (1987) found from a study of asbestos-rich serpentinitic sediments in northwestern Washington State that trace metals, including Ni, were associated with both the magnetic (mainly magnetite) and the nonmagnetic (mainly chrysotile) fractions; more than 50% of the Ni and the Cr present in the sediments were extracted by leaching with oxalic and citric acid. Short (1961), in a study of four soils derived from basalt, granite, granodiorite, and meta-andesite, in four different climatic locations in the United States, found Ni to accumulate relative to Al at each site.

Weathering of Eocene clastic sequences of carbonate-rich sediments derived from dolomite and serpentinite under highly oxidizing and alternating wet and dry conditions in northern Iraq had led to the association of 64% of the Ni with crystalline mineral phases (residual silicates and serpentine), about 10% with a citrate–dithionite–bicarbonate extractable fraction (mainly Fe), and 26% with an HCl-soluble fraction (carbonate minerals) (Dhannoun and Al-Dabbagh, 1990).

Within temperate zones, Ni is more readily lost upon weathering under restricted drainage than under unrestricted drainage (Tiller, 1958). It would seem that the breakdown of the ferromagnesian minerals is more rapid under restricted drainage probably because of permanent or persistent wetness, and there is more rapid transfer of Ni and other trace metals to the clay fraction (Mitchell, 1964). Under humid Mediterranean conditions with an annual rainfall of about 900 mm and where basic igneous rocks were transforming to mildly acidic soils with kaolinitic clays, Ni was not highly mobile and losses were relatively small (Navrot and Singer, 1976). In Tasmania during the formation of soils on dolerite in areas with annual rainfall from 475 to 1750 mm, the Ni concentrated in the clay fraction, and the losses, although relatively moderate and not related to rainfall, were greatest from surface horizons (Tiller, 1963), which suggests that the greatest changes had occurred in the initial stages of weathering. However, a study of core-stones and the rinds of weathering basalts by Eggleton *et al.* (1987) showed that whereas there was a loss of 40% of the original Mn, there was no loss of Ni. An earlier study by Wilson (1978) of one of the basalts studied by Eggleton *et al.* (1987) obtained similar results for Ni in the core-stone and rinds, but in soil remote from the weathering rock (where kaolinite had "replaced" montmorillonite) the estimates loss of Ni was of the order of 40 to 50%.

The weathering of serpentinite in Maryland gave rise to silt loams whose clay fractions were dominated by smectite and vermiculite; it appeared that under the conditions of soil formation the Ni concentration in the surface horizons had been decreased due to either leaching or dilution by the addition of aeolian material (Rabenhorst *et al.*, 1982). A similar observation was made by Schellmann (1989a) with respect to lateritic soils in a number of countries, but whether the foreign material was of alluvial–colluvial or aeolian origin is uncertain.

Upon the weathering of primary minerals it seems that the preferred octahedral site for Ni is in trioctahedral layer silicates substituting mostly for Mg. In the absence of these sites, Ni^{2+} substitutes for Fe^{2+} (e.g., magnetite) and possibly for Mn^{2+} in Mn oxides. In laboratory-prepared magnetites coprecipitated Ni was uniformly distributed in the crystals (Sidhu *et al.*, 1978). On partial oxidation to maghematite the Ni concentrated inside the crystals, but upon oxidation of all the Fe(II) to Fe(III) to form hematite, the Ni was ejected (Sidhu *et al.*, 1980). In synthetic goethites formed at high pH, the coprecipitated Ni was not uniformly distributed in the crystals and it was suspected to be present as inclusions of amorphous $Ni(OH)_2$ 2 to 30 nm in diameter (Gerth, 1990). Voskresenskaya (1987) argues that most of the Ni associated with low-grade lateritic Ni ores is present in goethite as inclusions of $Ni(OH)_2$. Also, Ni is present in the Mn oxide asbolane as layers of $Ni(OH)_2$ (Manceau *et al.*, 1987), there is evidence of the tendency of the Ni to "cluster" in Ni-rich domains in hydrous Ni layer silicates (Gerard and Herbillon, 1983; Manceau and Calas, 1985, 1986); and chloritization of vermiculite may involve the insertion of a Ni-rich brucite interlayer. The above suggests that of all the options available to Ni^{2+}, $Ni(OH)_2$ may be the preferred form of Ni in some weathering environments.

Perhaps many of the difficulties and uncertainties discussed in this section on weathering could be resolved by a more systematic and thorough study of the physical and chemical fractions of soils derived from igneous rocks, the associations of Ni with those fractions, and the influence of soil forming factors on the distribution of Ni between the fractions.

B. Sorption

The rapid removal of Ni from solution at low concentrations is due to sorption by any of the surfaces proffered by the solid phases of soil: amorphous Mattsonian-type precipitates; the layer silicates; organic matter; hydrous oxides of Si, Al, Fe, and Mn; and carbonates. Internal surfaces as well as external surfaces contribute to the range of surfaces available. Because of the range of surfaces that exist there is a continuum of sorption sites with various affinities for the species of Ni (Fig. 1). Although cationic species of Ni (Ni^{2+} and $NiOH^+$) appear to be the most strongly sorbed species, the coadsorption of Ni and a ligand either concurrently or sequentially may assume some significance under some conditions, but these have yet to be determined.

Traditionally it is considered that there are two main forms of sorbed ions: (*a*) nonspecifically sorbed ions (outer-sphere surface complexes), which are operationally defined as *readily exchangeable*, and (*b*) specifically sorbed ions (inner-sphere surface complexes), which are operationally defined as *nonexchangeable*. The experimental distinction between these two forms is made by measuring sorption in the absence and presence of an indifferent electrolyte, respectively, although in soil the electrolyte in the soil solution is rarely indifferent. It is also difficult to distinguish experimentally between forms associated with external surfaces and those associated with internal surfaces. The hysteresis of sorption–desorption isotherms describing specific adsorption studies may be due in part to the extreme and apparent reluctance of "lattice" forms to diffuse into the bulk solution. For lattice forms to form, Ni^{2+} must diffuse through the micropores in the crystal structure and either bond to surface groups lining the pores or replace similar ions in the structure. Presumably a small cation partially or completely stripped of its hydration shell would be absorbed more readily than a larger one when only size is considered, but the associated increase in charge intensity counters any improved mobility associated with either partial or total dehydration. The partial hydrolysis of divalent trace metal cations, which would favor inner-sphere complexation, decreases the degree of hydration and increases the charge intensity and thereby must be favored as the species most likely to be involved in solid-state diffusion of divalent trace metal cations such as Ni into soil minerals such as goethite (Bruemmer *et al.*, 1988). With the passage of time the scavenging of Ni that we observe in some secondary minerals must occur as a result of either coprecipitation or solid-state diffusion.

One might argue that the crystalline oxides of, say, Mn, which often contain relatively high concentrations of Ni, were formed at a time when the ambient concentrations of Ni were much higher. Or, one might imagine Ni being specifically adsorbed onto an accreting surface and ultimately being occluded in the lattice of the mineral. Unfortunately most sorption studies are relatively brief and usually are carried out under uniform conditions. Yet in soil the conditions are very different and the bulk solution is dynamic in its chemical composition because of the influence of changing temperature, soil water potential, microbial activity, etc. It is against this background that we must consider sorption and desorption studies of Ni.

Sorption and desorption studies of Ni species, predominantly Ni^{2+}, with the main soil constituents and with whole soils have been made (Table X), mainly in comparison with other trace metals (Table XI). In many adsorption studies the concentrations used are unrealistically high (10^3 to 10^6 times the concentrations found in soil solutions), even for polluted soils, and the times for reaction are unrealistically brief (as brief as 5 min) relative to plant growth and soil processes.

The sorption of Ni is strongly pH dependent for hydrous oxides and organic matter, but much less so for illite and kaolinite, and virtually absent for montmorillonite (Andersson, 1977). Kinniburgh *et al.* (1976)

Table X

Sorption Studies of Ni with Model Systems and with Soils

	A. Hydrous oxides
Fe	Kinniburgh *et al.* (1976); McKenzie (1980); Theis and Richter (1980); Schultz *et al.* (1987); Gerth and Brümmer (1983)
Al	Kinniburgh *et al.* (1976); Schulthess and Huang (1990)
Mn	McKenzie (1967, 1972, 1980); Grasselly and Hetényi (1968); Murray (1975); Bruemmer *et al.* (1988); Ooi *et al.* (1987)
Si	Theis and Richter (1980); Schulthess and Huang (1990)
	B. Layer silicates
Chlorite	Koppelman and Dillard (1977)
Illite	Andersson (1977); Koppelman and Dillard (1977)
Kaolinite	Andersson (1977); Koppelman and Dillard (1977); Mattigod *et al.* (1979); Angle (1982); Bansal (1982); Puls and Bohn (1988); Schulthess and Huang (1990)
Montmorillonite	Andersson (1977); Bowman and O'Connor (1982); Takahashi and Imai (1983); Puls and Bohn (1988); Schulthess and Huang (1990)
Sepiolite	Helios-Rybicka (1985)
	C. Zeolite
Mordenite	Schulthess and Huang (1990)
	D. Mixtures
Montmorillonite + δ-MnO_2	Traina and Doner (1985)
	E. Organic matter
Peat	Bloom and McBride (1979)
Humic acid	Gamble (1986)
Chelex 100	Figura and McDuffie (1977)
	F. Soils
	Andersson (1977); Biddappa *et al.* (1981); Bowman *et al.* (1981); Sadiq and Zaidi (1981); Harter (1983); O'Connor *et al.* (1983); Sadiq and Enfield (1984b); Tiller *et al.* (1984); Reddy and Dunn (1986); Anderson and Christensen (1988)

found that the selectivity sequence based on the pH at which 50% of the trace metal cation is adsorbed (pH_{50}) for a freshly precipitated Fe gel was Cu(pH 4.4) > Zn(5.4) > Ni(5.6) > Co(6.0), whereas for an Al gel it was Cu(4.8) > Zn(5.6) > Ni(6.3) > Co(6.5). The lower pH_{50} for the adsorption of Ni on the Fe gel indicates greater selectivity of the gel for Ni. The sorption sequences for the inorganic materials are very similar to one another (Table X) and are similar to the degree to which the metals are coprecipitated; for example, when Cd^{2+}, Cr^{3+}, Ni^{2+}, Pb^{2+}, and Zn^{2+} were coprecipitated with geothite under identical conditions (1:100 molar ratio

Table XI

Relative Orders of Sorption of Some Trace Metals including Ni onto Soil Materials[a]

Sorption order	Reference
A. Aluminium oxides	
Al gel: Cu(4.8)>Zn(5.6)>**Ni(6.3)**>Co(6.5)	Kinniburgh *et al.* (1976)
B. Iron oxides	
Fe gel: Cu(4.4)>Zn(5.4)>**Ni(5.6)**>Co(6.0)	Kinniburgh *et al.* (1976)
Goethite: Cu(4.2)>Co(5.9)>**Ni(6.1)**	McKenzie (1967)
Goethite: Zn(4.9)>**Ni(5.6)**>Cd(5.8)	Gerth and Brümmer (1983)
Geothite: Zn(5.0)>Cd(5.8)>**Ni(5.9)**	Bruemmer *et al.* (1988)
Hematite: Cu>Zn>Co>**Ni**	McKenzie (1980)
Goethite: Cu>Zn>Co>**Ni**	McKenzie (1980)
C. Kaolinite	
Cu(5.3)>Co(6.0)=**Ni(6.0)**>Zn(6.2)	Andersson (1977)
Cd(4.5)>Zn(5.4)>**Ni(5.8)**	Puls and Bohn (1988)
D. Manganese oxides	
Nodules: Cu(4.3)>Co(6.5)>**Ni(7.1)**	McKenzie (1967)
Birnessite, lithiophorite, hollandite: Cu>Co>**Ni**	McKenzie (1967)
Birnessite: Cu>Co>Zn>**Ni**	McKenzie (1980)
Cryptomelane: Cu>Co>Zn>**Ni**	McKenzie (1980)
Hausmannite: Cu>Zn>Co>**Ni**	McKenzie (1972)
Hydrous oxide: Cu>Co>Zn>**Ni**	Murray (1975)
Manganite: Cu>Zn>Co>**Ni**	McKenzie (1972)
Partridgeite—initial: Cu>Co=Zn>**Ni**	McKenzie (1972)
Partridgeite—aged: Co≫Cu>Zn>**Ni**	McKenzie (1972)
λ-MnO_2: Cu>Co>Zn>**Ni**	Ooi *et al.* (1987)
E. Montmorillonite	
Cd(4.7)>Zn(4.8)>**Ni(5.3)**	Puls and Bohn (1988)
Pb>Zn>**Ni**>Cd	Schulthess and Huang (1990)
F. Organic Matter	
Chelex 100: Cu(<2.5)>Cd(2.8)>**Ni**=Zn(3.3)>Co(3.8)	Figura and McDuffie (1977)
H^+-peat: Cu>**Ni**>Mn	Bloom and McBride (1979)
Humic acid: Cu>>Zn>**Ni**>Co>Mn	Rashid (1974)
Humic acid: Cu≫Fe(II)>Co=**Ni**=Zn>Mn	Bloom and McBride (1979)
Humic acid (pH 3.7): Cu>Cd=Zn=**Ni**>Co=Mn	Kerndorff and Schnitzer (1980)
Humic acid (pH 5.8): Cu>Cd>Zn>**Ni**>Co>Mn	Kerndorff and Schnitzer (1980)
Humic acid: Cu>Cd>**Ni**=Zn>Co>Mn	Gamble (1986)
Peat moss: Cu>**Ni**>Co>Zn>Mn	Rashid (1974)
G. Sepiolite	
Zn>Cd>**Ni**	Helios-Rybicka (1985)
H. Soils	
Allophane: Cu(4.9)>Zn=Cd(5.6)>Mn(6.0)>**Ni(6.4)**>Co(6.5)	Yamamoto (1984)
Illitic clay subsoil: Cu>Zn>**Ni**	Andersson (1977)
Volcanic soils: Cu>Zn>Cd>**Ni**	Biddappa *et al.* (1981)
Fe oxide dominant: Zn>**Ni**>Cd	Tiller *et al.* (1984)
Various: Cu>Zn>**Ni**	Harter (1983)

[a] Values in parentheses are pH_{50}.

of metal:Fe), the order of coprecipitation and percentage of metal incorporated was Cd(50.0) = Ni(50.0) $<$ Pb(53.8) $<$ Cr(66.0) $<$ Zn(69.2) (Francis and Dodge, 1990). Similar preferences have been found for Mn oxides and nodules (McKenzie, 1967), but with aging the relative order changes. For example, for partridgeite (Mn_2O_3) the order changed from Cu $>$ Co $>$ Zn $>$ Ni to Co $\gg$ Cu $>$ Zn $>$ Ni (McKenzie, 1972). The surface oxidation of the sorbed Co^{2+} to Co^{3+} by the oxide and the replacement of Mn(IV) by Co^{3+} are undoubtedly responsible for the change in order and account for the accumulation of Co in Mn oxides in soils and elsewhere.

A notable feature of the data in Table XI is that in the general order of sorption, Co and Zn are sorbed more favorably than Ni on inorganic surfaces and on soil, but the converse is true for organic matter. It would seem that in soils the inorganic phase of the soil colloids is the dominant phase in sorption reactions and that it is only in sands and peats that organic colloids would be the dominant phase.

Sorption not only is pH dependent but also depends on the concentration of the suspension, of the Ni itself, and of other species of metals and ligands and on time and temperature (Bruemmer *et al.*, 1988). Bowers and Huang (1987) found that the pattern of adsorption of Ni(II) by γ-Al_2O_3 was reversed by the presence of equimolar (10^{-4} M) EDTA such that adsorption was less strongly pH dependent and decreased with increasing pH; Fe(III) had no effect on the adsorption pattern of Ni-EDTA. However, although the same reversal of the adsorption pattern of Ni(II) by EDTA was observed at micromolar concentrations, in the presence of Fe(III) the effect of EDTA was effectively removed.

Also, because many sorption studies are carried out at high ratios of liquid volume to solid volume, their direct application to soils needs to be done with some, probably extreme, caution. For example, the pH_{50} for the adsorption of Cd (0.5 μM) onto an amorphous Fe gel decreased from about 7.5 to 6.0 when the total Fe concentration was increased 100-fold from 0.13 to 13 mM (Benjamin and Leckie, 1981); at 1 mM Fe the pH_{50} increased, respectively, from 6.3 to 7.3 when the concentration of Cd was increased 0.01 to 50 μM. Such results of sorption studies illustrate that it is difficult, and probably hazardous, to predict behavior in a multiphase system such as soil, particularly since the ratios of water to solids used in sorption studies are very different from those met with in soils, and the nature of all the organic ligands in the soil solution are not well characterized.

Sorption studies by Bloom and McBride (1979) with humic acid and peat found pH dependence similar to that of the hydrous oxides, but, perhaps rather surprisingly, they found little difference in preferences between the binding of Ca^{2+}, Mn^{2+}, and Ni^{2+} by peat and of Ca^{2+}, Co^{2+}, Mn^{2+}, Ni^{2+}, and Zn^{2+} by humic acid. For Cu^{2+} there was a marked and expected preference shown by both organic materials. The bonding was thought to be associated with the formation of inner-sphere complexes for Cu^{2+} and of outer-sphere complexes for Ni^{2+} and the other metal cations.

The sorption of Ni^{2+} onto kaolinite (Mattigod *et al.*, 1979) was shown to be very sensitive to the supporting electrolyte such that replacing NO_3^- with $SO_4^=$ and Na^+ with Ca^{2+} decreased adsorption, so that when $CaSO_4$ was the supporting electrolyte little Ni^{2+} was adsorbed at any concentration of Ni^{2+}. The combined tendency to form ion-pairs with $SO_4^=$ and to exchange with Ca^{2+} leads to the conclusion that in soil a mineral such as kaolinite, with its low cation exchange capacity (CEC), usually high crystallinity, and low specific area, would contribute little to the adsorption and behavior of Ni^{2+}.

The adsorption studies with mineral soils (Andersson, 1977; Harter, 1983) show an increasing tendency for the adsorption of Ni^{2+} to increase with increasing pH and for this pH dependency to be more pronounced in A horizons than B horizons, presumably because of the higher organic matter in the A horizons (Harter, 1983). However, in an organic soil (~80% organic matter) adsorption of both Ni^{2+} and Co^{2+} increased only slightly with increasing pH up to about pH 7, above which adsorption decreased strongly, whereas for Cu^{2+} the changes with pH were less marked (Andersson, 1977). It would seem that as the pH increases the solubility of trace metal cations such as Cu, Ni, and Zn is determined largely by the result of a competition between the increasing capacity and tendency of organic matter to form soluble complexes and the increasing tendency for cationic species to be sorbed onto soil colloids.

C. Soluble Complex Formation

The formation of soluble complexes between Ni^{2+} and organic ligands in the soil solution can be seen as a means by which Ni may be kept in solution when either a precipitating anion is added or the pH is increased. Comparative studies of the behavior of Cu and Zn in soils have shown that the total concentration of Cu in the soil solution is remarkably constant with increasing pH, whereas that for Zn decreases (Brümmer and Herms, 1983; Jeffery and Uren, 1983; Herms and Brümmer, 1984); the solubility of recently added Cu^{2+} (and Zn^{2+}) shows greater pH dependence than native Cu (Jeffery and Uren, 1983). It would seem that in the case of Cu, its increasing affinity for organic ligands with increasing pH, and possibly increasing activity of soluble organic ligands with increasing pH (Jahiruddin *et al.*, 1985), successfully opposes the increasing tendency for Cu^{2+} to be adsorbed as the pH increases. Because Zn^{2+} forms less stable complexes than Cu^{2+}, it virtually loses the battle to stay in solution, but not to the extent that its solubility decreases 100-fold for each unit increase in pH, as would be predicted from the solubility product of $Zn(OH)_2$. A number of studies have shown that the decrease in solubility of Zn per unit increase in pH is more likely 4-fold (Jeffery and Uren, 1983; Tiller, 1983; Herms and Brümmer, 1984). The concentrations and speciation of Ni in soil solutions of contaminated soils follow that of Zn and show strong pH dependence (Sanders *et al.*, 1986a,b).

Dudley *et al.* (1986), in a study of sluge-amended soil, found that soluble Ni (0.01 *M* $CaCl_2$ saturated paste extract) fluctuated in a seemingly random manner over 30 weeks of incubation at −33 kPa. Subsequent to application of GEOCHEM to the analytical data and gel separation, the results indicated that in the initial stages, when the pH and soluble organic C concentrations were high, most of the Ni was present as organic complexes, but after 30 weeks, when the pH had decreased and the soluble organic C was low, the species were predominantly inorganic (Dudley *et al.*, 1987).

Fletcher and Beckett (1987) determined the general affinity sequence for soluble organic matter derived from digested sewage sludge and found $H \gg Pb(II) > Cu(II) > Cd(II) > Ni(II) > Zn(II) \sim Fe(III) > Co(II) \sim Mn(II) > Ca > Mg$. They suggested that in sewage sludge the concentration of Ca and Mg would usually be high enough (0.01 *M*) to compete strongly with all metals present except Pb, Cu, and Cd.

The majority of complexes that form are probably labile, which means that the cation and the ligand associate and dissociate instantaneously. However, some complexes may be much less labile. For example, in a kinetic study of the species formed between Ni(II) and a fulvic acid, it was found that while the species formed at pH 4 and 5 were labile, some 40% of the Ni in the complex that had formed at pH 6.4 took 10 days to be separated from the fulvic acid by reaction with 4-(2-pyridylazo)resorcinol (Lavigne *et al.*, 1987). A similar result was obtained by Cabaniss (1990), who found that increasing the pH and decreasing the ionic strength decreased markedly the dissociation rate of Ni–fulvic acid complexes; increasing the molar ratio of fulvic acid to Ni(II) decreased the dissociation rate also but to a lesser extent than pH and ionic strength. The highly inert component observed by Lavigne *et al.* (1987) was not apparent in the study by Cabaniss (1990). These studies of Ni–fulvic acid complexes indicate the kinetic inertness of some organo-complexes, a factor that is rarely considered in discussions of nutrient behavior in soil solutions in relation to availability, although lability has been used as a basis for speciation of trace metals in aqueous environmental samples (Figura and McDuffie, 1980) and in soil solutions (Jeffery and Uren, 1983).

D. Waterlogging

The severity of waterlogging, as measured by a decrease in E_h is a function of temperature, time, and relative availability of electron donors (organic C) and electron acceptors (O_2, NO_3^-, Mn oxides, Fe oxides, $SO_4^=$, CO_2) (Ponnamperuma, 1972, 1984). Here mild waterlogging shall refer to the situation where O_2 and NO_3^- have disappeared and Mn is coming into solution, and a severe waterlogging refers to that where all the readily reducible Mn oxides have been reduced, Fe oxides are dissolving, and $S^=$ is appearing.

Because waterlogging can cause (*a*) reduction and solubilization of Fe and Mn oxides and (*b*) the formation of complex-forming acids, it is not surprising that mobilization of trace metals has been observed when waterlogging occurs (Ng and Bloomfield, 1962; Bloomfield, 1981); sometimes though there may be no change in the solubility but an increase in the extractability in a reagent such as acetic acid (Iu *et al.*, 1981). Francis and Dodge (1990) have shown that trace metals coprecipitated with goethite were solubilized upon anaerobic incubation with an anaerobic N_2-fixing species of *Clostridium*; the release was partly due to exchange and acidity but largely enhanced by the accompanied reduction of Fe(III) to Fe(II). Initially as O_2 and NO_3^- are disappearing the exchangeable forms would tend to be mobilized by the increase in H^+ activity, and during a mild waterlogging the next to be mobilized would be those associated with reactive oxides of Mn. Subsequently the difficult to reduce Mn oxides and the Fe oxides would dissolve and release their trace metals. At the same time the reduction reactions are consuming protons so that the pH after an initial decrease tends to increase, which in turn favors adsorption. Lattice forms associated with silicates would be unaffected by waterlogging no matter how severe. If waterlogging is severe and persists for more than several weeks then the metals may begin to precipitate as sulfides if $S^=$ is available (depends on the E_h and availability of $SO_4^=$ to be reduced to $S^=$), or they may be leached if percolation is going on, albeit slowly.

The conflicting reports of the effects of waterlogging on the behavior of Cu and Zn (Iu *et al.*, 1982) suggest that because of the numerous factors that influence the course of any waterlogging episode it is difficult to predict with any certainty what may happen to Ni. Katyal (1977) reported that the concentration of water-soluble Zn decreased continuously when the waterlogging of three soils proceeded for 12 weeks in the presence and absence of additional organic matter.

Upon the return of aerobic conditions, the reoxidation of the Fe and Mn oxides may produce more reactive oxides than those that existed prior to the waterlogging, so much so that the availability of ions strongly sorbed by

Table XII

Effect of Soil Drainage on the Extractability of Ni and Co from Soil and Their Concentrations in Red Clover and Rye Grass[a]

	Ni		Co	
	Free drainage	Poor drainage	Free drainage	Poor drainage
Acetic acid soluble (ppm)	1.3	3.4	1.3	2.9
Red clover (ppm)	2.0	5.9	0.16	1.4
Rye grass (ppm)	1.0	3.4	0.18	1.5

[a] From Mitchell and Burridge (1979) by courtesy of *Philosophical Transaction B*, The Royal Society.

these oxides is less than that prior to the waterlogging event; for example, the availability of phosphate to a dryland crop following paddy rice was decreased markedly (Willett, 1979, 1982). Even though reoxidation produces some conditions that one might expect to result in the coprecipitation and occlusion of Ni, the net effect of waterlogging and reoxidation in experiments by Ng and Bloomfield (1962) was to increase the extractability of Ni and other trace metals (Bloomfield, 1981).

In poorly drained soils Ni and Co are more readily extracted and apparently more available to plants than in well-drained counterparts (Table XII). In pedogenic terms the fate of Ni in weathering and soil formation would depend upon the availability of octahedral sites and the competition for them. In the poorly drained soils the primary location of octahedral sites is in the layer silicates, since the Fe and Mn oxides tend to develop only poorly or not at all. In strongly leaching environments the layer silicates tend to be absent and Ni tends to be associated with the oxides of Fe and Mn. The development of concretionary forms of Fe and Mn in poorly drained soils is an apparent contradiction here, but because they develop, by as yet unsatisfactorily explained mechanisms, within a medium in which the mobility of Ni and other divalent trace metals is high, they provide a sink for these metals (although if the metal is not captured and protected by the concretion, it is lost slowly by leaching).

E. Microbial

Because soil microorganisms are major determinants in the behavior of Mn in soils (Ghiorse, 1988) and may have an influence on Zn (Zamani *et al.*, 1984), it is possible that microbial activity may affect the behavior and availability of Ni. This proposition is plausible, since some bacteria have the ability to both accumulate the tolerate above-normal concentrations of Ni (Bordons and Jofre, 1987a,b); for example, 4 of 200 isolates of strains of aerobic bacteria isolated from soils and wastes were able to tolerate $NiCl_2$ concentrations between 7.5 and 20 m*M* (Schmidt and Schlegel, 1989). Also, the finding that in a typical digested sludge 85 to 95% of the Cu, Ni, and Zn was associated with bacterial detritus (MacNicol and Beckett, 1989) suggests that microbial biomass may at times affect the behavior of Ni in soils, particularly in those soils to which sewage sludge has been added. Although one might have thought that the inherent toxicity of Ni would preclude any major direct effect of soil microorganisms on the behavior of Ni, the converse, that of heavy metals on soil biomass, has been found (Brookes and McGrath, 1984; Brookes *et al.*, (1986). Studies of the effects of heavy metals on the urease activity in soils have been made (Tabatabai, 1977; Doelman and Haanstra, 1986), but it is impossible to determine whether the observed inhibition at high rates of addition are due to direct effects on urease or on the microbial population.

Babich and Stotzky (1982) found that in soil cultures the tolerance of various eubacteria, actinomycetes, yeasts, and filamentous fungi varied. Concentrations of Ni from 50 to 750 ppm were required for incipient inhibition of growth, and the toxicity was decreased by the addition of $CaCO_3$, kaolinite, and montmorillonite. Babich and Stotzky (1983) have reviewed the toxicity of Ni to microbes and have indicated that toxic activities of Ni might have detrimental effects on important soil processes such as mineralization of C and N, nitrification, soil enzymatic activity, and litter decomposition. Schellmann (1989a) observed that termites avoid nickeliferous areas although they are not totally absent from serpentine areas (Brooks, 1987), not all of which are nickeliferous.

F. Mobility

For Ni to be mobile in soil it must be in solution. The previous discussion on reactions has indicated that there are processes that tend to remove Ni from solution (sorption predominantly) and that lead to increases in solution (desorption and reduction of host oxides). The net effect determines the concentration and tendency of Ni to move. This section deals with some aspects of Ni mobility in soils.

Some indication of the likely movement of Ni in the short term may be gauged from leaching studies of ^{60}Co, ^{64}Cu, and ^{65}Zn in columns (16 cm) of virgin soils of neutral to alkaline pH (Jones *et al.*, 1957; Jones and

Belling, 1967). The isotopes were added to the surface in solution and leached with an equivalent application of 460 mm rain in 6 hr; dressings of $CuSO_4 \cdot 5H_2O$ (5 lb/acre), $CoSO_4 \cdot 6H_2O$ (1 lb/acre), $ZnSO_4 \cdot 7H_2O$ (7 lb/acre), superphosphate (112 lb/acre), extracts of lucerne (ca. pH 6), and water saturated with CO_2 were applied as treatments. In soils with a moderate CEC there was no movement with or without any of the treatments, but in soils with low CEC (sands, low organic matter) there was some movement for ^{60}Co and ^{65}Zn, increasing in the order $CoSO_4$ = $ZnSO_4$ < $CuSO_4$ < superphosphate < extract from lucerne < water saturated with CO_2, whereas for ^{64}Cu the order was the same except that the lucerne extract gave the same result as the water saturated with CO_2. By comparison, when EDTA was applied to a soil column in which there had been no movement of ^{60}Co, 67% of the ^{60}Co was leached from the column. The evidence from these experiments suggest that under the same conditions Ni would have shown similar behavior, with the effect of the water saturated with CO_2 suggesting that decreases in pH favor mobilization.

The pedogenic mobility of Ni as discussed earlier is favored by the development of partial or complete anaerobiosis accompanied by leaching, and conversely retention appears to be greatest under conditions that favor the development of soils dominated by oxides of Fe and Ti. The occurrence of Ni concentrations of 8 to 21 μg/liter in drainage waters (Böhmer, 1989) and similar concentrations of the same order in soil solutions (Bradford *et al.*, 1971) indicate that Ni is in solution and capable of being leached from the soil.

Short-term leaching studies to determine the mobility of elements in soils do not imitate well field conditions or the concomitant development of reactive phases such as oxides of Fe. For example, in most leaching columns the soil is often saturated and probably anaerobic in some parts, it is leached with what is a normal year's rainfall in a few days or weeks, and in some cases it is exposed to inordinately high concentrations of organic acids. High rates of application of water are of course appropriate to irrigation, particularly with wastewater and of sludge-amended soils. Because temporary waterlogging, often in the presence of additional quantities of readily biodegradable organic matter, may occur, it is under these conditions that the mobility of Ni is most likely to be greatest and some Ni may be leached to depth or into the groundwater.

The addition of readily biodegradable organic matter (glucose and dried berseem clover) to a red soil (pH 6.7, 24% clay) along with varying concentrations of Ni as $NiSO_4$ kept more Ni in a form extractable with NH_4^+ acetate than without it when measured at 20 days, but the effect had virtually disappeared by 80 days except at 100 ppm of Ni where the concentration was lower in the samples of soil that had been amended with the organic matter (Misra and Pande, 1974a).

One type of study of trace metal movement in soils involves the change in the distribution of the trace metals with depth as a result of amendment with either sewage or sewage sludge. After 80 years of irrigation of pastures with sewage at Werribee (Australia) onto loam soil overlying clay, Ni was added at about 80 kg ha^{-1}. Although there appeared to be little movement of the Ni below 20 cm there was some evidence that there had been movement of a small proportion to at least 60 cm (Evans *et al.*, 1979). Similarly, Aboulroos *et al.* (1989) investigated the accumulation of trace metals upon irrigation of sandy soils with sewage effluent from Cairo city for more than 70 years, and they found that the total Ni concentration had increased from 14 to 125 ppm in the 0- to 30-cm layer and from 9 to 64 ppm in the 30- to 60-cm layer. There may have been movement below these depths but no analyses were made.

After 8 years of annual incorporations of two different sewage sludges into the top 20 cm of a loam soil at Berkeley in California at annual rates up to 225 t ha^{-1}, there was no significant movement of any metals out of the zone of incorporation (Williams *et al.*, 1987). Similarly, Chang *et al.* (1984) found after 6 years of incorporation of composted sludges and liquid sludge into two soils (Greenfield sandy loam and Domino loam) at rates up to 41, 792, 491, 200, 624, and 1928 kg ha^{-1} Cd, Cr, Cu, Ni, Pb, and Zn, respectively, that 90% of the metals remained in the top 15 cm and there was slight but statistically significant movement into the 15- to 30-cm layer, but no movement was evident below this latter depth.

Surface application of sewage sludge for 4 years to grassland on two freely draining soils, a sandy loam and a calcareous loam, led to movement of all metals studied to a depth of 10 cm, but 86 and 83%, respectively, of the Ni was recovered in the top 5 cm and 100 and 94% in the top 7.5 cm (Davis *et al.*, 1988). The order of mobility on the sandy loam was Cu $>$ Zn $>$ Mo $>$ Cd $>$ Ni $>$ Pb $=$ Cr, whereas on the calcareous loam it was Mo $>$ Cd $=$ Cu $=$ Pb $>$ Cr $=$ Ni $=$ Zn. The authors believe that movement was due mainly to infiltration of sludge particles and incorporation by the activity of earthworms, a belief which is supported by the unusual relatively high mobility of Cu on the sandy loam, although the greater mobility of Mo on the calcareous soil suggests that solubility and leaching also had an influence.

Other studies of mobility involve columns of soil, some of which employ high rates of application of Ni (e.g., 622 and 794 kg ha^{-1}) in solutions of $Ni(NO_3)_2$ and leaching. The results appear to be difficult to interpret other than to gauge some sort of relative mobility under the conditions of the experiment and appear to have very little general applicability to common situations. Similar experiments have been carried out where sewage sludge was applied to light-textured soils and leached with 5 m of water over 25 months (Emmerich *et al.*, 1982a) and 1.7 m over 13 months (Welch and Lund, 1982). In the former study there was no significant movement of trace metals including Ni below the depth of incorporation, yet the soil solution concentrations varied from 10 to 620 μg/liter (Emmerich *et al.*, 1982b). Welch and Lund (1982) found that about 14% of the Ni applied in the sludge had moved below the depth of incorporation (15 cm) in soils containing from 1 to 3% clay and for soils with 12 and 16% clay the movement below 15 cm was 4.8 and 3.5%, respectively. The concentrations of Ni in the leachate ranged from 3100 to 15,200 μg/liter, which is considerably greater than those reported by Emmerich *et al.* (1982b), who observed little movement (Emmerich *et al.*, 1982a).

From the discussion of weathering and the above, it appears that Ni is not completely immobile and that the extent of movement of Ni down through the profile will depend on (*a*) the forms and quantities applied; (*b*) the amount of water passing through the profile; (*c*) the soil properties, with pH, clay, and organic matter content being the most significant; and (*d*) the propensity of the soil to produce soluble organic ligands capable of forming soluble complexes with Ni.

G. Fractionation and Speciation

The determination and characterization of the nature, forms, and relative proportions of an element such as Ni in environmental samples such as soils and sediments are essential steps in providing the necessary information for sound management decisions, whether they be recommendations of fertilizer rates or reclamation procedures for polluted land. The terms fractionation and speciation have been used, respectively, with regard to solid samples and aqueous samples including soil solutions, and here the terms will be used in that context even though speciation has been used sometimes with regard to adsorbed species.

Apart from a basic desire to understand soil processes, fractionation and speciation are means of assessing bioavailability and changes associated with the long-term reactions involved in weathering and determining answers to questions such as "what are the active forms and the less reactive forms?" In the case of heavy metals such as Ni, fractionation studies may be used as a guide to the likely success of encouraging and accelerating the less active forms to form at the expense of the more active forms (i.e., reversion; see under Reversion), a strategy which may be employed in the reclamation of polluted soils.

1. Soil Solution

Speciation has not been applied often to soil solutions, probably because of the analytical difficulties involved in determining activities of many of the elements and their species. In recent years the analytical capability has improved markedly, which has allowed the determination not only of trace metals in solution but also of many of the ligands present, including organic molecules. There is considerable current interest in the speciation of trace metals and Al because of the perceived importance of species in relation to bioactivity.

Two approaches have been used in the speciation of trace metals that may be present in soil solutions. The oldest is that where the species are *operationally defined* on the basis of, for example, size (<0.45 μm) and then on the tendency of the total species present to react with anion or cation exchange resins or in response to the presence of an electrochemical stimulus (e.g., anodic stripping voltammetry). The more recent type of physicochemical speciation involves first of all the characterization of the solution with respect to the total concentrations of the elements and ligands present, and bulk properties such as pH; then the distribution of the species present are calculated with computer-based equilibrium models (Sposito, 1983). The success of these computer models depends very heavily on the availability of quality data, especially the equilibrium constants, and Martell *et al.* (1988) provide discussion on these aspects of metal complexes and metal speciation.

All of these methods require the isolation of the soil solution, an exercise that has its own problems, among which are the poor buffering capacity of the soil solution and the lower partial pressure of CO_2 in the atmosphere to which the solution is exposed. The considerable amount of analytical chemistry involved in the complete characterization of the solutions is a limiting factor to progress, and, at present, temporal changes in speciation have not been attempted.

The first operational speciation of trace metals in soil solutions was carried out by Hodgson *et al.* (1965, 1966) who used solvent extraction to separate the complexed and uncomplexed forms. More recently Jeffery and Uren (1983) devised a scheme based on those for natural waters (Batley and Florence, 1976; Figura and McDuffie, 1980), where anodic stripping voltammetry was used in conjunction with a chelating resin to determine the lability of the trace metal species in soil solutions as affected by liming. Very few attempts have been made to speciate Ni in soil solutions, but an ion-exchaange equilibrium method indicated that in the displaced soil solution of three Danish soils most Ni was present as Ni^{2+} at pHs lower than 6, whereas at pH 7 and thereabouts about 20% of the Ni was present as complexes (Sanders *et al.*, 1986b). These estimates agree

Table XIII

Equilibrium Speciation of Ni in Two Soils upon the Addition of Wastewater as Predicted by GEOCHEM[a]

	Hanford fine sandy loam	Holtville silty loam
pH	5.40	7.90
% adsorbed	82.1	99.1
M_{TS}[b]	5.76	8.02
$[Ni^{2+}]$[b]	5.91	9.02
Principal aqueous species[c]	Ni^{2+}, $NiSO_4^0$, Org	$NiCO_3^0$, $NiHCO_3^+$, Ni^{2+}, $NiB(OH)_4^+$

[a] Sposito (1983).
[b] Expressed as −log of molarity.
[c] Listed in order of decreasing importance.

reasonably well with computer-generated speciation for Ni such as that referred to earlier by Sposito and Bingham (1981). The same computer program (GEOCHEM) was used to predict the principal aqueous species of trace metals in two soils upon the addition of wastewater (Table XIII), the general results of which were a high degree of adsorption, the absence of precipitation as stoichiometric compounds, and the predominance of inorganic species except for Cu.

The usefulness of speciation by both methods has yet to be realized in practical terms, but their use together seems to be of paramount importance in improving our understanding of the role that the various species play in soil chemistry and plant nutrition.

2. Solid Phase

Fractionation of the solid phase of soils may be on the basis of physical properties, such as particle size, density, and magnetic susceptibility, or selective chemical dissolution, or various combinations. The recent interest in the pollution of soils and sediments has led to an upsurge in the use of fractionation schemes, which in turn have given rise to a number of reviews that not only reiterate the objectives of such schemes but also spell out the shortcomings and the need for better schemes (Lake *et al.*, 1984; Pickering, 1986; Beckett, 1989; Calvet *et al.*, 1990). No attempt shall be made here to review the voluminous literature, nor to comment on factors such as soil-to-solution ratios, temperature, shaking time, and residual volumes, which affect extraction efficiency, but comment shall be made on aspects that appear to apply specifically to the fractionation of Ni in soils.

Very few schemes attempt to separate the water-soluble forms and more commonly they are included in the readily exchangeable fraction.

Although the latter fraction may appear to be the easiest of all the fractions to determine, the choice of the exchanging cation can influence markedly the quantity extracted. For example, Crooke (1956) found that a 1% solution of $ZnSO_4$ extracted 294 ppm Ni from a serpentine soil while a 0.5% K_2SO_4 removed only 17 ppm. Also, Soon and Bates (1982) found that when an extraction with 0.125 *M* Cu(II) acetate (shaken for 16 hr) followed a 1 *M* NH_4^+ acetate (pH 7 and shaken 2 hr), it extracted seven times as much Ni from a range of soils, some of which were polluted with trace metals. As Leeper (1978) has pointed out there appears to be a range of degrees of exchangeabilities of these most reactive forms that are of considerable significance in availability and reversion.

Another difficulty associated with the estimation of exchangeable forms of trace metals with electrolytes and some other extractants (e.g., Na_2H_2 EDTA) is that often the pH of extraction is lower than the natural pH of the soil. Not only will the higher activity of H^+ facilitate desorption of trace metal cations, but also it will cause reduction of reactive oxides of Mn and thus the release of trace metals associated with them. The extent of this reduction will depend on such factors as the decrease in pH, the reducibility of the Mn oxides, and the reducing power of the organic matter. To overcome this difficulty with respect to Mn it is necessary to prevent the decrease in pH; one can argue in favour of a need to increase the pH. The latter may lead to the removal from solution of the trace metal via adsorption or precipitation, but this may be blocked with the inclusion of a chelate such as EDTA or DTPA. Beckwith (1956) designed a reagent of 1.0% NaCa HEDTA in 1.0 *M* ammonium acetate at pH 8.3 to overcome these difficulties with respect to Mn, and it is possible that such a reagent is appropriate for other trace metals, including Ni.

After the water-soluble plus exchangeable fraction has been extracted, the next step is commonly to attempt to remove organic forms; these are forms that presumably range from inner-sphere surface complexes to highly stable forms such as porphyrin complexes. Either oxidation of the organic matter with reagents such as H_2O_2 and NaOCl or dispersion with pyrophosphate is used, but unfortunately they all can cause the dissolution of reactive oxides of Mn (Anderson, 1963; Lavkulich and Wiens, 1970; Uren *et al.*, 1988). The H_2O_2 can reduce Mn oxides, the NaOCl oxidizes the oxides to permanganate, and the pyrophosphate peptizes organic matter, which in turn, despite the higher pH, reduces the reactive oxides of Mn. The reactive oxides of Mn are obviously a problem and it is simpler to remove them with a mild reducing agent at a neutral pH; if a low pH is used at this stage the reduction of active oxides of Fe and the dissolution of acid-soluble oxides of Fe and Al and other compounds (e.g., $CaCO_3$) may occur.

Once the reactive oxides of Mn are removed, the destruction of organic matter with an oxidant at high pH such as NaOCl can proceed without complication. If readsorption during previous steps was blocked and desorption is not a possibility upon acidification, then the removal of $CaCO_3$, if present, can proceed. Then commonly follows attempts to remove amorphous oxides of Fe with a reducing agent, and then crystalline oxides of Fe with a yet more severe reducing agent; each of these two reducing steps will dissolve the less reactive oxides of Mn, Al associated with the Fe oxides, and of course all the trace metals associated with all the oxides that are dissolved. The residue after such a sequence of extractions is thought to be largely silicates but little interest has been shown in them even though they frequently contain the bulk of most trace metals in nonpolluted soils. Although understandably the emphasis is placed on the most reactive forms, there is a need to determine the forms of Fe, Al, Ti, and Si, which are the major constituents of the less reactive fractions.

Validation of any fractionation scheme is currently impossible because all phases of soil cannot be quantitatively identified and for that reason the schemes will continue to be empirical for some time yet. However, since quantitative phase analysis (QPA) by Reitveld analysis of X-ray and neutron diffraction patterns of a mixture of six minerals has been successful (Howard *et al.*, 1988), it is possible that QPA could be applied successfully at least to the residual fractions where presumably the most highly crystalline phases are found. Selective dissolution associated with X-ray diffraction has been used to characterize Mn oxides (Tokashiki *et al.*, 1986), a job made difficult because of the poor crystallinity of the oxides, their low concentration, their diffuse X-ray diffraction patterns, and the coincidence of diagnostic peaks. Support for any scheme's validity should be provided by its application to well-defined model soils and tests for readsorption (Nirel and Morel, 1990). A limited study of readsorption of some trace metals onto sediments indicated that no readsorption of Ni occurred from 1 *M* Na acetate at pH 5 and from acidic H_2O_2 (Belzile *et al.*, 1989). There was no readsorption of As, Cd, Cu, and Zn from these and other extractants, but in one case with Pb there was some readsorption.

Because Ni is associated with Mn and Fe the possibility of physical fractionation on the basis of density and magnetic susceptibility at first appears to be a promising means by which presumably simpler mixtures of minerals could be separated and investigated by chemical and physical methods. For example, when finely ground dolerite was fractionated using bromoform and a magnet, the total concentration of Ni in the separates was in the order magnetic fraction $\gg$ pyroxenes $>$ feldspars (Tiller, 1959). However, Suzuki *et al.* (1971) found that of both coarse and fine sand fractions of a serpentine soil, the nonmagnetic light fractions had the highest concentrations of Ni (Table IX). Another study showed that the <2 μm fraction of Domino soil (15% <2 μm) had a Ni concentration of 47 ppm, whereas in coarser fractions concentrations were uniformly lower, at about 28 ppm (Essington and Mattigod, 1990). After 8 years of sludge additions at a rate of 90 Mt ha^{-1} yr^{-1}, which increased the Ni concentration of the top 15 cm threefold, the concentration of Ni in the <2-μm fraction (5.2%) had increased fivefold to 217 ppm, and to a weighted average of 130 ppm in the coarser fractions. When the <2-μm and 2- to 45-μm fractions were separated on the basis of density, the lightest fractions (<2.10 g cm^{-3}) had the highest concentrations of Ni. A similar study of a sludge-amended soil (Ducaroir *et al.*, 1990) showed the concentration of Ni in the <2-μm fraction to be highest in the densimetric fraction of 2.02 g cm^{-3}; other trace metals (Cd, Cr, Mn, Fe, Pb, and Zn) except for Cu had their highest concentrations in the 2.18 g cm^{-3} fraction. The concentration of Cu increased with decreasing density, a trend which was thought to be associated with increasing organic matter concentration. These results are consistent with data for sewage sludge studied by Bergman *et al.* (1979), who found that 75% of the trace metals Cd, Cu, Pb, Ni, and Zn, were in the low-density and organic-rich fractions, but high concentrations of all metals were found at high densities (2.9 g cm^{-3}), whereas distinct minima were observed around a density of 2.6 g cm^{-3} (mainly quartz and feldspar).

If these results are typical of unamended soils as well as sludge-amended soils, and there is no reason to doubt that they are not, then it may mean that the admixture of the phases in soil is so complex that satisfactory fractionation of all phases, organic and inorganic, amorphous and crystalline, is currently no more than a dream. The adherence of organic matter and oxides onto surfaces is undoubtedly the main reason for this unsatisfactory situation. However, one is hopeful that, given more effective dispersion of the phases (without redistribution); separation into greater numbers of fractions on the basis of size, density, and magnetic susceptibility; and greater sensitivity in quantitative phase analysis, the desired level of fractionation will be achieved.

IV. AVAILABILITY

A. Essentiality

The essentiality of Ni as a plant nutrient in soybeans (Eskew *et al.*, 1983, 1984), chickpeas (Eskew *et al.*, 1984), and temperate cereals (Brown *et al.*, 1987a,b) has been illustrated, but its essential functions in higher plants other than in urease have yet to be established (Brown *et al.*, 1990). The occurrence of Ni deficiency in the field is unlikely, but if deficiency were to occur naturally then it would be most likely to arise on those soils where severe Co, Cu, and Zn deficiency occurs, such as on the pumice soils of New Zealand and infertile sands in Australia. Although the liming of these soils, where acidic, would increase the likelihood of deficiency markedly, the fertilizers that must be used on these infertile soils to overcome both the major and the minor element deficiencies would contain enough Ni to satisfy the requirement for Ni. Mishra and Kar (1974) cited several instances where the application of Ni has improved the growth of plants, but it is very often difficult to discern the cause for the observed stimulation, particularly where the fungicidal effects of the Ni may have played a role (Welch, 1981). For example, it is difficult to provide a satisfactory explanation of the result of an experiment by Arnon (1937) who found that Ni at 0.05 ppm in nonaerated solution cultures doubled the yield of barley when N was supplied as NH_4^+, but decreased the yield by 25% when the source of N was NO_3^-.

B. Toxicity

The phytotoxic effects of Ni have been known for a long time (Vanselow, 1966; Mishra and Kar, 1974). Apart from a decrease in growth, the symptoms of Ni toxicity include chlorosis, stunted root growth, and sometimes brown intervenal necrosis and symptoms specific to the plant species. It is usual to attribute the chlorosis to Fe deficiency induced by antogonism from the Ni^{2+}, but the Fe deficiency may simply be the result of the poor root growth, which is a common, if not predisposing, cause in many situations where Fe deficiency occurs, since active root growth is essential for the acquisition of Fe by plants growing in soil (Uren and Reisenauer, 1988).

In an experiment where the roots of oats grew between filter papers moistened with a solution of $NiSO_4$, the root growth of oat seedlings was inhibited by 2 ppm Ni; yet no improvement in root growth was observed in response to liming when oat seedlings were grown in a serpentine soil which contained before liming about 80 ppm water-soluble plus exchangeable Ni (Siow, 1990). The tolerance of root growth to toxic concentrations of Ni is markedly increased by both Mg and Ca (Gabbrielli and Pandolfini, 1984; Robertson, 1985) and the absence of these cations from the filter paper culture must be a major reason for the inhibition of root growth at 2 ppm Ni. Another and as yet not commonly recognized factor may be associated with the difference between liquid and solid media where it has been observed that beans and soybeans will tolerate, respectively, much higher concentrations of Mn (Kohno and Foy, 1983) and Al (Horst *et al.*, 1990) in sand culture than in solution culture.

In the research of Siow (1990) referred to previously, in a separate experiment the liming of the serpentine soil increased the growth of oats and caused the typical foliar symptoms of Ni toxicity of the oats to disappear. Thus, it would seem that interference in internal metabolism that leads to the foliar expression of toxicity occurs at lower concentrations of Ni in soil-grown plants than those concentrations that inhibit root growth.

The tolerance of plants to Ni has been studied in species that colonize either contaminated soils near smelters (e.g., Rauser and Winterhalder, 1985) or serpentine areas (e.g., Gabbrielli *et al.*, 1990). Tolerance derives from the ability of some species to limit the uptake and translocation of Ni and in others by the accumulation in nontoxic forms in leaves. Plants endemic to serpentine areas employ both strategies, but despite the extensive research on the problem of infertility on serpentine soils a definite or single cause has yet to be established (Brooks, 1987).

The concentrations of Ni associated with Ni toxicity vary widely. For example, Patterson (1971) reported toxicity symptoms in spring wheat at 8 ppm but no yield loss in oats at 90 ppm. Similarly, Bolton (1975) reported no yield loss in oats at 147 ppm while Khalid and Tinsley (1980) observed slight chlorosis in perennial rye grass at 50 ppm without any loss in yield.

C. Uptake

The main forms and reactions that are thought to be important in determining the mobility and availability of Ni to plants are illustrated in Fig. 4. Emphasis has been placed on Ni^{2+} since it is the form most likely to predominate in the soil solution, to participate in sorption reactions, and to be taken up by plants (Mishra and Kar, 1974; Tinker, 1986).

The soluble forms in the soil solution are the most free to move toward actively absorbing regions of the roots of plants; the formation of uncharged and anionic complexes would facilitate diffusion, but at the point of absorption the Ni^{2+} is probably separated from ligands associated with it, in the same way it seems to occur for Cu^{2+} and EDTA (Goodman and Linehan, 1978). A system devised by Ratkovic and Vucinic (1990) to measure the uptake of paramagnetic ions by intact plant roots using H^+ NMR relaxation could be employed in investigations of the uptake of different forms of Ni^{2+}.

Much discussion assumes that Ni^{2+} is actively taken up in much the same way as any divalent metal cation of an essential element such as Zn^{2+} or

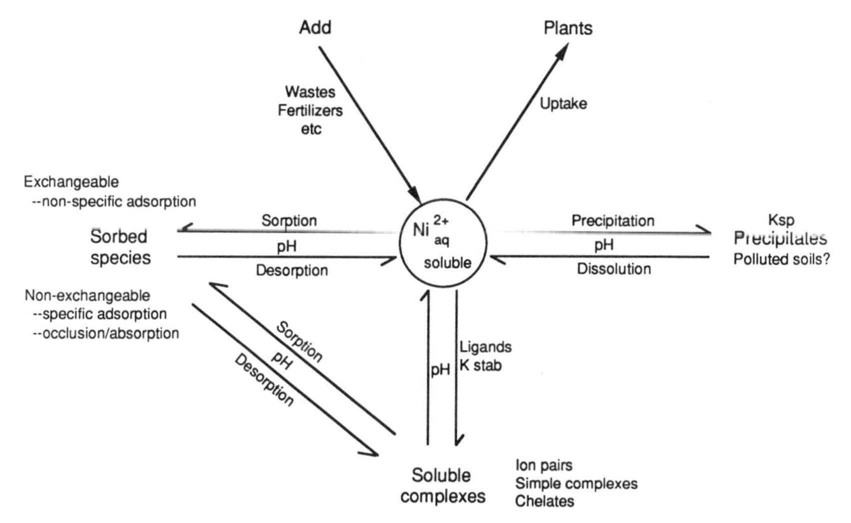

Figure 4. Forms and reactions affecting the mobility and availability of Ni in soils.

Mn^{2+}. Little work has been done on the site and mechanism of uptake of Ni by roots but evidence from work with excised barley roots suggests that Ni^{2+} is absorbed actively across the plasmalemma (Körner *et al.*, 1986). Similarly the uptake of Ni from solution cultures by oat seedlings appeared to be under metabolic control at concentrations of Ni too low to be toxic (Aschmann and Zasoski, 1987). Cataldo *et al.* (1978) have suggested that uptake of Ni^{2+} occurs in soybeans at the same carrier site as for Cu^{2+} and Zn^{2+}.

1. Solution Culture Systems

In solution cultures, the uptake of Ni increases with increasing concentration in the solution up to the point where toxicity decreases growth and the capacity for uptake is curtailed due to damage to the roots. Various foliar symptoms of toxicity have been reported, but the most consistently reported are akin to iron chlorosis (Hewitt, 1948). In one study with corn referred to previously (Robertson, 1985), the toxicity, as measured by a decrease in root growth, was most effectively alleviated by increasing the Mg^{2+} concentration up to a concentration at which the Mg^{2+} itself inhibited root growth; the Mg^{2+} inhibition was removed by additional Ca^{2+}. Crooke and Inkson (1955) found for oats that Ca decreased the toxic effects of Ni more effectively than Mg or K, but P increased them. The latter effect of P was thought to be the result of the P decreasing the availability of Fe in the culture system by precipitation; the Ni toxicity then revealed itself. It should be pointed out that these culture solution systems are perhaps not quite as simple as some would believe. In the case of the P × Fe interaction not only may there be precipitation but also the reducibility of the Fe oxyhydroxide is decreased by the adsorption of phosphate (Willett and Cunningham, 1983), and the trace metals such as Cu and Zn are adsorbed onto the precipitate. The adsorption appears to be readily reversible but strongly pH dependent at the pH's used in solution cultures (Uren and Edwards, 1984).

The uptake of Ni in solution cultures (Hunter and Vergnano, 1952) and sand cultures (Crooke *et al.*, 1954) has been shown to increase with increasing pH, which is in agreement with the observed uptake of other trace metal cations as the pH is increased in solution cultures containing clay in suspension (Epstein and Stout, 1951). The effect of chelates on the uptake of Ni from solution cultures also appears to be the same as that on other trace metal cations where uptake is unaffected or slightly enhanced at low molar ratios of ligand to metal, but as the ratio is increased the uptake is decreased (Brown *et al.*, 1960, 1961). At toxic concentrations of Ni, a 1:1 molar ratio of EDTA to Ni decreased the toxic effects of Ni and the uptake of Ni from solution culture by mustard and tomato plants (DeKock, 1956; DeKock and Mitchell, 1957). Brown *et al.* (1987b) noted that high concentrations of chelating agents such as EDTA may induce symptoms of Ni deficiency, particularly when the supply of Fe is low.

Solution culture systems in some respects are not good imitations of soil systems. In the latter, longitudinal and radial gradients in chemical, microbial, and physical characteristics develop along roots (Uren and Reisenauer, 1988), whereas in solution culture systems they may develop (sand cultures) or may not be allowed to (flowing cultures).

2. Soil

In soil, where the activity of the species in the soil solution is influenced markedly by the the sorption properties of the various solid phases, the behavior and uptake of Ni are determined by bulk properties such as texture, organic matter content, and pH. For soils of the same pH, mechanical composition, and organic matter content, the nature of the adsorbing surfaces (as determined by composition, specific surface area, and accessibility to roots) become the major determinants of availability.

Changes which increase the solubility of Ni in soils will usually increase both mobility and the availability of Ni to plants, although in some situations the increase in mobility may not necessarily lead to an increase in uptake by plants because the cause of the enhanced mobilization (e.g., waterlogging) may not be favorable to plant growth. For example, high concentrations of NaCl may increase the mobility of Ni^{2+} (Doner, 1978), but the concentrations of salt are high enough to be toxic to most plants. Such work is of relevance in heavily polluted environments such as waste dumps or perhaps where sediments carried by fresh water mix with saline water in estuaries (Rohatgi and Chen, 1975) or where saline ground water invades nonsaline soils.

Studies on the availability of Ni to plants in soils have been made on serpentine soils, normal soils, and normal soils to which sewage sludge has been added. Many experiments have been carried out to test the availability of Ni added either as salts or in sludge and to test the effect of pH.

In many of these experiments a wide range of rates of application of Ni have been used (Table XIV) and the question arises "What are realistic rates of addition of Ni?" A grain crop of 5 t/ha with a concentration of 1 ppm Ni contains 5 g of Ni. If deficiency were to exist then rates similar to or less than those required to cure Cu or Zn deficiency, say, 1 to 5 kg/ha every 5 to 10 years, would be appropriate. At the other extreme, rates of

Table XIV

Some Examples of Rates of Addition of Ni in Experiments and in Practice

Situation	Application rate (kg ha^{-1} yr^{-1})	Reference
Dust and rain	0.002–0.2	This chapter
Superphosphate (100 Kg ha^{-1} yr^{-1})	<0.03	This chapter
Normal manuring	<0.12	Andersson (1977)
Suggested loadings (all countries)		Webber *et al.* (1984)
Range	0.045–2.3	
Median	0.5	
Sewage disposal—irrigated pasture	1	Evans *et al.* (1979)
Sludge—market garden	6	McGrath (1984)
Sludge—field crops	0–7	Sanders *et al.* (1986b)
Sludge—field crops	0–9	Rappaport *et al.* (1987)
Sludge—pots	~15	Dijkshoorn *et al.* (1981, 1983a)
Sludge—field plots	22	Vlamis *et al.* (1978)
Salts—pots	0–1000	Halstead (1968)
Salts—pots	0–960	MacLean and Dekker (1978)
Salts—pots	0–1000	Wallace *et al.* (1977a)
Salts—pots	360	Taylor and Griffin (1981)
Salts—pots	13	Dijkshoorn *et al.* (1983b)
Salts—columns	290	Doner *et al.* (1982)
Salts—columns	622	Tyler and McBride (1982)
Salts—columns	794	Biddappa *et al.* (1982)
Sludge—columns	236	Emmerich *et al.* (1982a,b)

addition to just avoid toxicity may be of the order of one-quarter to one-eighth of the "Zn equivalent" (560 kg/ha), i.e., 140 to 70 kg/ha. Although all of the Ni might be added at once in a spill or with a highly contaminated sludge, such amounts are in practice more likely to have been added over a long period of, say, 20 to 30 years or more.

A review of heavy metal additions in sewage sludge to agricultural land (Matthews, 1984) indicates that the recommended limits may be as high as 140 kg/ha over a 30-year period with up to one-fifth, i.e., 28 kg/ha, being added in one application (UK); this is to be compared with the USDA recommendation of maximum cumulative applications to privately owned land, which increase from 50 to 200 kg Ni/ha as the CEC increases from <5 to >15 meq%, respectively. In practice then the annual rates of addition of Ni are likely to be of the order of 5 kg/ha or less, and although we should be aware of what happens at high rates (50–1000 kg/ha) we should base our theories of normal behavior on those experiments where seemingly normal rates of Ni have been applied.

Since most additions of Ni to soils are not always going to be as salts, but often associated with complex materials, one needs to be aware that in some cases the addition of trace metals in the form of salts may lead to a greater uptake of trace metals than when equivalent amounts are added in complex materials such as sewage sludge (Dijkshoorn *et al.*, 1981; Smilde, 1981; Kiekens *et al.*, 1984). However, when Cunningham *et al.* (1975) compared metal-enriched sludge with equivalent amounts of metals added directly to soil, they found Ni concentrations to be higher in the plants grown on soil amended with the enriched sludge, whereas concentrations of Cr, Cu, and Zn were higher in the plants grown on the salt-amended soil. In contrast, Sorteberg (1981) reported that the concentrations of Cu, Ni, and Zn were higher in oats grown on sludge-amended soil than on salt-amended soils. Obviously, the nature of the sludge and the forms and concentrations of the metals present strongly influence the availability of the metals when the sludge is applied to land. Sewage sludges are complex mixtures of sand, silt, and clay; organic matter in various degrees of decomposition and biodegradability; and secondary sulfides and carbonates; such variability in terms of composition and metal concentrations in relative and absolute terms makes it difficult to generalize on the availability of trace metals when sludge is applied to land.

Of all soil properties, pH is the major determinant of trace metal behavior and availability (Leeper, 1978). Nickel is no exception and the effect of increasing the pH by the addition of lime is to consistently decrease the availability of Ni (e.g., Table XV). In this particular example, symptoms of toxicity were observed where greater than 20 ppm Ni had been added to the soil at pH 5.1, and where 160 ppm had been added at pH

Table XV

The Effect of pH and Ni Additions on the Concentration (ppm) of Ni in Spring Wheat[a]

Soil pH	Ni added to soil (ppm)						
	0	5	10	20	40	80	160
5.1	2.5	4.5	3.0	**8.0**[b]	**10.0**	**74.0**	—
5.5	2.2	2.5	3.7	4.7	6.5	**17.2**	**105.0**
6.5	1.0	0.75	2.2	2.0	2.75	3.0	**8.25**
7.5	—	0.5	0.5	—	0.75	1.25	3.0

[a] From Patterson (1971) by courtesy of the publisher.
[b] Concentrations in bold type are those of plants exhibiting toxicity symptoms.

6.5; no symptoms of toxicity were observed at pH 7.5 at rates of addition up to 160 ppm. The effect of increasing pH is due to an increasing affinity for the solid phase and it should be emphasized that this effect overrides the increasing ability evident in sand cultures of plants to take up Ni as pH increases (Crooke *et al.*, 1954).

Another point of concern in the interpretation of plot and pot experiments arises where lime has been added. Although the the general conclusion that liming decreases the availability of Ni is reasonably the most apt, the interpretation of a response to lime should be made with some caution. Because uptake is the product of yield and concentration, a decrease in the uptake of Ni on the application of lime may be due to (*a*) lower solubility of Ni through the adsorption or precipitation of Ni^{2+}, (*b*) competition for uptake from Ca^{2+} (or Mg^{2+}), or (*c*) increased growth due to increased availability of a deficient nutrient (e.g., Mo or, particularly, N) or to a decrease in the toxicity of an element other than Ni (e.g., Al or B).

In serpentine soils the addition of lime usually causes an increase in growth and a decrease in the concentration of Ni in the plant (Crooke, 1956; Halstead, 1968; Halstead *et al.*, 1969). Since sodium carbonate had the same effect (Crooke, 1956), the possibility in this case of a positive effect of Ca^{2+} can be ruled out. Similar effects of lime are observed with both normal and sludge soils but even though a Ca^{2+} effect is much less likely than on serpentine soils, only experiments using noncalcic forms of alkali and Ca^{2+} salts can possibly resolve this apparent difficulty.

The effect of P on the availability of Ni to oats in sand cultures was to increase the uptake of Ni and to increase the severity of symptoms of Ni toxicity (Crooke and Inkson, 1955). However, Nicholas and Thomas (1954) found the increasing rates of P application as NaH_2PO_4 in soil had no effect on the uptake of Ni or on symptoms of Ni toxicity in tomatoes. Crooke (1956) found that a mixed fertilizer containing superphosphate had no effect on the extractability of added Ni in neutral ammonium acetate from an acidic soil. P added to four soils at a rate of 500 ppm P as $CaH_4(PO_4)_2 \cdot 2H_2O$ increased the extractability of added Ni (0–500 ppm), decreased the pH slightly, and increased the concentration and uptake of Ni in oats, but only once appeared to induce toxic effects (Halstead *et al.*, 1969). Pezzarossa *et al.* (1990), upon the addition of superphosphate containing 60 ppm Ni added at rates up to 3640 kg ha^{-1}, found no effect on Ni uptake by lettuce; the concentrations of EDTA (pH 4.65) extractable Ni in the substrate were not affected significantly by the treatments.

MacLean and Dekker (1978) found with corn that P added to a loam (pH 6.3) as $CaH_4(PO_4)_2 \cdot 2H_2O$ at a rate of 50 ppm doubled the yield but had no clear effect on Ni (Table XVI). In this experiment the greater toxicity of Ni over that of Zn is observed, but curiously there was no decrease in yield due to the Zn in the plants without additional P, whereas in those plants with the high rate of P the high Zn treatments caused a 20% decrease in yield. These experiments with a P-responsive soil indicate that at realistic rates of addition of Ni there appears to be no Ni × P interaction of any significance and that in practice it would be difficult to carry out an experiment that is not confounded by changes in pH and Ca status. Whether P has an antagonistic effect on Ni availability in low Ni situations, as with Zn (Lindsay, 1972), has yet to be investigated.

The ability of manganese oxides to sorb Ni and other trace metals is likely to decrease the availability of Ni, but perhaps not to the same extent

Table XVI

Effect of P and Rates of Ni and Zn on Yield, Concentration, and Uptake of Ni and Zn in Corn Grown on Grenville Loam[a]

Rate of metals added (μg/g)	P added (μg/g)	Mean yield (g/pot)		Mean concentration (μg/g)		Mean uptake (μg/pot)	
		Ni	Zn	Ni	Zn	Ni	Zn
0	0	5.6	5.6	0.6	48	3.4	269
60	0	3.8	4.7	1.6	146	6.1	686
240	0	2.0	5.7	21.9	397	43.8	2263
480	0	0.5	5.7	78.1	709	39.1	4041
0	50	12.7	12.7	0.6	28	7.6	356
60	50	9.9	12.2	1.5	60	14.9	732
240	50	2.4	11.5	17.6	292	42.2	3358
480	50	0.3	10.0	166.1	436	49.8	4360

[a] From MacLean and Dekker (1978) by courtesy of the Agricultural Institute of Canada.

as Co (Adams *et al.*, 1969). McKenzie (1978) found that birnessite and cryptomelane added at concentrations of 1000 and 3000 ppm Mn to three red-brown earths had no effect on the Ni uptake in one soil (pH 4.7), but significantly decreased the uptake of Ni in another. In the third soil, cryptomelane was effective, but birnessite had no effect. For Pb and Co, the uptake was effectively decreased by both oxides on all soils, whereas for Cu and Zn the oxides had little influence. If the soils had been stored longer before the plants were grown or had the Mn oxides been precipitated *in situ*, then the effect of the Mn oxides may have been greater.

Because of the association of Ni with Fe and Mn oxides one expects to observe an increase in the availability of Ni upon waterlogging much the same as that for Co (Adams and Honeysett, 1964). The increase in Ni extractability and uptake by ryegrass and red clover in poorly drained soils (Mitchell and Burridge, 1979) was noted earlier (Table XI). In studies of short-term decreases in soil redox potential, Brown *et al.* (1989) found little effect on the concentration of Ni in the soil solution of one soil and its amended counterpart. In another soil naturally richer in Ni there was a marked increase, but the increase in solubility was only reflected in the Ni uptake by oats and not for ryegrass.

The role of organic matter in the availability of trace metals has been emphasized by many, but in this author's opinion the reputation is overrated. There appears to be an uncritical willingness to assume that all organic matter has the chelating abilities of synthetic compounds such as EDTA, whereas only a small proportion of organic ligands in soil may have such capability; in reality the complexing ability of the bulk of organic matter is on a par with oxalate. The success of curing Cu deficiency on calcareous peats in contrast to the failure of soil-applied Mn to the same soils is a good illustration of the contrasting effects of complexing of Cu by organic matter and the microbial oxidation of Mn to insoluble oxides. Also, the finding (Bloom and McBride, 1979) that Ca^{2+}, Co^{2+}, Mn^{2+}, Ni^{2+}, and Zn^{2+} formed only outersphere complexes with humic acid, and that Cu^{2+} formed inner sphere complexes, supports the contention that the bulk of organic matter in soil binds most cations with only moderate strength when compared with inorganic forms.

Similarly, much has been made of the possible role of synthetic chelates in modifying trace metal behavior and availability in soils, but even so there is no consistent pattern of effect on uptake. Norvell (1972) derived stability diagrams for 10 Ni chelates and concluded that DTPA, HEDTA, EDTA, and CDTA and to a lesser extent NTA and EDDHA, most effectively increased the solubility of Ni in soil, whereas the remainder, including citrate and oxalate, had little ability in this regard. When Na-EDTA was added at a high rate of 1.0 g kg^{-1} to a soil contaminated with heavy metals, the uptake of Cu, Fe, Pb, and Zn by perennial ryegrass was increased, but the uptake of Ni decreased markedly (Albasel and Cottenie, 1985). On another soil contaminated with Zn, DTPA had inconsistent effects on the uptake of Zn by Italian ryegrass and barley. In other studies DTPA, NTA, and EDTA, addded at rates of 50 and 100 $\mu g\ g^{-1}$ soil, increased the uptake of Ni by bush beans two- to threefold (Wallace *et al.*, 1977a,b), and Fe-DTPA (5 $\mu g\ g^{-1}$) increased the yield and uptake of Ni by bush beans grown on some serpentine soils from California (Wallace *et al.*, 1977c). The lack of consistency in the effects of chelates may be due, among many factors, to the rates of application used. In a comparative study by Willaert and Verloo (1988) of the uptake of Ni from a sandy loam treated with 50 ppm Ni as $NiSO_4$, Ni-EDTA, Ni-DTPA, and Ni-NTA, it was found that $NiSO_4$ was toxic to spinach and yet the chelated forms increased both the yield and the concentration of Ni in the plants. The concentrations of Ni in the plants were 49, 240, 102, and 118 ppm, respectively, all high enough to be toxic.

The data of Sorteberg (1978) and that in Table XVII serve to emphasize several points about the availability of Ni in soil. The severity of toxicity, which was in the order peat> sandy soil > clay, was practically eliminated by the heavy liming; it is unfortunate that no information on soil pH's was given other than that the heavy liming was sufficient to bring the pHs

Table XVII

Relative Yields (Grain + Straw) of Oats in Three Soils to which Ni[a] and Lime Had Been Added[b]

	1973	1974	1975	1976	Uptake 1973–1976 % of added Ni
Peat					
Low lime	27	48	88	81	5.4
High lime	86	103	98	115	5.1
Sandy soil					
Low lime	37	58	77	81	3.2
High lime	107	101	82[c]	98	1.5
Clay					
Low lime	41	51	55	78	1.9
High lime	100	103	98	97	0.5
Mean Ni concentration in grain (ppm)					
Low lime	110	92	74	58	
High lime	45	47	35	27	

[a] 250 mg/pot ≈ 100 kg/ha.
[b] From Sorteberg (1978) by courtesy of the publisher.
[c] Mn deficiency present.

between 6 and 7. Notwithstanding the lack of pH data, the availability of Ni was greatest on the peat and least on the clay, and the recovery of the added Ni was greatest on the peat and least on the clay. Also, the extractability of Ni from the soils reflected the relative toxicities and uptake of Ni, and similar results were obtained in sludge-amended soils (Sorteberg, 1981). The decrease in availability and severity of toxicity is a common phenomenon with additions of trace metals to soils and will be discussed later.

Of all the factors discussed previously, pH is the most important in determining the availability of Ni to plants. The pH is significant because it determines the solubility of Ni through its effects on the sorptive capacity of the inorganic and organic colloids. Few other factors appear to be of similar significance.

D. Estimation of Availability

From the previous discussion on the forms and reactions of Ni in soils and the factors influencing its availability, it can be easily appreciated that the behavior and availability of Ni are fairly close to those of other trace metals such as Zn. It is not surprising then that no special reagent has been used in an attempt to estimate the availability of Ni in soils. Sillanpää and Lakanen (1969) estimated the availability of Ni in normal soils using acidic ammonium acetate and timothy as the test plant. The correlation coefficients between extractable Ni concentrations and the concentration of Ni in the plant were 0.705, 0.609, and 0.787 for clay, sandy, and organic soils, respectively, and for all soils (216) the correlation coefficient was 0.734.

Misra and Pande (1974b) evaluated nine extractants on 32 alluvial alkaline soils of Uttar Pradesh, and the correlation coefficients between extractable Ni concentrations and uptake by sorghum grown in pots increased in the order 0.02 *M* EDTA (−0.24) < 0.1 *M* $K_2P_2O_7$ (−0.14) < 1% citric acid (0.13) < 0.1 *N* H_2SO_4 (0.40) < 0.1 *N* HNO_3 (0.41) < 1.0 *M* NH_4^+ acetate (0.57) < 0.5 *M* acetic acid (0.71) < 0.1 *N* HCl (0.74) < Grigg's reagent–acidic NH_4^+ oxalate (0.95).

In comparison, Haq *et al.* (1980) extracted 46 soils with varying degrees of contamination with metals. The nine extractants gave linear correlations between Ni concentrations in Swiss chard and extractable Ni concentrations with r^2 values from 0.05 to 0.42; for example, for EDTA the r^2 was 0.17 and for DTPA it was 0.18. The r^2 values were improved dramatically by the inclusion of other so-called independent variables in the regression equations. Where pH was the only independent variable other than extractable Ni concentration, 0.5 *M* acetic acid gave an r^2 of 0.82, which was increased even further when organic matter (%) and clay (%) were included. Quantitatively the acetic acid extracted on average 10 and 20 times less Ni than EDTA and DTPA, respectively, and about 20 times more Ni than that in the tops of the Swiss chard; in comparison water extracted about 3 times the amount of Ni in the tops of the Swiss chard. Water then provided quantitatively the most realistic estimate of the amount taken up but it was poorly correlated with plant Ni and the inclusion of five independent variables in the regression equation were required to achieve an r^2 of 0.73. Other extractants such as $CaCl_2$ and $CuCl_2$ have been evaluated and as with others the r^2 values were improved, to 0.84 and 0.93, respectively, by the incorporation of pH, clay content, and organic matter content in the regression equations (Horst *et al.*, 1988).

Adams and Alloway (1988) investigated the effect of temperature (5 to 38.5°C) on the extractability of trace metals (Cu, Mn, Ni, Pb, and Zn) from five soils by five reagents. The results were not consistent, although extractable Ni concentration, which tended to increase with increasing temperature, was the most consistent; the extractable Zn concentrations exhibited anomalous trends.

The DTPA soil test was developed by Lindsay and Norvell (1978) to assess the availability of trace metals, Zn, Fe, Mn and Cu, on near-neutral and calcareous soils. O'Connor (1988) has drawn attention to the inappropriate application of the DTPA soil test to situations outside those for which it was designed. Acidic soils and sludge-amended soils are two such examples where the test often fails, although in the former case and as noted above the regression coefficients are often improved by the inclusion of a pH term in the regression equation.

On the basis of the limited evidence available, it would seem that an extractant containing Cu^{2+} provides a better estimate of availability than most extractants, although it, like all the others, benefits from the inclusion of pH in the regression equation.

E. Reversion

The term reversion is used to describe the decrease in availability of an element, essential or nonessential, with time after its addition to soil (Leeper, 1970, 1978); it originates from the tendency of phosphatic fertilizers in soils of neutral to alkaline pH to revert by successive stages of incongruent dissolution of Ca phosphates to an apatite, the original form in rock phosphate. Its wider application to include phosphate added to acidic soils and other elements has meant that the term is now used both to describe an agronomic situation and to imply a mechanism of slow recrystallization, or at most a mechanism of decreasing solubility of inorganic compounds only (Leeper, 1978). In the case of Ni, as there must be with other trace metals, there is a limit to how much can be added, since in serpentine soils rich in Ni the reversion in those soils is inadequate to protect the plants from toxic concentrations of Ni.

An example of the decrease in the availability of Ni is given in Table XVII, where the reversion is illustrated by the decreasing toxicity of the Ni in the lightly limed soils and by the decreasing concentrations of Ni in the oat grain. These changes in availability were reflected in changes in the extractability of Ni from the soils, and similar observations were recorded for Ni added to soil in sewage sludge (Sorteberg, 1981).

Berrow and Burridge (1990), in a study of the persistence of metal residues in soil treated with sewage sludge 17 years previously, found that in the top 15 cm the acetic acid extractable concentrations of Cd, Cu, Ni, and Zn increased in the early years after the additions were made and since then there has been a gradual decrease, but the concentrations still remain well above those of the untreated soil. In experiments such as these with high rates of Ni added along with biodegradable organic matter, it is likely that the apparent rate of reversion may be due to movement out of the layer that has been sampled. Apparent errors may arise from changes in the bulk density; for example, the bulk density, which may be low initially due to the high rates of addition of organic matter, may after several years increase such that upon sampling to the same depth the soil lower in the profile with a lower concentration of metal is mixed with the contaminated soil from above. Plant analysis is obviously an essential component of any reversion study.

In soils contaminated with Ni or other heavy metals it is important to know whether reversion occurs, and if it does, by what means can reversion be stimulated and what rate of reversion is likely to be achieved. Some estimates of apparent rates of reversion of essential trace elements have been made in comparison with that of P (Table XVIII). With the exception of Mn, the apparent rates of reversion are 40 to 400 times less than that of P. The rates of reversion of Cu and Zn, those of trace elements most similar to Ni, are of the order of 0.2 kg ha^{-1} yr^{-1} in situations where they are marginally deficient and where one might expect rapid reversion.

In contrast, the EEC recommended a tolerance rate of addition of 2 kg ha^{-1} yr^{-1} over 10 years up to a limit of 30 ppm in soil; the latter would be reached in 120 years at a rate of 0.25 kg ha^{-1} yr^{-1} and 240 years at a rate of 0.125 kg ha^{-1} yr^{-1}, which are the estimated rates of reversion of the essential trace metals of Zn and Cu, respectively. These values should be compared with the estimated gains of Ni of 0.002 to 0.2 kg ha^{-1} yr^{-1} in rain and 0.03 kg ha^{-1} yr^{-1} as a contaminant in fertilizer and with leaching losses of Ni of 0.009 to 0.071 kg ha^{-1} yr^{-1}.

Table XVIII

Estimates of the Apparent Rates of Reversion of Trace Elements Relative to P

Element	Nutrient concentration		Application rate		Effectiveness (years)	Apparent reversion rate	
	ppm	Relative	kg ha^{-1}	Relative		kg ha^{-1} yr^{-1}	Relative
P	2000	1	10.0	1	1	10.0	1
B	4	1/500	0.5	1/20	2	0.25	1/40
Co	0.05	1/40,000	0.25	1/40	5	0.05	1/200
Cu	4	1/500	1.25	1/8	10	0.125	1/80
Mn	40	1/50	5.0	1/2	1	5.0	1/2
Mo	1	1/2000	0.05	1/200	5	0.01	1/1000
Zn	20	1/100	1.25	1/8	5	0.25	1/40

Note. Estimates are based on data from Sherrell (1983), Simpson (1983), Murphy and Walsh (1972), Burridge *et al.* (1983), Reuter (1975), Gartrell (1981), and Hosking *et al.* (1986) Removal of products not deducted.

The mechanisms of reversion have not been studied in much detail and probably little progress will be made until the fractionation schemes for Ni and other trace metals are validated. Sorption, dilution by leaching and mixing, and removal of metals in harvested plant and animal products are likely to be important contributors to reversion, but whether Ni enters lattice forms and can be encouraged to do so is open to investigation. However, some caution is necessary since any treatments intended to decrease the mobility of Ni in less soluble and less reactive forms may interfere with the availability of other trace metals. Fortunately, in most situations where Ni is a serious contaminant, other trace metals are also.

V. CONCLUSIONS

The forms of Ni in soil are diverse and range from those of high activity (water-soluble and exchangeable forms) to ones that have no reactivity in the short term (the life of a plant). The reactions of Ni are dominated by sorption reactions on soil colloids whose nature, relative proportions, and concentrations are important. Soil pH plays a major role in these reactions and thus influences the availability of Ni to plants quite markedly; few other factors appear to be as significant as pH. Deficiency of Ni in plants is unlikely to be a practical problem, but toxicity is more likely.

Interest in the occurrence of Ni toxicity is dominated by research on serpentine soils (ancient Ni) and soils to which large quantities of Ni have been added in wastes (modern Ni), particularly in the forms of sewage and sewage sludge. Much of our understanding of the forms and reactions of Ni in soils comes from experiments dealing with the chemistry and biology of these two extreme situations. Needless to say, caution is essential when attempting to determine what is normal.

Future research on Ni in soil could fruitfully pursue, for example, the role of microbes in influencing the behavior of Ni, the fractionation of Ni in studies of weathering and reversion of active forms of Ni to less active ones, and the speciation of Ni in the soil solution and the extent and cause of temporal changes. For plants there is also scope to investigate the forms of Ni taken up by roots, the functions of Ni in plants and the changes induced by toxic concentrations in relation to the expression of symptoms (root and shoot growth, foliar symptoms, etc.), and the mechanisms for resistance to Ni toxicity. The bio-inorganic chemistry of Ni is a new challenge, whereas an old challenge, the serpentine factor, remains a source of perpetual fascination.

References

Aboulroos, S. A., Holah, Sh. Sh., and Badawy, S. H. (1989). *Z. Pflanzenernähr.* **152,** 51–55.

Adams, S. J., and Alloway, B. J. (1988). *Environ. Technol. Lett.* **9,** 695–702.

Adams, S. N., and Honeysett, J. L. (1964). *Aust. J. Agric. Res.* **15,** 357–367.

Adams, S. N., Honeysett, J. L., Tiller, K. G., and Norrish, K. (1969). *Aust. J. Soil Res.* **7,** 29–42.

Albasel, N., and Cottenie, A. (1985). *Soil Sci. Soc. Am. J.* **49,** 386–390.

Anderson, A. J., Meyer, D. R., and Mayer, F. K. (1973). *Aust. J. Agric. Res.* **24,** 557–571.

Anderson, J. U. (1963). *Clays Clay Miner.* **10,** 380–388.

Anderson, P. R., and Christensen, T. H. (1988). *J. Soil Sci.* **39,** 15–22.

Andersson, A. (1977). *Swedish J. Agric. Res.* **7,** 7–20.

Andrews, R. K., Blakeley, R. L., and Zerner, B. (1988). *In* "Metal Ions in Biological Systems" (H. Sigel, ed.), Vol. 23. pp. 165–284, Dekker, New York.

Angle, J. S. (1982), *Soil Sci. Soc. Am. J.* **46,** 1119–1121.

Anonymous (1987). "Annual Report." Macaulay Institute for Soil Research.

Arnon, D. I. (1937), *Soil Sci.* **44,** 91–113.

Aschmann, S. G., and Zasoski, R. J. (1987). *Physiol. Plant.* **71,** 191–196.

Babich, H., and Stotzky, G. (1982). *Environ. Poll.* **29,** 303–315.

Babich, H., and Stotzky, G. (1983). *Adv. Appl. Microbiol.* **29,** 195–265.

Baes, C. F., and Mesmer, R. E. (1976). "The Hydrolysis of Cations." Wiley, New York.

Bansal, O. P. (1982). *J. Soil Sci.* **33,** 63–71.

Barrie, L. A., Lindberg, S. E., Chan, W. H., Ross, H. B., Arimoto, R., and Church, T. M. (1987). *Atmos. Environ.* **21,** 1133–1135.

Batley, G. E., and Florence, T. M. (1976). *Anal. Lett.* **9,** 379–388.

Beckett, P. H. T. (1989). *Adv. Soil Sci.* **9,** 143–176.

Beckwith, R. S. (1956). *Aust. J. Agric. Res.* **6,** 299–307.
Belzile, N., Lecomte, P., and Tessier, A. (1989). *Environ. Sci. Technol.* **23,** 1015–1020.
Benjamin, M. M., and Leckie, J. O. (1981). *J. Colloid Interface Sci.* **79,** 209–221.
Bergman, S. C., Ritter, C. J., Zamierowski, E. E., and Cothern, C. R. (1979). *J. Environ. Qual.* **8,** 416–422.
Berrow, M. L., and Burridge, J. C. (1990). *Int. J. Environ. Anal. Chem.* **39,** 173–177.
Biddappa, C. C., Chino, M., and Kumazawa, K. (1981). *J. Environ. Sci. Health B* **16,** 511–528.
Biddappa, C. C., Chino, M., and Kumazawa, K. (1982). *Plant Soil* **66,** 299–316.
Bloom, C. R., and McBride, M. B. (1979). *Soil Sci. Soc. Am. J.* **43,** 687–692.
Bloomfield, C. (1981). *In* "The Chemistry of Soil Processes" (D. J. Greenland and M.H.B. Hayes, eds), pp. 463–504. Wiley, London.
Böhmer, M.-B. (1989). *Arch. Acker-Pflanzenbau Bodenkd. Berlin* **33,** 475–482.
Bolton, J. (1975). *Environ. Pollut.* **9,** 295–304.
Bordons, A., and Jofre, J. (1987a). *Environ. Technol. Lett.* **8,** 495–500.
Bordons, A., and Jofre, J. (1987b). *Enzyme Microb. Technol.* **9,** 709–713.
Bosio, N. J., Hurst, V. J., and Smith, R. L. (1975). *Clays Clay Miner.* **23,** 400–403.
Bowers, A. R., and Huang, C. P. (1987). *Water Res.* **7,** 757–764.
Bowman, R. S., and O'Connor, G. A. (1982). *Soil Sci. Soc. Am. J.* **46,** 933–936.
Bowman, R. S., Essington, M. E., and O'Connor, G. A. (1981). *Soil Sci. Soc. Am. J.* **45,** 860–865.
Boyle, R. W., and Robinson, H. A. (1988). *In* "Metal Ions in Biological Systems" (H. Sigel, ed.), Vol. 23, pp. 1–29. Dekker, New York.
Bradford, G. R., Bair, F. L., and Hunsaker, V. (1971). *Soil Sci.* **112,** 225–230.
Bradford, G. R., Page, A. L., Lund, L. J., and Olmstead, W. (1975). *J. Environ. Qual.* **4,** 123–127.
Bremner, J. M., and Mulvaney, R. L. (1978). *In* "Soil Enzymes" (R. G. Burns, ed.), pp. 149–196. Academic Press, New York.
Brindley, G. W., and de Souza, J. V. (1975). *Clays Clay Miner.* **23,** 11–15.
Brindley, G. W., and Hang, P. T. (1973). *Clays Clay Miner.* **21,** 27–40.
Brookes, P. C., and McGrath, S. P. (1984). *J. Soil Sci.* **35,** 341–346.
Brookes, P. C., Heijnen, C. E., McGrath, S. P., and Vance, E. D. (1986). *Soil Biol. Biochem.* **18,** 383–388.
Brooks, R. R. (1987). "Serpentine and its Vegetation: A Multidisciplinary Approach" Croom Helm, London.
Brown, J. C., Specht, A. W., and Resnicky, J. W. (1961). *Agron. J.* **53,** 81–85.
Brown, J. C., Tiffin, L. O., and Holmes, R. S. (1960). *Plant Physiol.* **35,** 878–886.
Brown, P. H., Welch, R. M., Cary, E. E., and Checkai, R. T. (1987a). *J. Plant Nutr.* **10,** 2125–2135.
Brown, P. H., Welch, R. M., and Cary, E. E. (1987b). *Plant Physiol.* **85,** 801–803.
Brown, P. H., Dunemann, L., Schulz, R., and Marschner, H. (1989). *Z. Pflanzenernähr. Bodenk.* **152,** 85–91.
Brown, P. H., Welch, R. M., and Madison, J. T. (1990). *Plant Soil* **125,** 19–27.
Bruemmer, G. W., Gerth, J., and Tiller, K. G. (1988). *J. Soil Sci.* **39,** 37–52.
Brümmer, G., and Herms, U. (1983). *In* "Effects of Accumulation of Air Pollutants in Forest Ecosystems" (B. Ulrich and J. Pankrath, eds.), pp. 233–243. Reidel, Dordrecht.
Burns, R. G. (1965). *Nature* **205,** 999.
Burns, R. G. (1970). "Mineralogical Applications of Crystal Field Theory." Cambridge Univ. Press.
Burns, R. G. (1976). *Geochim. Cosmochim. Acta* **40,** 95–102.

Burridge, J. C., Reith, J. W. S., and Berrow, M. L. (1983). *In* "Trace Elements in Animal Production and Veterinary Practice" (N. F. Suttle, R. G. Gunn, W. M. Allen, K. A. Linklater, and G. Wiener, eds.), pp. 77–85. Occ. Pub. No. 7, British Society Animal Production.

Cabaniss, S. E. (1990). *Environ. Sci. Technol.* **24,** 583–588.

Calvet, R., Bourgeois, S., and Msaky, J. J. (1990). *Int. J. Environ. Anal. Chem.* **39,** 31–45.

Campbell, D. J., and Beckett, P. H. T. (1988). *J. Soil Sci.* **39,** 283–298.

Cataldo, D. A., Garland, T. R., and Wildung, R. E. (1978). *Plant Physiol.* **62,** 563–565.

Chang, A. C., Warneke, J. E., Page, A. L. and Lund, L. J. (1984). *J. Environ. Qual.* **13,** 87–91.

Cheshire, M. V., Berrow, M. L., Goodman, B. A., and Mundie, C. M. (1977). *Geochim. Cosmochim. Acta* **41,** 1131–1138.

Childs, C. W. (1975). *Geoderma* **13,** 141–152.

Childs, C. W., and Leslie, D. M. (1977). *Soil Sci.* **123,** 369–376.

Colin, F., Noack, Y., Treseases, J.-J., and Nahon, D. (1985). *Clay Miner.* **20,** 93–113.

Cotton, F. A., and Wilkinson, G. (1966). "Advanced Inorganic Chemistry," 2nd ed. Wiley, New York.

Cotton, F. A., and Wilkinson, G. (1980). "Advanced Inorganic Chemistry," 4th ed. Wiley, New York.

Cotton, F. A., and Wilkinson, G. (1988). "Advanced Inorganic Chemistry," 5th ed. Wiley, New York.

Crooke, W. M. (1956). *Soil Sci.* **81,** 269–276.

Crooke, W. M., Hunter, J. G., and Vergnano, O. (1954). *Ann. Appl. Biol.* **41,** 311–324.

Crooke, W. M., and Inkson, R. H. E. (1955). *Plant Soil* **6,** 1–15.

Crowther, D. L., Dillard, J. G., and Murray, J. W. (1983). *Geochim. Cosmochim. Acta* **47,** 1399–1403.

Cunningham, J. D., Keeney, D. R., and Ryan, J. A. (1975). *J. Environ. Qual.* **4,** 460–462.

Dalton, D. A., Evans, H. J., and Hanus, F. J. (1985). *Plant Soil* **88,** 245–258.

Davis, R. D. (1979). *J. Sci. Food Agric.* **30,** 937–947.

Davis, R. D., Carlton-Smith, C. H., Stark, J. H., and Campbell, J. A. (1988). *Environ. Pollut.* **49,** 99–115.

Dawson, B. S. W., Fergusson, J. E., Campbell, A. S., and Cutler, E. J. B. (1985). *Geoderma* **35,** 127–143.

Decarreau, A., Colin, F., Herbillon, A., Manceau, A., Nahon, D., Paquet, H., Trauth-Badaud, D., and Trescases, J. J. (1987). *Clays Clay Miner.* **35,** 1–10.

DeKock, P. C. (1956). *Ann. Bot.* **20,** 133–141.

DeKock, P. C., and Mitchell, R. L. (1957). *Soil Sci.* **84,** 55–62.

Dhannoun, H. Y., and Al-Dabbagh, S. M. A. (1990). *Chem. Geol.* **82,** 57–68.

Dijkshoorn, W., Lampe, J. E. M., and van Broekhoven, L. W. (1981). *Plant Soil* **61,** 277–284.

Dijkshoorn, W., Lampe, J. E. M., and van Broekhoven, L. W. (1983a). *Neth. J. Agric. Sci.* **31,** 181–188.

Dijkshoorn, W., Lampe, J. E. M., and van Broekhoven, L. W. (1983b). *Fert. Res.* **4,** 63–74.

Dillard, J. G., Crowther, D. L., and Murray, J. W. (1982). *Geochim. Cosmochim. Acta* **46,** 755–759.

Dillard, J. G., Crowther, D. L., and Calvert, S. E. (1984). *Geochim. Cosmochim. Acta* **48,** 1565–1569.

Dixon, J. B. (1989). *In* "Minerals in the Soil Environment" (J. B. Dixon and S. B. Weed, eds.), 2nd ed., pp. 393–403. American Society of Agronomy, Madison, Wisconsin.

Dixon, N. E., Blakely, R. L., and Zerner, B. (1980). *Can. J. Biochem.* **58,** 469–473.

Dixon, N. E., Gazzola, C., Blakely, R. L., and Zerner, B. (1976). *Science* **191,** 1144–1150.
Doelman, P., and Haanstra, L. (1986). *Biol. Fertil. Soils* **2,** 213–218.
Doner, H. E. (1978). *Soil Sci. Soc. Am. J.* **42,** 882–885.
Doner, H. E., Traina, S. J., and Pukite, A. (1982). *Soil Sci.* **133,** 369–376.
Ducaroir, J., Cambier, P., Leydecker, J.-P., and Prost, R. (1990). *Z. Pflanzenernähr. Bodenk.* **153,** 349–358.
Dudley, L. M., McNeal, B. L., and Baham, J. E. (1986). *J. Environ. Qual.* **15,** 188–192.
Dudley, L. M., McNeal, B. L., Baham, J. E., Coray, C. S., and Cheng, H. H. (1987). *J. Environ. Qual.* **16,** 341–348.
During, C. (1972). "Fertilisers and Soils in New Zealand Farming," 2nd ed. Government Printer, Wellington, New Zealand.
Eggleton, R. A., Foudoulis, C., and Varkevisser, D. (1987). *Clays Clay Miner.* **35,** 161–169.
Emmerich, W. E., Lund, L. J., Page, A. L., and Chang, A. C. (1982a). *J. Environ. Qual.* **11,** 174–178.
Emmerich, W. E., Lund, L. J., Page, A. L., and Chang, A. C. (1982b). *J. Environ. Qual.* **11,** 182–186.
Epstein, E., and Stout, P. R. (1951). *Soil Sci.* **72,** 47–65.
Eskew, D. L., Welch, R. M., and Cary, E. E. (1983). *Science* **222,** 621–623.
Eskew, D. L., Welch, R. M., and Norvell, W. A. (1984). *Plant Physiol.* **76,** 691–693.
Essington, M. E., and Mattigod, S. V. (1990). *Soil Sci. Soc. Am. J.* **54,** 385–394.
Evans, K. J., Mitchell, I. G., and Salau, B. (1979). *Prog. Water Technol.* **11,** 339–352.
Farago, M. E., and Cole, M. M. (1988). *In* "Metal Ions in Biological Systems" (H. Sigel, ed.), Vol. 23, pp. 47–90. Dekker, New York.
Figura, P., and McDuffie, B. (1977). *Anal. Chem.* **49,** 1950–1953.
Figura, P., and McDuffie, B. (1980). *Anal. Chem.* **52,** 1433–1439.
Filby, R. H., and van Berkel G. J. (1987). *In* "Metal Complexes in Fossil Fuels" (R. H. Filby and J. F. Branthaver, eds.), ACS Symposium Series No. 344, pp. 2–39. ACS Washington, D.C.
Fish, R. H., Reynolds, J. G., and Gallegos, E. G. (1987). *In* "Metal Complexes in Fossil Fuels" (R. H. Filby and J. F. Branthaver, eds.), ACS Symposium Series No. 344, pp. 332–349. ACS, Washington, D.C.
Fleet, M. E., and Stone, W. E. (1990). *Contrib. Miner. Petrol.* **105,** 629–636.
Fletcher, P., and Beckett, P. H. T. (1987). *Water Res.* **21,** 1163–1172.
Foster, M. D. (1963). *Clays Clay Miner.* **12,** 70–95.
Francis, A. J., and Dodge, C. J. (1990). *Environ. Sci. Technol.* **24,** 373–378.
Gabbrielli, R., and Pandolfini, T. (1984). *Physiol. Plant.* **62,** 540–544.
Gabbrielli, R., and Pandolfini, T., Vergnano, O., and Palandri, M. R. (1990). *Plant Soil* **122,** 271–277.
Gamble, D. S. (1986). *In* "The Importance of Chemical "Speciation" in Environmental Processes" (M. Bernhard, F. E. Brinkman, and P. J. Sadler, eds.), pp. 217–236. Springer-Verlag, Berlin.
Gartrell, J. W. (1981). *In* "Copper in Soils and Plants" (J. F. Loneragan, A. D. Robson, and R. D. Graham, eds.), pp. 313–349. Academic Press, New York.
Gerard, P., and Herbillon, A. J. (1983). *Clays Clay Miner.* **31,** 143–151.
Gerth, J. (1990). *Geochim. Cosmochim. Acta* **54,** 363–371.
Gerth, J., and Brümmer, G. (1983). *Fresenius Z. Anal. Chem.* **316,** 616–620.
Ghiorse, W. C. (1988). *In* "Manganese in Soils and Plants" (R. D. Graham, R. J. Hannam, and N. C. Uren, eds.), pp. 75–85. Kluwer Academic, Dordrecht.
Gilkes, R. J., and McKenzie, R. M. (1988). *In* "Manganese in Soils and Plants" (R. D. Graham, R. J. Hannam, and N. C. Uren, eds.), pp. 23–35. Kluwer Academic, Dordrecht.
Golightly, J. P., and Arancibia, O. N. (1979). *Can. Miner.* **17,** 719–728.

Goodman, B. A., and Linehan, D. J. (1978). *In* "The Soil–Root Interface" (J. L. Harley and R. S. Russell, eds.), pp. 67–82. Academic Press, San Diego/London.
Grasselly, G., and Hetényi, M. (1968). *Acta Mineral. Petrogr.* **18,** 85–98.
Halstead, R. L. (1968). *Can. J. Soil Sci.* **48,** 301–305.
Halstead, R. L., Finn, B. J., and MacLean, A. J. (1969). *Can. J. Soil Sci.* **49,** 335–342.
Haq, A. U., Bates, T. E., and Soon, Y. K. (1980). *Soil Sci. Soc. Am. J.* **44,** 772–777.
Harter, R. D. (1983). *Soil Sci. Soc. Am. J.* **47,** 47–51.
Hausinger, R. P. (1987). *Microbiol. Rev.* **51,** 22–42.
Helios-Rybicka, E. (1985). *Clay Miner.* **20,** 525–527.
Herms, U., and Brümmer, G. (1984). *Z. Pflanzenernähr. Bodenk.* **147,** 400–425.
Hewitt, E. J. (1948). *Nature* **161,** 489–490.
Hiller, D. A., Brümmer, G., and Ackermand, D. (1988). *Z. Pflanzenernähr. Bodenk.* **151,** 47–54.
Hodgson, J. F., Geering, H. R., and Norvell, W. A. (1965). *Soil Sci. Soc. Am. Proc.* **29,** 665–669.
Hodgson, J. F., Lindsay, W. L., and Trierweiler, J. F. (1966). *Soil Sci. Soc. Am. Proc.* **30,** 723–726.
Holtzclaw, K. M., Keech, D. A., Page, A. L., Sposito, G., Ganje, T. J., and Ball, N. B. (1978). *J. Environ. Qual.* **7,** 124–127.
Horst, H., Brüne, H., Sauerbeck, D., and Hein, A. (1988). *Landwirtsch. Forsch.* **41,** 286–296.
Horst, W. J., Klotz, F., and Szulkiewicz, P. (1990). *Plant Soil* **124,** 227–231.
Hosking, W. J., Caple, I. W., Halpin, C. G., Brown, A. J., Paynter, D. I., Conley, D. N., and North-Coombes, P. L. (1986). "Trace Elements for Pastures and Animals in Victoria." F. D. Atkinson, Government Printer, Melbourne.
Howard, C. J., Hill, R. J., and Sufi, M. A. M. (1988). *Chem. Aust.* **55,** 367–369.
Hunter, J. G., and Vergnano, O. (1952). *Ann. Appl. Biol.* **39,** 279–284.
Hutchinson, T. C. (1981). *In* "Effect of Heavy Metal Pollution On Plants" (N. W. Lepp, ed.), Vol 1, pp. 171–211. Applied Science Publishers, London.
Iu, K. L., Pulford, I. D., and Duncan, H. J. (1981). *Plant Soil* **59,** 327–333.
Iu, K. L., Pulford, I. D., and Duncan, H. J. (1982). *Plant Soil* **66,** 423–427.
Jahiruddin, M., Livesey, N. T., and Cresser, M. S. (1985). *Commun. Soil Sci. Plant Anal.* **16,** 909–922.
Jeffery, J. J., and Uren, N. C. (1983). *Aust. J. Soil Res.* **21,** 479–488.
Johnston, W. R., and Proctor, J. (1981). *J. Ecol.* **69,** 855–869.
Jones, G. B., and Belling, G. B., (1967). *Aust. J. Agric. Res.* **18,** 733–740.
Jones, G. B., Riceman, D. S., and McKenzie, J. O. (1957). *Aust. J. Agric. Res.* **8,** 190–201.
Katyal, J. C. (1977). *Soil Biol. Biochem.* **9,** 259–266.
Kerndorff, H., and Schnitzer, M. (1980). *Geochim. Cosmochim. Acta* **44,** 1701–1708.
Khalid, B. Y., and Tinsley, J. (1980). *Plant Soil* **55,** 139–144.
Kiekens, L., Cottenie, A., and van Landschoot, G. (1984). *Plant Soil* **79,** 89–99.
Kimura, H. S., and Swindale, L. D. (1967). *Soil Sci.* **104,** 69–76.
Kinniburgh, D. G., Jackson, M. L., and Syers, J. K. (1976). *Soil Sci. Soc. Am. J.* **40,** 796–799.
Kohno, Y., and Foy, C. D. (1983). *J. Plant Nutr.* **6,** 677–693.
Koppelman, M. H., and Dillard J. G. (1977). *Clays Clay Miner.* **25,** 457–462.
Körner, L. E., Møller, I. M., and Jensén, P. (1986). *Physiol. Plant.* **68,** 583–588.
Krauskopf, K. B. (1979). "Introduction to Geochemistry," 2nd ed. McGraw–Hill, New York.
Ladd, J. N. (1985). *In* "Soil Organic Matter and Biological Activity" (D. Vaughan and R. E. Malcolm, eds.), pp. 175–221. Martinus Nijhoff/Dr. W. Junk Publishers, Dordrecht.
Lake, D. L., Kirk, P. W. W., and Lester, J. N. (1984). *J. Environ. Qual.* **13,** 174–183.

Lavigne, J. A., Langford, C. H., and Mak, M. K. S. (1987). *Anal. Chem.* **59,** 2616–2620.
Lavkulich, L. M., and Wiens, J. H. (1970). *Soil Sci. Soc. Am. Proc.* **34,** 755–758.
Leeper, G. W. (1970). "Six Trace Elements in Soil." Melbourne University Press, Carlton.
Leeper, G. W. (1978). "Managing the Heavy Metals on the Land." Dekker, New York.
Le Riche, H. H. (1968). *J. Agric. Sci.* **71,** 205–208.
Le Riche, H. H., and Weir, A. H. (1963). *J. Soil Sci.* **14,** 225–235.
Lindsay, W. L. (1972). *In* "Advances in Agronomy," (N. C. Brady, ed.), Vol. 24, pp. 147–186. Academic Press, San Diego.
Lindsay, W. L., and Norvell, W. A. (1978). *Soil Sci. Soc. Am. J.* **42,** 421–428.
Llorca, S. (1987). *C. R. Acad. Sci. Paris Ser. 2* **304,** 15–18.
MacLean, A. J., and Dekker, A. J. (1978). *Can. J. Soil Sci.* **58,** 381–389.
MacNicol, R. D., and Beckett, P. H. T. (1985). *Plant Soil* **85,** 107–129.
MacNicol, R. D., and Beckett, P. H. T. (1989). *Water Res.* **23,** 199–206.
Makitie, O. (1962). *J. Sci. Agric. Soc. Finland* **34,** 91–95.
Manceau, A., Busek, P. R., Miser, D., Rask, J., and Nahon, D. (1990). *Am. Miner.* **75,** 490–494.
Manceau, A., and Calas, G. (1985). *Am. Miner.* **70,** 549–558.
Manceau, A., and Calas, G. (1986). *Clay Miner.* **21,** 341–360.
Manceau, A., and Calas, G. (1987). *Clay Miner.* **22,** 357–362.
Manceau, A., Calas, G., and Decarreau, A. (1985). *Clay Miner.* **20,** 367–387.
Manceau, A., Llorca, S., and Calas, G. (1987). *Geochim. Cosmochim. Acta* **51,** 105–113.
Martell, A. E., Motekaitis, R. J., and Smith, R. M. (1988). *Environ. Toxicol. Chem.* **7,** 417–434.
Mason, B., and Moore, C. B. (1982). "Principles of Geochemistry," 4th ed. Wiley, New York.
Matthews, P. J. (1984). *CRC Crit. Rev. Environ. Control* **14,** 199–250.
Mattigod, S. V., Gibali, A. S., and Page, A. L. (1979). *Clays Clay Miner.* **27,** 411–416.
Mattigod, S. V., Sposito, G., and Page, A. L. (1981). *In* "Chemistry in the Soil Environment" (R. H. Dowdy, ed.), pp. 203–221. Amer. Soc. Agron., Madison.
McBride, M. B. (1989). *Adv. Soil Sci.* **10,** 1–56.
McGrath, S. P. (1984). *In* "Trace Substances in the Environment" (D. D. Hemphill, ed.), pp. 242–252. University of Missouri.
McKenzie, R. M. (1967). *Aust. J. Soil Res.* **8,** 97–106.
McKenzie, R. M. (1972). *Geoderma* **8,** 29–35.
McKenzie, R. M. (1975). *Aust. J. Soil Res.* **13,** 177–188.
McKenzie, R. M. (1977). *Geol. Geochem. Manganese* **1,** 259–269.
McKenzie, R. M. (1978). *Aust. J. Soil Res.* **16,** 209–214.
McKenzie, R. M. (1980). *Aust. J. Soil Res.* **18,** 61–73.
Mishra, D., and Kar, M. (1974). *Bot. Rev.* **40,** 395–452.
Misra, S. G., and Pande, P. (1974a). *Plant Soil* **40,** 679–684.
Misra, S. G., and Pande, P. (1974b). *Plant Soil* **41,** 697–700.
Mitchell, R. L. (1945). *Soil Sci.* **60,** 63–69.
Mitchell, R. L. (1964). *In* "Chemistry of the Soil" (F. E. Bear, ed.), 2nd ed., pp. 320–368. Reinhold, New York.
Mitchell, R. L., and Burridge, J. C. (1979). *Philos. Trans. R. Soc. London B* **288,** 15–24.
Mobley, H. L. T., and Hausinger, R. P. (1989). *Microbiol. Rev.* **53,** 85–108.
Murphy, L. S., and Walsh, L. M. (1972). *In* "Micronutrients in Agriculture" (J. J. Mortvedt, P. M. Giordano, and W. L. Lindsay, eds.), pp. 347–387. Soil Sci. Soc. Amer., Madison.
Murray, J. W. (1975). *Geochim. Cosmochim. Acta* **39,** 505–519.
Murray, J. W., Healy, T. W., and Fuerstenau, D. W. (1968). *Adv. Chem. Ser.* **79,** 74–81.

Nahon, D. B., and Colin, F. (1982). *Am. J. Sci.* **282,** 1232–1243.
Navrot, J., and Singer, A. (1976). *Soil Sci.* **121,** 337–345.
Newman, A. C. D., and Brown, G. (1987). *In* "Chemistry of Clays and Clay Minerals" (A. C. D. Newman, ed.), Mineralogical Society Monograph No. 6, pp. 2–128. Longmans Scientific and Technical, Harlow.
Ng, S. K., and Bloomfield, C. (1962). *Plant Soil* **16,** 108–135.
Nicholas, D. J. D., and Thomas, W. D. E. (1954). *Plant Soil* **5,** 182–193.
Nielsen, F. H., Reno, H. T., Tiffin, L. O., and Welch, R. M. (1977). *In* "Geochemistry and the Environment" Vol. 2, pp. 40–53. National Academy of Sciences, Washington, D.C.
Nirel, P. M. V., and Morel, F. M. M. (1990). *Water Res.* **24,** 1055–1056.
Norrish, K. (1975). *In* "Trace Elements in Soil–Plant–Animal Systems" (D. J. D. Nicholas and A. R. Egan, eds.), pp. 55–81. Academic Press, San Diego.
Norvell, W. A. (1972). *In* "Micronutrients in Agriculture" (J. J. Mortvedt, R. M. Giordano, and W. L. Lindsay, eds.), pp. 115–138. Soil Sci. Soc. Amer., Madison.
O'Connor, G. A. (1988). *J. Environ. Qual.* **17,** 715–718.
O'Connor, G. A., Essington, M. E., Elrashidi, M., and Bowman, R. S. (1983). *Soil Sci.* **135,** 228–235.
Ooi, K., Miyai, Y., and Katoh, S. (1987). *Solvent Extraction Ion Exchange* **5,** 561–572.
Ostwald, J., and Dubrawski, J. V. (1987). *Miner. Mag.* **51,** 463–466.
Page, A. L., and Chang, A. C. (1979). *Phytopathology* **69,** 1007–1011.
Patterson, J. B. E. (1971). *In* "Trace Elements in Soils and Crops," MAFF Tech. Bull. No. 21, pp. 193–207. HMSO.
Pezzarossa, B., Malorgio, F., Lubrano, L., Tognoni, F., and Petruzzelli, G. (1990). *Commun. Soil Sci. Plant Anal.* **21,** 737–751.
Pickering, W. F. (1986). *Ore Geol. Rev.* **1,** 83–146.
Ponnamperuma, F. N. (1972). *In* "Advances in Agronomy" (N. C. Brady, ed.), Vol. 24, pp. 29–96. Academic Press, San Diego.
Ponnamperuma, F. N. (1984). *In* "Flooding and Plant Growth" (T. T. Kozlowski, ed.), pp. 9–45. Academic Press, San Diego.
Pratt, P. F., Bair, F. L., and McLean, G. W. (1964). *Soil Sci. Soc. Am. Proc.* **28,** 363–365.
Puls, R. W., and Bohn, H. L. (1988). *Soil Sci. Soc. Am. J.* **52,** 1289–1292.
Purves, D. (1985). "Trace-Element Contamination of the Environment," Rev. ed. Elsevier, Amsterdam.
Quirke, J. M. E. (1987). *In* "Metal Complexes in Fossil Fuels" (R. H. Filby and J. F. Branthaver, eds.), ACS Symposium Series No. 344, pp. 74–83. ACS, Washington, D.C.
Rabenhorst, M. C., Foss, J. E., and Fanning, D. S. (1982). *Soil Sci. Soc. Am. J.* **46,** 607–616.
Rankama, K., and Sahama, T. G. (1950). "Geochemistry." Univ. of Chicago Press.
Rappaport, B. D., Martens, D. C., Reneau R. B., Jr., and Simpson, T. W. (1987). *J. Environ. Qual.* **16,** 29–33.
Rashid, M. A. (1974). *Chem. Geol.* **13,** 115–123.
Ratkovic, S., and Vucinic, Z. (1990). *Plant Physiol. Biochem.* **28,** 617–622.
Rauser, W. E., and Winterhalder, E. K. (1985). *Can. J. Bot.* **63,** 58–63.
Reddy, M. R., and Dunn, S. J. (1986). *Environ. Pollut. Ser. B* **11,** 303–313.
Reuter, D. J. (1975). *In* "Trace Elements in Soil–Plant–Animal Systems" (D. J. D. Nicholas and A. R. Egan, eds.), pp. 291–324. Academic Press, New York.
Robertson, A. I. (1985). *New Phytol.* **100,** 173–189.
Rohatgi, N., and Chen, K. Y. (1975). *J. Water Pollut. Control Fed.* **47,** 2298–2316.
Sadiq, M., and Enfield, C. G. (1984a). *Soil Sci.* **138,** 262–270.
Sadiq, M., and Enfield, C. G. (1984b). *Soil Sci.* **138,** 335–340.
Sadiq, M., and Zaidi, T. H. (1981). *Water Air Soil Pollut.* **16,** 293–299.

Sanders, J. R., Adams, T. McM., and Christensen, B. T. (1986a). *J. Sci. Food Agric.* **37,** 1155–1164.
Sanders, J. R., McGrath, S. P., and Adams, T. McM. (1986b). *J. Sci. Food Agric.* **37,** 961–968.
Schellmann, W. (1986). *Chem. Erde.* **45,** 39–52.
Schellmann, W. (1989a). *Chem. Geol.* **74,** 351–364.
Schellmann, W. (1989b). *Miner. Deposita* **24,** 161–168.
Schmidt, T., and Schlegel, H. G. (1989). *FEMS Microbiol. Ecol.* **62,** 315–318.
Schreier, H., Omueti, J. A., and Lavkulich, L. M. (1987). *Soil Sci. Soc. Am. J.* **51,** 993–999.
Schulthess, C. P., and Huang, C. P. (1990). *Soil Sci. Soc. Am. J.* **54,** 679–688.
Schultz, M. F., Benjamin, M. M., and Ferguson, J. F. (1987). *Environ. Sci. Technol.* **21,** 863–869.
Schwertmann, U., and Latham, M. (1986). *Geoderma* **39,** 105–123.
Shacklette, H. T., Hamilton, J. C., Boerngen, J. G., and Bowles, J. M. (1971). "Elemental Composition of Surficial Materials in the Conterminous United States," Geol. Survey Paper 574-D. U.S. Govt. Printing Office, Washington, D.C.
Shannon, R. D., and Prewitt, C. T. (1969). *Acta Crystallogr. B* **25,** 925–946.
Sherrell, C. G. (1983). *N. Z. Agric. Sci.* **17,** 217–221.
Short, N. M. (1961). *J. Geol.* **69,** 534–571.
Sidhu, P. S., Gilkes, R. J., and Posner, A. M. (1978). *J. Inorg. Nucl. Chem.* **39,** 1953–1958.
Sidhu, P. S., Gilkes, R. J., and Posner, A. M. (1980). *Soil Sci. Soc. Am. J.* **44,** 135–138.
Sidhu, P. S., Sehgal, J. L., Sinha, M. K., and Randhawa, N. S. (1977). *Geoderma* **18,** 241–249.
Sillanpää, M., and Lakanen, E. (1969). *J. Sci. Agric. Soc. Finland* **41,** 60–65.
Simpson, K. (1983). "Soil." Longmans, London.
Siow, A. K. F. (1990). "The Behaviour of Nickel in Soils and its Toxicity to Plants." M. Agr. Sc. thesis, La Trobe University.
Slingsby, D. R., and Brown, D. H. (1977). *J. Ecol.* **65,** 597–618.
Smilde, K. W. (1981). *Plant Soil* **61,** 3–14.
Soon, Y. K., and Bates, T. E. (1982). *J. Soil Sci.* **33,** 477–488.
Sorteberg, A. (1978). *J. Sci. Agric. Soc. Finland* **50,** 317–334.
Sorteberg, A. (1981). *J. Sci. Agric. Soc. Finland* **53,** 1–15.
Sposito, G. (1983). *In* "Applied Environmental Geochemistry" (I. Thornton, ed.), pp. 121–170. Academic Press, San Diego/London.
Sposito, G. (1984). "The Surface Chemistry of Soils." Oxford Univ. Press.
Sposito, G., and Bingham, F. T. (1981). *J. Plant Nutr.* **3,** 35–49.
Suzuki, S., Mizuno, N., and Kimura, K. (1971). *Soil Sci. Plant Nutr.* **17,** 195–198.
Swaine, D. J. (1962). "The Trace-Element Content of Fertilizers," Technical Communication No. 52. Commonwealth Agricultural Bureau.
Tabatabai, M. A. (1977). *Soil Biol. Biochem.* **9,** 9–13.
Takahashi, Y., and Imai, H. (1983). *Soil Sci. Plant Nutr.* **29,** 111–122.
Taylor, R. M. (1987). *In* "Chemistry of Clays and Clay Minerals" (A. C. D. Newman, ed.), pp. 129–201, Mineralogical Soc. Monograph No. 6. Longman Scientific and Technical, Harlow.
Taylor, R. W., and Griffin, G. F. (1981). *Plant Soil* **62,** 147–152.
Taylor, R. M., and McKenzie, R. M. (1966). *Aust. J. Soil Res.* **4,** 29–39.
Thauer, R. K., Diekert, G., and Schönheit, P. (1980). *Trends Biochem. Sci.* **5,** 304–306.
Theis, T. L., and Richter, R. O. (1980). *Adv. Chem. Series* **189,** 73–96.
Tiller, K. G. (1958). *J. Soil Sci.* **9,** 223–241.
Tiller, K. G. (1959). *Proc. R. Soc. Tas.* **93,** 153–158.

Tiller, K. G. (1963). *Aust. J. Soil Res.* **1,** 74–90.
Tiller, K. G. (1983). *In* "Soils: An Australian Viewpoint, Division of Soils, CSIRO," pp. 365–387. CSIRO/Academic Press, San Diego.
Tiller, K. G. (1989). *Adv. Soil Sci.* **9,** 113–142.
Tiller, K. G., Gerth, J., and Brümmer, G. (1984). *Geoderma* **34,** 17–35.
Tinker, P. B. (1986). *J. Soil Sci.* **37,** 587–601.
Tokashiki, Y., Dixon, J. B., and Golden, D. C. (1986). *Soil Sci Soc. Am. J.* **50,** 1079–1084.
Traina, S. J., and Doner, H. E. (1985). *Clays Clay Miner.* **33,** 118–122.
Tyler, L. D., and McBride, M. B. (1982). *Soil Sci.* **134,** 198–205.
Uren, N. C., Asher, C. J., and Longnecker, N. E. (1988). *In* "Manganese in Soils and Plants" (R. D. Graham, R. J. Hannam, and N. C. Uren, eds), pp. 309–328. Kluwer Academic, Dordrecht.
Uren, N. C., and Edwards, L. B. (1984). *Plant Soil* **81,** 145–149.
Uren, N. C., and Reisenauer, H. M. (1988). *Adv. Plant Nutr.* **3,** 79–114.
Vanselow, A. P. (1966). *In* "Diagnostic Criteria for Plants and Soils" (H. D. Chapman, ed.), pp. 302–309. University of California Division of Agricultural Sciences.
Vlamis, J., Williams, D. E., Fong, K., and Corey, J. E. (1978). *Soil Sci.* **126,** 49–55.
Voskresenskaya, N. T. (1987). *Geochem. Int.* **24,** 97–107.
Wallace, A. (1982). *Soil Sci.* **133,** 319–323.
Wallace, A., Romney, E. M., Cha, J. W., Soufi, S. M., and Chaudhry, F. M. (1977a). *Commun. Soil Sci. Plant Anal.* **8,** 757–764.
Wallace, A., Romney, E. M., Alexander, G. V., Soufi, S. M., and Patel, P. M. (1977b). *Agron. J.* **69,** 18–20.
Wallace, A., Romney, E. M., and Kinnear, J. E. (1977c). *Commun. Soil Sci. Plant Anal.* **8,** 727–732.
Webber, M. D., Kloke, A., and Tjell, J. Chr. (1984). *In* "Processing and Use of Sewage Sludge" (P. L'Hermite and H. Ott, eds.), pp. 371–386. Reidel, Dordrecht.
Welch, J. E., and Lund, L. J. (1982). *J. Environ. Qual.* **16,** 403–410.
Welch, R. M. (1981). *J. Plant Nutr.* **3,** 345–356.
Wells, N. (1960). *J. Soil Sci.* **11,** 409–424.
Wiewiora, A., and Szpila, K. (1975). *Clays Clay Miner.* **23,** 91–96.
Willaert, G., and Verloo, M. (1988). *Plant Soil* **107,** 285–292.
Willett, I. R. (1979). *Plant Soil* **52,** 373–383.
Willett, I. R. (1982). *Aust. J. Soil Res.* **20,** 131–138.
Willett, I. R., and Cunningham, R. B. (1983). *Aust. J. Soil Res.* **21,** 301–308.
Williams, D. E., Vlamis, J., Pukite, A. H., and Corey, J. E. (1987). *Soil Sci.* **143,** 124–131.
Williams, P. C. (1967). *Nature* **214,** 628.
Williams, R. J. P. (1981). *Philos. Trans. R. Soc. London B* **294,** 57–74.
Williams, R. J. P. (1987). *Coord. Chem. Rev.* **79,** 175–193.
Wilson, M. J., and Berrow, M. L. (1978). *Chem. Erde* **37,** 181–205.
Wilson, R. E. (1978). "Mineralogy, Petrology and Geochemistry of Basalt Weathering." B. Sc. honours thesis, La Trobe University.
Wolfenden, E. B. (1965). *Geochim. Cosmochim. Acta* **29,** 1051–1062.
Woolhouse, H. W. (1983). *Plant Physiol. New Ser. C* **12,** 245–300.
Yamamoto, K. (1984). *Bull. Natl. Inst. Agric. Sci. Ser. B* (36), 171–232.
Yamasaki, S., Yoshino, A., and Kishita, A. (1975). *Soil Sci. Plant Nutr.* **21,** 63–72.
Zamani, B., Knezek, B. D., and Dazzo, F. B. (1984). *J. Environ. Qual.* **13,** 269–273.
Zantua, M. I., and Bremner, J. (1977). *Soil Biol. Biochem.* **9,** 135–140.

Electrochemical Techniques for Characterizing Soil Chemical Properties

T. R. Yu

Institute of Soil Science, Academia Sinica,
Nanjing, China

I. INTRODUCTION

Electrochemical techniques are a category of methods used for the study of the composition and properties of a medium based on electrochemical reactions occurring at the electrode–solution interface. They are developed on the basis of the integration of electrochemistry and analytical chemistry and have been applied to many research fields, including soil science.

Most electrochemical techniques are simple in operation. Generally only one to a few minutes are required to make a determination. In many cases it is possible to evaluate a soil property *in situ* by simply inserting two electrodes into the soil. This should be of particular value for characterizing soils, which are dynamic systems under natural conditions. It is no wonder that some of these techniques, for instance, determinations of pH and electrical conductance, have become routine in many soil laboratories. However, it was in the last two decades that the application of electrochemical techniques to soil science became increasingly widespread, stimulated apparently by the rapid progress in electroanalytical chemistry.

In this article, the principles and applications of most important electrochemical techniques in soil science will be reviewed.

II. FUNDAMENTALS OF ELECTROCHEMICAL TECHNIQUES

A. Classification of Electrochemical Techniques

The basic principle of electrochemical techniques is to convert one chemical parameter of a medium into an electrical parameter by an appropriate device, usually an electrode or an electrode pair, and then to measure this electrical parameter using an instrument. As in the case of electrical measurements, in electrochemical techniques four parameters can be utilized. The electrode potential may be a function of the composition of the solution. Techniques based on the measurement of this potential are called potentiometry. When an external voltage is applied to an electrode, an electrical current may flow through the circuit. Techniques utilizing this current–voltage relationship are called voltammetry. One can also measure the electrical resistance (conductance) between two electrodes in the medium, and such a technique is called conductometry. Sometimes the total quantity of electricity involved in an electrochemical reaction is measured, and techniques based on this principle are called coulometry. In soil science, this last technique is not commonly used and will not be dealt with in this chapter.

In all these techniques, the electrode is the site at which electrochemical reactions take place and is the key element in electrochemical analyses. Therefore, in the following sections, the electrode potential at the electrode–solution interface when no current flows is first examined, and then the effect of current on electrode potential, electrode polarization, is discussed.

B. Origin of Electrode Potential

When two substances are in contact, an interface at the site of contact is formed. These two substances, called two phases, can be solid–solid, solid–liquid, or liquid–liquid. Because of the difference in chemical composition between the two phases, a material transfer occurs at the interface. If the transferred material is charged (electrons, ions, dipoles), a difference in electrical potential between the two phases results.

Suppose that a silver metal is dipped in a water solution of silver nitrate. The metal contains silver ions and electrons, and the solution contains silver ions and nitrate ions together with hydrogen ions and hydroxyl ions. At the interface, silver ions can transfer from the silver metal to the solution and can also transfer in an opposite direction. Within a given time interval the number of silver ions transferred from the solution to the silver metal is proportional to the number of that ion in solution. Under ordinary conditions the number of silver ions transferred from the solution to the silver metal is always larger than that transferred in the opposite direction. As a result, the metal is positively charged and the solution negatively charged, and a potential difference between the two sides of the interface is produced.

If an inert electrode, such as a platinum electrode, is dipped in a solution containing ferric and ferrous ions, electrons can be liberated from Fe^{2+} ions and captured by the platinum metal, or leave the platinum metal and be captured by Fe^{3+} ions. The rate of liberation of electrons from Fe^{2+} ions is dependent on the Fe^{2+} concentration in solution, and the capture of electrons by Fe^{3+} ions depends on the Fe^{3+} concentration. Therefore, at the beginning, the rate of electron transfer to platinum is determined by the concentrations of these two ion species. Since under ordinary conditions the rate of electron transfer from platinum to solution is larger than that of the reverse transfer, the platinum electrode would be positively charged and the solution negatively charged. The production of a potential difference between the two phases would retard further transfer of electrons, until a dynamic equilibrium is established.

In the above examples, the transfer of ions and electrons between a metal and a solution is considered. For many ion-selective electrodes, the main component is not a metal, but another solid or an organic liquid immiscible with water. When these electrodes are in contact with water there is also a charge transfer between the electrode and water. For charged organic liquids, the charge carriers are ions. For ionic solids, such as glass and lanthanum fluoride, the charge carriers are ions together with a small portion of electrons.

From the previous discussions it follows that, when an interface is produced by the contact of two phases, a charge transfer would occur between the two sides of the interface, if the substances constituting the interface contain mobile charged components (electrons, ions). This charge transfer results in an uneven charge distribution between the two phases, and thus a potential difference. For substances that contain dipoles or are capable of producing induced dipoles, an interfacial potential difference can also be produced by the orientation of dipoles. Therefore, the presence of an interfacial potential difference, in short, an interface potential, is a common phenomenon in nature. Utilization of the electrode potential constitutes the basis of electroanalytical chemistry.

C. Nernst Equation

From the previous discussions, it is known that the interface potential of an electrode is related to both the nature of the electrode material and the quantity of the relevant substance M in solution. This relationship can be quantitatively expressed by the Nernst equation:

$$E = E^0 + \frac{\mathrm{RT}}{zF} \ln a_{\mathrm{M}}{}^{z+} \tag{1}$$

For the electrode reaction

$$aA + bB + ze \rightleftharpoons cC + dD, \tag{2}$$

the generalized Nernst equation is

$$E = E^0 + \frac{RT}{zF} \ln \frac{A^a B^b}{C^c D^d}. \tag{3}$$

The Nernst equation is the basis of potentiometry. It is essential to understanding the meaning and uses of the parameters in the equation and each parameter in the Nernst equation will be discussed. The parameter

E^0 is the standard potential, i.e., the electrode potential under standard conditions where the activities of the reactants and products of the electrode reaction are unity. This is an important parameter in electrochemically characterizing the system. For most inorganic systems the E^0 value has been precisely determined or calculated. If the E^0 value of an electrode is not known, it may be determined by measuring the electrode potential in a series of solutions with known ion activities and extrapolating the electrode potential to unity activity. The R is the gas constant, and its numerical value is 8.325 J mol^{-1} K^{-1}, T is absolute temperature, and F is the Faraday constant. Its numerical value is 96496 C/g equivalent.

For electrodes, the term (RT/zF) ln is generally called the potential response coefficient. It corresponds to the slope of the straight line of the electrode potential–ion activity relationship. In practical use, it is often called the response slope or the Nernst slope. Its numerical value is temperature-dependent. The z is the number of electrons involved in the electrode reaction. It can be seen from Eq. (1) that for electrodes responsive to divalent ions the response slope is only one-half that for electrodes responsive to monovalent ions. Therefore, the larger the numerical value of z, the less sensitive the electrode potential is to the change in ion activity.

The a is the activity of the ions participating in the electrode reaction. According to the Nernst equation, the electrode potential is related to the logarithm of a. Therefore, it is possible to determine the ion activity varying within a range of several orders of magnitude. On the other hand, because 10-fold change in ion activity induces the change in electrode potential by only about 59 mV for monovalent ions and about 29 mV for divalent ions, a high precision in potential measurement is required.

The electrode responds to ion activity rather than to ion concentration. This point is of significance in chemistry, because in the calculation of many equilibrium constants the required parameter is ion activity, but not ion concentration. In particular, when an ion species is in dynamic adsorption–dissociation equilibrium in a soil system, what is of interest is just the actual concentration (activity) of that ion species in the diffuse double layer and the free solution, which cannot be determined by any other method. This is one of the reasons potentiometry has been applied so widely in soil chemical studies.

D. Membrane Potential

In a membrane electrode, two sides of the membrane are in contact with solutions. Therefore, two interface potentials arise. Also, for some membranes there may be a diffusion potential within the membrane caused by the migration of charge carriers. Because the interface potential is related to ion activity in a Nernstian manner, in the simple case where the diffusion potential within the membrane can be neglected, the membrane potential would be

$$E = \frac{RT}{zF} \ln \frac{a_1}{a_2}, \tag{4}$$

where a_1 and a_2 are the ion activities in the two solutions, respectively.

Generally, a membrane electrode is constructed of a membrane, an inner solution with a constant composition, and an inner reference electrode. Because the inner interface potential and the potential of the inner reference electrode are constants, the measured potential of the membrane electrode is only a function of the outer interface potential, which is related to the ion activity in the outer (test) solution following the Nernst equation.

E. Electrode Polarization

1. Definition

In the discussion so far it was assumed that a dynamic equilibrium at the electrode–solution interface has been established. Under dynamic equilibrium conditions the rates of charge transfer carried by ions or electrons to and from the electrode surface are equal, so that no net charge transfer and thus no net current occurs at the interface. The interface potential so established is an equilibrium potential. On the other hand, if a current is allowed to flow through the interface, the original equilibrium interface potential is disturbed. The electrode so disturbed is said to be polarized, and the phenomenon is called electrode polarization.

2. Causes of Electrode Polarization

In an electrochemical cell consisting of two electrodes and a solution, current may originate from various sources. In a galvanic cell, electrochemical reactions can occur spontaneously. To an electrolytic cell, electrons can be introduced by an externally applied voltage. In a cell for the determination of ion activities by potentiometry, there is always a small current coming from the terminals of the millivolt meter. Therefore, for practical electrochemical cells the presence of a current and thus the occurrence of electrode polarization are frequently encountered.

The potential of an electrode is determined by the concentration (activity) of charged particles (ions, electrons) at the electrode surface. When a dynamic equilibrium is attained, the concentration of ions or electrons at the electrode surface is constant. Then, the electrode potential should also be constant.

When an electric current flows through the electrode, some electrochemical reactions may occur at the electrode surface. For example, when a current flows through a copper electrode dipped in a solution containing copper ions with the direction that the current flows from the solution to the copper electrode, owing to the discharrge of copper ions, the concentration of copper ions at the electrode surface would be reduced. As a result, the electrode potential would be lower than its equilibrium value. The difference in electrode potential between the two cases is called overpotential. In this example, the overpotential is called concentration overpotential. Concentration polarization is the basis of voltammetry.

When ion deposition or gas evolution occurs during an electrode reaction, because a series of energy barriers must be overcome during the transfer of hydrated ions from the solution to the electrode surface and the conversion of the ions to the final form, some activation energies are required. It is only under conditions in which the electrode potential is more positive (for anodic reaction) or more negative (for cathodic reaction) than the equilibrium potential that the electrode reaction can occur. The phenomenon related to the activation energy is called activation polarization.

During some electrode reactions, an oxide or other substance may be produced at the electrode surface. This results in the production of a resistance and thus an IR drop during the flow of current. The overpotential related to this resistance is called ohmic overpotential.

Under a given experimental condition, the total overpotential is the sum of the three overpotentials. The magnitude of this overpotential is of significance in practical electrochemical measurements, as shall be shown in later sections.

3. Factors Affecting Overpotential

For a given electrode, several factors may affect the magnitude of overpotential (Bockris and Reddy, 1970). Overpotential is closely related to the nature of the electrode material, because for different metals the exchange current density may differ by a factor of several orders of magnitude. Exchange current density is the current density at the electrode–solution interface when the rate of electronation is equal to that of deelectronation. It denotes the rate of electron exchange between the electrode and the solution at an equilibrium state. According to the Butler–Volmer equation, as the exchange current density increases, the overpotential of the electrode decreases.

For a practical electrode, overpotential is determined mainly by the relative magnitude of net current density flowing through the electrode surface and the exchange current density.

If the rate of electrode reaction is not determined by the transfer of electrons at the electrode–solution interface but by the rate of migration of reacting substances to or from the electrode surface, concentration polarization is the dominant factor affecting the overpotential.

4. Some Practical Considerations

In practical electrochemical analyses, sometimes it is desired to make the electrode sufficiently polarized, and sometimes it is necessary that no electrode polarization occur.

In potentiometry, it is necessary to keep the potential of both the indicator electrode and the reference electrode at their equilibrium potential, which can be established only under unpolarized conditions. In conductometry, electrode polarization should also be avoided.

For reference electrodes, the choice and design of the electrode are important. Silver–silver chloride electrodes and calomel electrodes have a high exchange current density. Because exchange current density is dependent on ion concentration, it is usual practice to use a high chloride concentration for these electrodes.

During the passage of current, the current density at the electrode is inversely proportional to the surface area of the electrode. Therefore, it would be desirable to have an electrode with as large an surface area as practical. This is why the silver–silver chloride electrode constructed for the determination of reducing substances in the field and designed by Ding *et al.* (1982) had a very large surface area. The effective surface area of a platinum electrode can be increased by electrolytic deposition of fine grains of platinum on the electrode surface. Some silver–silver chloride electrodes are fabricated by electroplating a layer of silver on platinum or silver wire to increase the surface area.

Under certain conditions, the possible effect of a minute but unavoidable current on the electrode potential cannot be overlooked. For example, when the oxidation–reduction potential of an aerobic soil is determined with a microplatinum electrode, if it is assumed that the surface area of the electrode is 0.01 cm^2 and the current in the measuring circuit is 10^{-11} A, because the exchange current density of platinum is about 10^{-10} A/cm^2, it can be calculated from the Butler–Volmer equation that the overpotential may be as large as 100–200 mV. Thus, the measuring error caused by electrode polarization makes the results meaningless. In this case, the current in the measuring circuit should be kept at 10^{-13} A or less, or an electrode with a larger surface area should be used.

Conversely, in voltammetry it is required that the working electrode be highly polarizable, so that the electrode potential can be easily controlled by externally applied voltage as desired. For this purpose, in addition to the use of working electrodes with a surface area as small as possible so that the current density at the electrode is high, the choice of electrode material is a key point. Because the exchange current density of mercury is extremely low, mercury electrodes are easily polarizable and are suitable as working electrodes in voltammetry (polarography). On the contrary, the overpotential for the evolution of both hydrogen gas and oxygen gas at carbon electrodes is high. Hence these electrodes are suitable as the working electrode for both cathodic reduction and anodic oxidation. This is one of the reasons graphite electrode or glassy carbon electrodes are used for the determination of reducing substances in soils described in Section VI.B.

III. POTENTIOMETRY: DETERMINATION OF SINGLE ION ACTIVITIES

A. Ion-Selective Electrodes

1. Types

An ion-selective electrode is one whose potential selectively responds to a given ion species. According to the kind of electroactive material constructing the electrode membrane, ion-selective electrodes can be classified as glass, solid-state, or liquid-state.

The most commonly used glass electrode is the pH electrode. The chief components of pH-selective glass are SiO_2 and Li_2O or Na_2O. The SiO_2 forms the network of the glass. When the glass membrane is immersed in water, a thin hydrated layer is formed at the membrane surface. In this layer there are mobile lithium ions together with a small amount of hydrogen ions to neutralize the negative charge of immobile silicate radicals. These lithium ions can exchange with hydrogen ions in solution, and at equilibrium an interface potential is established at the membrane surface.

The properties of the glass electrode are determined by the composition of the glass. Some typical glass electrodes are given in Table I. In making

Table I

Composition and Properties of Some Glass Electrodes

Use	Name	Composition[a]	Linear pX range	Remarks
		Low temperature		
pH	Jena 9050	$LBaUS_{30-5-2}$	1–11	$R \sim 50$ MΩ
	ISS[b] 6990	$LCULaS_{30-4.5-2.1-0.5}$	1–11	$R \sim 50$ MΩ
		General purpose		
pH	Corning 015	$NCS_{21.4-6.4}$	1–9.5	$R \sim 100$ MΩ
	Beckman E-2	$LBaS_{25-8}$	1–13	$R \sim 300$ MΩ
	L&N 399	$LCsBaLaS_{28-2-5-2}$	1–13	$R \sim 300$ MΩ
	ISS 5910B	$LCsBaLaNdS_{28-2-4-2-1}$	1–13	$R \sim 100$ MΩ
pNa	Orion 94-11	NAS_{11-18}	1–6	$K_{Na,K}$ 1×10^{-3}
	EIL GEA 33c	$LAS_{26-12.5}$	1–5	$K_{Na,K}$ $10^{-2} - 10^{-3}$
	ISS 7042	$NATlGeS_{20-10-0.5-1}$	1–5	$K_{Na,K}$ 10^{-2}
pK	Beckman 78137	NAS_{27-4}	1–4	$K_{K,Na}$ 0.1, K_{K,NH_4} 0.3

[a] Symbols: L, Li_2O; Ba, BaO; U, UO_2; S, SiO_2; C, CaO; La, La_2O_3; N, Na_2O; Cs, Cs_2O; Nd, Nd_2O_3; A, Al_2O_3; Tl, Tl_2O_3; Ge, GeO_2. Composition in mole percentage; for example, $LCULaS_{30-4.5-2.1-0.5}$ glass is 30% Li_2O, 4.5% CaO, 2.1% UO_2, 0.5% La_2O_3, and 62.9% SiO_2.

[b] Institute of Soil Science.

electrodes, the choice of glass is important. When making spear-shaped or flat-shaped electrodes or when the electrode is to be used at a low temperature, Jena 9050 or ISS 6990 glasses are suitable. However, these glasses have a relatively high sodium error at high pH. Therefore, when making bulb-shaped electrodes for general-purpose use, ISS 5910B, Beckman E-2, and L&N 397 glasses are more suitable.

If part of the silicon in the glass is replaced by aluminum, the electrode will also respond to sodium ions. This is the sodium ion-selective electrode. By properly choosing the composition of the glass, electrodes with a reasonable selectivity to potassium ions over sodium ions (Table I) can also be made.

Solid-state membrane electrodes may be made of single crystals, such as lanthanum fluoride for fluoride ion-selective electrodes, or of polycrystalline fine grains, such as silver chloride–silver sulfide compressed into a disc form for chloride ion-selective electrodes. The essence of the electrode is that the membrane be composed of a compound containing the cation or anion species to be determined and that it have a very low solubility, so that the potential of the electrode is determined by the activity of the relevant ions in solution. Generally, the solubility of the solid membrane determines the detection limit of the electrode.

In liquid-state membrane electrodes, the electroactive material, which is insoluble in water, is dissolved in an organic solvent that is immiscible with water, so that a clear-cut interface between the liquid and the test solution is formed. Two kinds of electroactive materials are currently in use. One is a negatively or positively charged ion exchanger, such as dialkyl phosphoric acid for calcium ion-selective electrodes and tetradodecyl ammonium for nitrate ion-selective electrodes. Another is a neutral carrier, such as valinomycin for potassium ion-selective electrodes. In the former case the exchanger forms a sparingly soluble compound with the ions to be determined, whereas in the latter case the neutral carrier forms a complex with the ions. The detection limit of this kind of electrode is determined primarily by the solubility of the compound or the stability constant of the complex.

2. Construction

Depending on one's purpose, various forms and sizes of ion-selective electrodes can be constructed. For glass electrodes, the most commonly used form is a bulb. For solid-state membrane electrodes, the flat membrane is generally fitted onto one end of a supporting body with an adhesive. Because the liquid-state electroactive substance is mobile, it must be filled in a porous ceramic disc or incorporated with an inert matrix such as PVC to form a membrane.

The interface potential of the electrode must be introduced to the measuring instrument through a lead. Usually, an inner reference electrode is employed for this purpose. Figure 1 shows the typical construction for a miniature liquid-state electrode that can be used to study ion diffusion in soils. The inner reference solution, which is sometimes immobilized by an agar–agar gel, contains an ion species, such as K^+, that is selectively sensed by the electrode, and chloride ions. Because the inner interface potential between the membrane and the inner reference solution and the potential of the Ag–AgCl electrode are constant, any change in the overall electrode potential reflects the change in interface potential between the outer surface of the liquid-state membrane and the test solution.

In some types of solid-state membrane electrodes the inner reference electrode and the inner solution are omitted. For example, when making $AgCl–Ag_2S$ discs one side of the disc is molded with a layer of metallic silver, which can be directly connected to the measuring instrument through a metal lead. Attempts have been made to construct similar connections for glass electrodes and liquid-state membrane electrodes. However, they have not been very successful, because it is difficult to keep the E^0 value of the electrode constant.

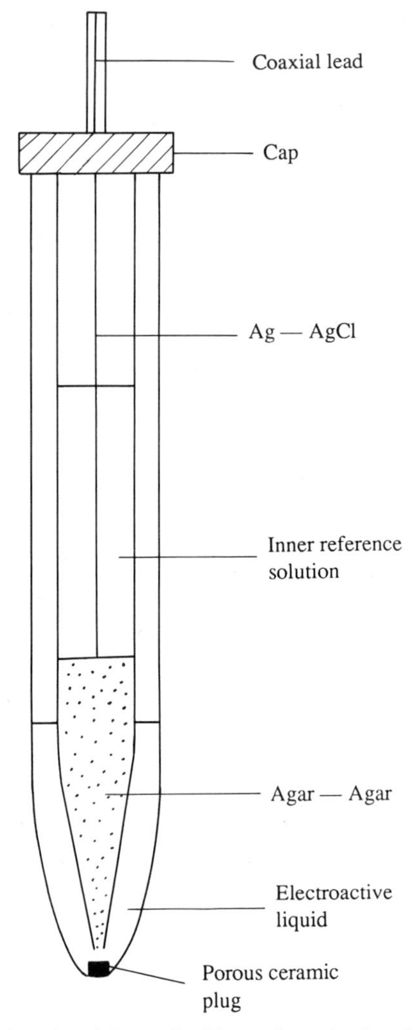

Figure 1. Construction of a miniature liquid-state ion-selective electrode (from Wang and Yu, 1989a, with permission of publisher).

In soil science research, for instance when *in situ* determinations are made, the ruggedness of the electrode is sometimes an important consideration. This is not difficult to achieve for solid-state membrane electrodes. For example, the chloride ion-selective electrodes that are used in the determination of the activity product of NaCl in field soils (discussed in Section V.E) may be made by fabricating an $AgCl–Ag_2S$ pellet into a spear shape, and then fitting it to one end of plastic or glass tubing with the aid of an insulating adhesive. For glass electrodes, the spear-shaped or flat-shaped ones can also be fabricated as desired (Su, 1991).

3. Selectivity

In principle, an ion-selective electrode should selectively respond to the activity of only one ion species. In practice, however, in addition to the primary ions, it may also respond to other ions, although to a lesser extent.

In the presence of an interfering ion species, the electrode potential is related to the two ion species according to the Nicolsky–Eisenman equation,

$$E = E^0 + \frac{RT}{z_A F}\ln(a_A + K_{A,B}\, a_B{}^{z_A/z_B}), \tag{5}$$

where a_A and a_B are the activities of the primary ion and the interfering ion, respectively, and z_A and z_B are their valency. The $K_{A,B}$ parameter is the potentiometric selectivity coefficient, which is a measure of the selectivity of the electrode to ions B relative to ions A. It is not a constant, but varies with experimental conditions and the ratio between the two ions. In Table II, approximate potentiometric selectivity coefficients for two kinds of ion-selective electrodes, each made with two kinds of electroactive materials with respect to some commonly encountered ions in soils, are given for illustration.

Table II

Approximate Potentiometric Selectivity Coefficients of Potassium Ion-Selective and Calcium Ion-Selective Electrodes Made of Different Electroactive Materials

Primary ion	Electroactive material	$K_{I,X}$
	Dibenzo-30-crown 10	Na^+ 10^{-3}
K^+		NH_4^+ 10^{-1}
		Ca^{2+} 10^{-4}
		Mg^{2+} 10^{-4}
	Valinomycin	Na^+ 10^{-4}
K^+		NH_4^+ 10^{-2}
		Ca^{2+} 10^{-4}
		Mg^{2+} 10^{-5}
Ca^{2+}	Dialkyl phosphate (exchanger)	Na^+ 10^{-2}
		K^+ 10^{-2}
		Mg^{2+} 10^{-2}
Ca^{2+}	ETH 1001 (neutral carrier)	Na^+ 10^{-4}
		K^+ 10^{-4}
		Mg^{2+} 10^{-5}

The potentiometric selectivity coefficient is a very important parameter in evaluating, choosing, and using ion-selective electrodes in soil science studies. It may help in judging how large an error would result from interference from other ions. For potassium ion-selective electrodes, for example, although those made of crown compounds are still in use in some laboratories due to the ready availability of the electroactive material and the relatively low resistance of the electrode, in soil science research the interference by ammonium ions may introduce a significant error. By contrast, electrodes made of valinomycin are practically interference-free for most soil measurements. For calcium ion-selective electrodes, those made of ETH 1001 are far superior to those made of dialkyl phosphoric acid, particularly when interferences of hydrogen ions and magnesium ions are of concern. Even for the former type of electrode the interference of hydrogen ions may be a problem for some strongly acid soils of the tropics, because the pCa value may be as high as 5–6, while the pH is generally 4–4.5. In this case, electrodes made of ETH 129, which have a lower $K_{\mathrm{Ca,H}}$ value, may be preferable to use. For sodium ion-selective glass electrodes, the $K_{\mathrm{Na,H}}$ is as high as 250. Therefore, the pH value of the test solution must be adjusted to at least 3 units higher than the expected pNa value.

If the concentration of interfering ions is constant, an electrode with a rather larger potentiometric selectivity coefficient to interfering ions can still be used under some conditions. Ji and Wang (1978) have successfully used a miniature glass electrode for studying the release and diffusion of ammonium ions from granulated ammonium bicarbonate in a neutral soil containing some sodium and potassium ions.

4. Detection Limit

Within a certain pX range, where X is the activity of the ions to be determined, the electrode potential changes linearly with the change in pX in a Nernstian or nearly Nernstian manner. Beyond a certain limit, the straight line curves gradually as the pX is increased, and eventually the electrode potential becomes independent of pX. The intersection point between the extrapolated linear portion of the E–pX curve and the horizontal line is defined as the detection limit of the electrode.

Generally speaking, the lower the solubility of the electroactive material, the lower the detection limit of the electrode. For example, the detection limit of chloride ion-selective electrodes made of AgCl with a K_{sp} of 1.6×10^{-10} is about $5 \times 10^{-5}\ M\ \mathrm{Cl}^-$. On the other hand, electrodes made of $\mathrm{Hg_2Cl_2}$ with a K_{sp} of 2×10^{-18} have a detection limit of $2 \times 10^{-7}\ M\ \mathrm{Cl}^-$. For a given kind of electrode, other factors, such as the surface conditions of the membrane, may also affect the detection limit.

This is the reason the electrode surface must be cleaned frequently, particularly when it is used in soil suspensions where clogging by colloidal particles often occurs.

Quite often the detection limit of an electrode is much higher than the expected value as calculated from the solubility of the electroactive material. For example, according to the solubility product of Ag_2S (pK_{sp} 51), the detection limit of a sulfide ion selective electrode should be about a pS^{2-} of 17. In practice, however, the lowest linear range of the electrode is usually about a pS^{2-} of 6. This occurs because it is very difficult to prepare a calibration solution with a low ion activity, due to contamination by reagent chemicals and water, adsorption on vessels, and the effect of atmospheric air, etc. A profitable way is to use pX buffer solutions, similar in principle to pH buffer solutions, for calibration. Actually some authors have found that, using this technique, the lowest linear range of calcium ion-selective electrodes was about a pCa of 8, in contrast to the generally observed value of a pCa of 5 when calibrated with $CaCl_2$ solutions. This has important practical implications in soil science research. Owing to adsorption–desorption and complexation–dissociation reactions, a soil acts as a buffer system with respect to many ions. Therefore, in practical determinations a measured pX value corresponding to one or more units larger than the lowest value on the straight line of the conventional calibration curve does not necessarily mean there is an error, provided that the electrode has attained a true steady potential. This extrapolation method has been applied to determinations of pS^{2-} in paddy soils (Pan *et al.*, 1982) and pCa in soils (Yu *et al.*, 1989a,b).

5. Response Time and Potential Stability

Strictly speaking, the response time of an electrode should be the time required for the attainment of an equilibrium and thus a steady electrode potential. However, this time interval is frequently very long. Therefore, for practical purposes, the IUPAC has recommended that the response time be defined as the time interval from the moment of contact of the electrode with the test solution, or the moment of a sudden change in concentration, to the moment when the potential deviates from the steady potential by 1 mV.

In the course of the establishment of an equilibrium electrode potential, two main processes are involved. The ions in the solution must migrate to the electrode surface. Then, a new chemical equilibrium would be established at the electrode surface through ion exchange, dissociation of the electrode material, or combination of the electroactive sites with the ions. Therefore, in addition to the kind of electroactive material, a variety of factors can affect the response time of an electrode. Generally speaking, the response time of glass electrodes is short, whereas that of liquid-state membrane electrodes is comparatively long.

Ordinarily, the response time of most ion-selective electrodes, which take 0.5–5 min, is sufficiently short to meet the requirements of studies in soil science. However, two extreme cases may cause some problems. Because the lower the ion activity the longer the response time, it may require a time interval of up to an hour to get a nearly steady electrode potential, if the ion activity is very low. An example is when sulfide ions are determined in reduced soils where the pS^{2-} value may be as high as 15 (Pan, 1985). Similarly, when the pH of upland soils is determined *in situ*, because the soil is unsaturated with water, a rather long time interval is needed to get a reliable pH reading (Cang and Yu, 1981). Another case is when the electrode is used to monitor fast chemical reactions, such as the rate of cation exchange on clay minerals (Malcolm and Kennedy, 1969). Here, the response time of the electrode may be critical for getting reliable results.

The potential of glass electrodes and solid-state membrane electrodes is quite stable, if the electrode surface does not contain contaminants. For liquid-state membrane electrodes, a slow drift in potential may occur, due to the gradual change in composition of the electrode surface caused by its dissolution in water. Therefore, frequent calibration is required. It is advisable to obtain fresh electrode surface by some mechanical means after a certain period of usage so that the electrode potential can be restored to its original value. This is particularly important in soil science research.

B. Measuring Circuit

The measuring circuit for the determination of single ion activities consists of four parts: the test solution, an indicator (ion-selective) electrode, a reference electrode, and a high-input millivolt meter. A simplified circuit is shown in Fig. 2. In the circuit, two important parameters, E and R, are of concern.

In the measuring circuit, there are a number of potentials connected in series. The E_{mo} is the interface potential at the outer surface of the electrode membrane and is a function of the activity of relevant ions in the test solution. The E_{mi} is the interface potential at the inner surface of the electrode membrane. The E_i is the potential of the inner reference electrode of the ion selective electrode. The E_r is the potential of the reference electrode. Generally, E_{mi}, E_i, and E_r are considered constant, if the temperature does not change. Usually E_D, a diffusion potential or liquid-

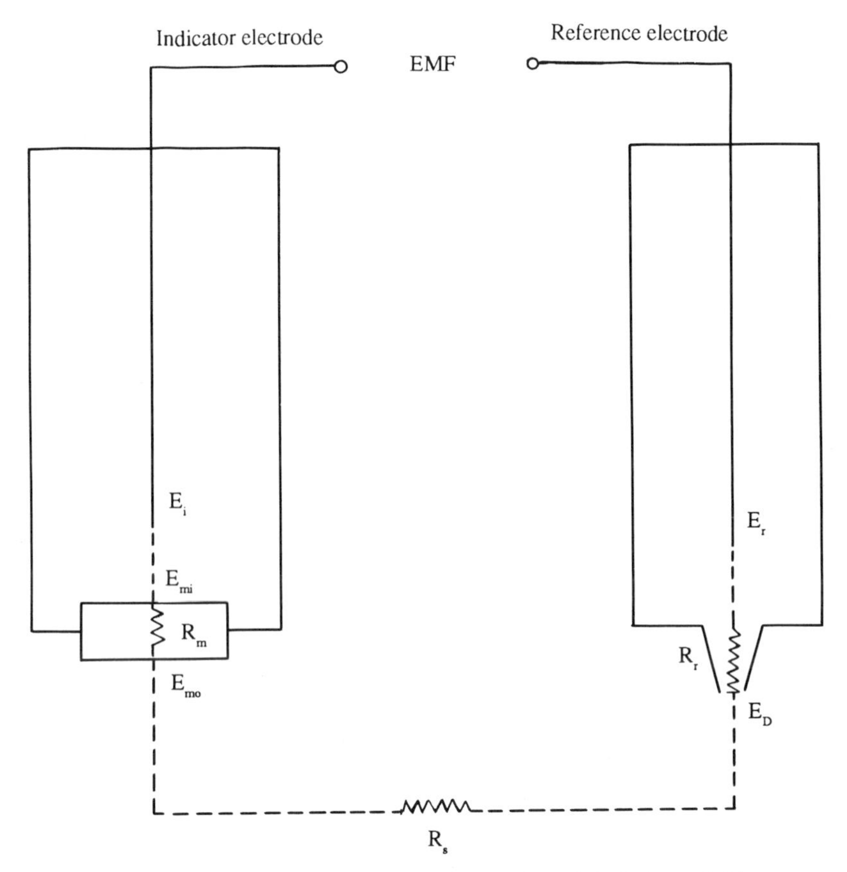

Figure 2. Measuring circuit for the determination of single ion activities.

junction potential at the interface between the salt bridge of the reference electrode and the test solution, is assumed to be constant and small in magnitude. Thus, the measured changes in potential difference between the two electrodes, or the EMF of the whole cell, would only reflect changes in E_{mo}.

An essential requirement in potentiometric measurements is that the potential difference between the indicator electrode and the reference electrode is wholly input to the millivolt meter. In actual cases, however, the situation is not so simple. Most of the problems are associated with the high resistance of the electrode membrane, R_{m}. For glass electrodes, R_{m} is of the order of $10^8\ \Omega$ and for the miniature type, $10^9\ \Omega$. Occasionally, there may be a high resistance at the tip of the reference electrode. When the pH of acid upland soils is determined *in situ*, the resistance R_{s} may also be very high. Therefore, in order to have a precision of 0.1% in potential measurements, the input resistance of the millivolt meter should be at least 10^{12} Ω.

During measurements, there is always a small but changing current flowing through the circuit. Because the inner reference electrode of the ion-selective electrode and the reference electrode are of the nonpolarizable type, i.e., silver–silver chloride electrode and/or calomel electrode, electrode polarization is not a problem. However, the IR drop on the high resistances may cause a deviation in the measured potential from the true E_{mo} value. For example, suppose that R_m is 500 MΩ and the current is 10^{-11} A. Then, the IR drop would be 5 mV. Because the resistances and the current may change with experimental conditions, such as temperature, a measuring error may result. Therefore, when potentiometric measurements are made using electrodes of high resistance, a millivolt meter with a small input current should be used. Moreover, because the resistance of the electrode is high, adequate insulation and proper shielding in the circuit should also be provided.

C. Measurements in Soil Systems

1. Potentiometry as an Analytical Technique

Most of the research conducted on the application of potentiometry to soil science has involved the determination of ion concentrations. Cations that have been determined include H, Na, K, NH_4, Mg, Ca, Al, Cu, and Cd, and anions include NO_3, NO_2, F, Cl, Br, CN, S, and SO_4. This research integrates the principles of soil chemistry and analytical chemistry. A review on this subject has been given elsewhere (Yu, 1985). In this chapter, only some general considerations will be discussed.

The first step in the determination of ion concentrations is to extract the ions from the soil into a solution phase, or to convert an element from a combined form into an ionic form and then to extract it. If the ions are readily soluble, such as nitrate in soils of temperate regions, they can be extracted with water. However, more often an extracting solution must be employed. In this respect, some important points should be considered. The extractant should be highly efficient in extracting the ions to be determined. Thus, for cations, extractants composed of polyvalent cations generally have a stronger extracting power than those composed of monovalent cations. The ions constituting the extractant must have a small potentiometric selectivity coefficient vis á vis the ion-selective electrode. For example, because the $K_{NO_3,X}$ value of the nitrate ion-selective electrode is about 10^{-2} for Cl^- and about 10^{-4} for SO_4^{2-}, apparently sulfate is a better extractant than chloride. Sometimes the impurities in the reagent may be a problem. For instance, for the extraction of exchangeable potassium, calcium chloride can be an ideal extractant, because calcium is a common ion species in soils, it has a strong replacing power, and it will enable one to obtain a clear solution that is desirable for centrifugation or filtration. However, because the concentration of the extracting solution is as high as 0.5 or 1 *M*, it is frequently found that the content of potassium in commercially available reagent $CaCl_2$ is so high that the lower limit of the linear portion of the calibration curve is only about pK 4. This makes it difficult to determine potassium in soils with a low exchangeable potassium content. This is why Ji (1991) chose 0.5 *M* barium chloride solution as the extractant for potassium. Many determinations using ion-selective electrodes may be affected by interferences from hydrogen or hydroxyl ions. Therefore, it would be advisable to keep the extracting solution buffered within a certain pH range. For example, an aluminum sulfate solution buffered at pH 3 with H_2SO_4 may be a suitable extractant for nitrate. Sometimes another substance may be added to the extracting solution to eliminate the interference from other ions. Examples are sulfanilamide or sulfamic acid for nitrite interference in nitrate determinations (Mahendrappa, 1969), silver sulfate for chloride interference in nitrate determinations (Onken and Sunderman, 1970), and sodium citrate for aluminum and iron interferences in fluoride determinations.

After extraction of the ions into solution, several techniques can be utilized to determine the ion concentration. The simplest one is to match the measured electrode potential with those in a series of standard solutions. Because the electrode responds to ion activity rather than to ion concentration, and in most cases one wants to know the ion concentration, a correction factor using the Debye–Hückel equation is made. A more accurate technique is the known addition method, in which a known volume of a standard solution is added to the test solution. The unknown concentration can be calculated from the difference in electrode potential before and after the addition of the standard solution, using an equation derived from the Nernst equation, if the potential response slope, S, of the electrode is known. If it is not known, a third electrode potential is measured after a twofold dilution with water, and a corresponding equation can be utilized for the calculation. If the volume of the test solution is limited, an alternate technique in which the test solution is added to a standard solution can be applied. If a still more accurate result is required, the potentiometric titration technique is a good choice. In this technique, the ions to be determined are titrated with a reagent that can react with these ions through precipitation, neutralization, complexation, oxidation–reduction, etc., and the change in electrode potential is monitored to find the equivalence point. Because in this case what is of concern is the relative magnitude in the change of electrode potential during the titration and the accuracy is determined chiefly by the measurement of volume, many interfering effects can be eliminated.

For some ions that cannot be measured using an ion-selective electrode, an indirect technique, in which the test solution is titrated with a reagent that is selectively sensed by the electrode, can be employed. For example, sulfate can be titrated with a barium chloride solution or a lead nitrate solution, using a barium ion-selective electrode or a lead ion-selective electrode, respectively. Likewise, aluminum ions can be titrated with sodium fluoride.

Ammonium ions cannot be determined with an ion-selective electrode directly, because with either the glass-type electrode or the liquid-state-membrane-type electrode the $K_{NH_4,X}$ values to some other ions, particularly to potassium ions, are too high. However, ammonium ions can be easily converted to ammonia by adjusting the solution to a pH of >11. The ammonia diffuses into a thin layer of ammonium chloride solution through a gas-permeable membrane. The pH of the ammonium chloride solution is a function of the amount of dissolved ammonia and can be measured with a pH electrode. Such a device, called an ammonia sensor, has been widely used in soil analyses, because the detection limit can reach 10^{-6} M ammonium, and the determination is almost interference-free in soils.

2. Studies on Interactions between Ions and Soil Particles

Because of the coulombic force between cations and negatively charged soil particles, only a portion of the cations adsorbed by soil can dissociate in solution. These dissociated cations are present in both the diffuse layer and the free solution. Any attempt to separate this portion of cations from those adsorbed by the soil and then to determine them by chemical methods would invariably lead to a disturbance of the original chemical equilibrium. On the other hand, ion-selective electrodes, which can be inserted in the soil paste directly, just respond to the activity of cations in the solution phase. This is the unique feature of using ion-selective electrodes in soil chemistry studies. C. E. Marshall pioneered this research field.

Marshall determined ion activities with clay membrane electrodes. The membrane is prepared (Marshall, 1942) by evaporating a suspension of the colloidal fraction of a suitable clay, such as bentonite or beidellite, to a membrane of 0.2–0.3 mm thickness. Then, it is subjected to a high temperature. It was empirically found that membranes heated to 623 K responded to mono- and divalent cations, whereas those heated to 763 K responded only to monovalent cations. Modern theories of ion-selective electrodes would suggest that this type of membrane is poor from both selectivity and measuring range considerations. However, Marshall and co-workers conducted a series of fruitful studies with such electrodes in the 1940s and the 1950s. Their results have been summarized (Marshall, 1953, 1964). Many subsequent studies using modern ion-selective electrodes (Yu, 1985) have yielded results that qualitatively supported some of the conclusions drawn by Marshall and his co-workers. That is, the dissociation of cations is closely related to the kind of soil or clay due to differences in surface charge properties and geometrical configuration. For a given adsorbent, dissociation is strongly dependent on the valence of the cations, being generally less than 5% for divalent cations and 20–35% for monovalent cations. For a given cation species, the dissociation usually increases with the percentage cation saturation and the water content of the system; accompanying ions may affect the dissociation through competitive adsorption or through other interactions with the cations that are being studied.

The behavior of anions in soils has not been extensively studied using ion-selective electrodes. One main reason for this may be that most of the research on soils assumed that soils were predominantly negatively charged, and thus did not appreciably adsorb anions. In recent years, there has been great interest in the theoretical and practical aspects of positive charge in soils. A limited number of studies suggest that the interactions of these soils with some anions are not as simple as was previously thought. It is generally felt that chloride and nitrate are nonspecifically adsorbed anions and that adsorption of these anions on soils involves only coulombic forces. However, with the use of ion-selective electrodes, Wang *et al.* (1987) observed that variable charge soils adsorbed more chloride ions than nitrate ions. This stronger adsorption of chloride made the soil less positive in electrokinetic potential. They suggested that, in addition to coulombic attraction, a specific force was involved in the adsorption of chloride ions. Unpublished data of Hu and Yu showed that during chloride adsorption the soil released more hydroxyl ions than during nitrate adsorption. After adsorption, the soil became more negative as revealed by more adsorption of sodium ions in the former case than in the latter case. Wang and Yu (1989a,b,c), studying the diffusion of ions with miniature ion-selective electrodes, were able to determine the effects of pH, quantity of diffusing source, accompanying ions, and addition of ferric oxides on the difference in diffusion coefficient between chloride ions and nitrate ions in terms of the interactions of these anions with the surface charge of the soil during their diffusion. They assumed a stronger specific force for chloride ions than for nitrate ions. It appears that mechanisms for anion reactions with soils may be elucidated using ion-selective electrodes.

In determining ion activities in soils using ion-selective electrodes, a theoretical question still remains to be clarified: how does the electrode respond to ions in various parts of the electrical double layer surrounding the soil particles and those in the intermicellar solution? A related practical question is do the measured ion activities in a soil suspension differ from those in the free solution? According to the concept of Marshall (1964), the measured ion activities represent a geometrical space average, assuming an equal mobility for the ions in the diffuse layer and in the intermicellar solution. The explanation of the "suspension effect" by Pallman (1930) and Wiegner (Wiegner and Pallman, 1930), which has largely been replaced by the liquid-junction potential theory discussed in Section IV, can still be supported by some recent experimental data. Perhaps more elaborate research will aid in the elucidation of this complicated question.

3. Problems Associated with *In Situ* Determinations

Because ion activities represent the effective concentration of ions in a soil system under dynamic conditions and are of significance in relating soil properties to plant growth, it is desirable to determine them *in situ*. However, because under most field conditions the pores of the soil are filled with both water and air, while the prerequisite for an ion-selective electrode to faithfully respond to ion activities is that the whole electrode membrane must be in contact with the solution, this may create some problems for *in situ* determinations. Some authors, (Chapman *et al.*, 1940; Huberty and Haas, 1940; Haas, 1941; Bailey, 1943), in attempting to determine soil pH *in situ* with glass electrodes in the early 1940s, encountered these problems. In recent years, Cang and Yu (1981), using a glass electrode, reexamined this question and gave it an interpretation in terms of a diffusion process.

The situation is illustrated in Fig. 3. When the water content of the soil is insufficiently high, the electrode surface will contact the soil solution only at AA′ and BB′ where the interface potentials are all equal to E_1. At the point A′B′, the interface potential will be E_2, which reflects the hydrogen ion activity of the water used in washing the electrode prior to ion determination. Therefore, as a whole, the electrode potential will be a mixed potential representing a geometric average of numerous E_1 and E_2 values.

However, the mixed potential arising from the above microscopic state cannot remain constant. Because the hydrogen ion activity at AA′ or BB′

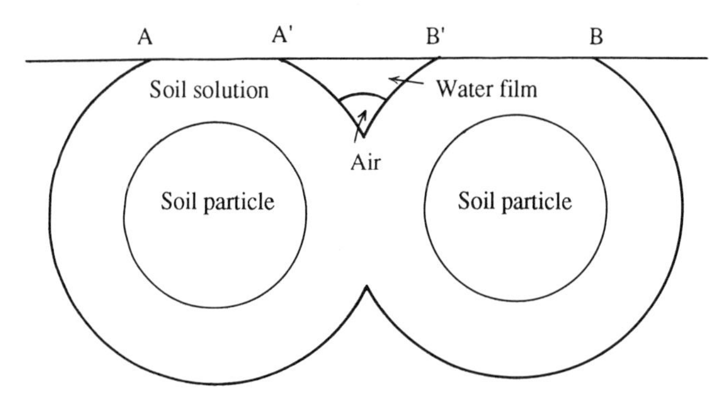

Figure 3. Diagram illustrating the contact between glass electrode membrane and water-unsaturated soil (from Yu, 1985, with permission of publisher).

is different from that at A′B′, the ions will diffuse between the two solutions until all the water films contacting the electrode membrane have the same composition representing the soil solution. This results in an equilibrium electrode potential to yield the real pH of the soil. Thus, in principle, the determination of soil pH *in situ*, even for soils unsaturated with water, is possible. Of course, the time required for the attainment of an equilibrium electrode potential is dependent on a variety of factors, such as the ratio of the volume of the soil solution to that of the soil air, the size of the pores, and the pH of the soil. With care, the determination of soil pH can be made within 3–5 min after insertion of the glass electrode into the soil with an error less than 0.2 pH unit for soils with water contents commonly encountered in the field.

For other types of glass electrodes and solid-state membrane electrodes, the situation is similar. Zhang *et al.* (1979) found that under ordinary field conditions a steady electrode potential can be obtained 2–10 min after inserting a spear-shaped sodium ion-selective electrode or a chloride ion-selective electrode into the soil, with the required time being dependent on the water content and the NaCl content of the soil. Gentle swirling of the electrode during insertion could shorten the required time to obtain a steady electrode potential.

If the soil is very dry, an alternative is to add several drops of water around the electrode, and then to wait for 2–3 min before measurement. In this case, the soil moisture condition around the electrode surface would be similar to that of a drained soil after rainfall.

Repeated determinations in soils may lead to the contamination of the electrode surface or the clogging of the micropores of some solid-state membrane electrodes, causing increased time for the establishment of the equilibrium electrode potential. Therefore, the electrode surface must be rubbed frequently with tissue paper or fine emery paper.

IV. PROBLEMS WITH LIQUID-JUNCTION POTENTIAL

In the measuring circuit for the determination of single ion activities shown in Fig. 2, it is generally assumed that the interface potential E_D (E_j) at the liquid-junction between the salt bridge solution of the reference electrode and the test solution, called the diffusion potential or liquid-junction potential, is low and constant in magnitude and thus can be neglected. This assumption is based on the consideration that for the salt bridge solution, KCl, the transference numbers of K^+ and Cl^- ions are nearly equal, and therefore at both sides of the interface there is little surplus negative or positive charge. In actual cases, however, this assumption may not be strictly true, because interface potential is an intensity parameter not a capacity parameter, and a minor excess of electric charge on one side of the interface may cause an interface potential. In particular, if the mobilities of the cations and the anions in the test solution differ markedly, for instance when H^+ or OH^- ions are involved, the counter-diffusion of these ions to the KCl solution may cause a significant liquid-junction potential.

The situation becomes more complicated if the salt bridge is in contact with a colloidal system, such as a soil suspension. Here the charged clay particles may affect the mobilities of K^+ and Cl^- ions, and consequently a liquid-junction potential would result. Jenny *et al.* (1950) first noticed this phenomenon, and subsequent work by other researchers (Bower, 1961; Tschapek, 1961; Xuan and Yu, 1964; Clark, 1966; Panicolau, 1970; Olsen and Robbins, 1971; Peinemann and Helmy, 1973) supported the observations of Jenny *et al.* (1950). Overbeek (1953, 1956) discussed this theoretically.

The liquid-junction potential is dependent on a variety of factors, including the composition and concentration of the salt bridge solution, the electric charge carried by the exchanger particles, and the exchangeable cation status of the solid. In Table III, the effects of the charge properties of the exchanger material and the kind of exchangeable ions on the sign and the magnitude of the liquid-junction potential are shown. It can be seen that the behavior of aluminum ions is quite peculiar. For kaolinite

Table III

Liquid-Junction Potential between Saturated KCl Solution and Exchangers[a]

Exchanger	Exchangeable ion	E_j (mV)
Montmorillonite	Na	−28.3
	H	−25.9
	Ca	−6.5
	Al	+2.4
Cation exchange resin (sulfonic)	H	−149.3
	Al	+7.1
Cation exchange resin (carbonic)	H	−43.1
Anion exchange resin	Cl	+51.3

[a] From Xuan and Xu (1964) by courtesy of the publisher.

with a low charge density, the liquid-junction potential is quite low. For acid soils, the liquid-junction potential may amount to 20 mV in certain cases. Thus, an error of as large as 0.3 pX unit may be introduced by the liquid-junction potential if it is neglected.

It is not necessary for there to be contact between the salt bridge and the charged soil particles directly to induce such a liquid-junction potential. Ji and Yu (1985) found that soil particles could affect the salt bridge even if the latter was several millimeters from the soil particles. The liquid-junction potential developed in two stages. A large part of the potential developed within 1–2 s. Then, there was a slow potential drift until a steady potential was reached. When the salt bridge was moved quickly to a distance beyond the effect of the force field exerted by the soil, there was also a slow lag in potential dissipation following an abrupt fall in potential. If the steady liquid-junction potential was plotted against the distance between the surface of the salt bridge and the soil, a potential profile shown in Fig. 4 could be obtained. From Fig. 4 it can be seen that for the Lantosol soil that carried a net positive charge, the sign of the liquid-junction potential was always positive, whereas for the Yellow-brown soil carrying a net negative charge the sign of the potential was always negative. When the soil was mixed with pure silica powder in different proportions to decrease the volume charge density of the soil mass, the potential decreased with the increase in the proportion of silica mixed. The cause of this distance effect is not exactly known.

From the previous discussion, it appears that the liquid-junction potential is a troublesome and complicated problem in potentiometry involving

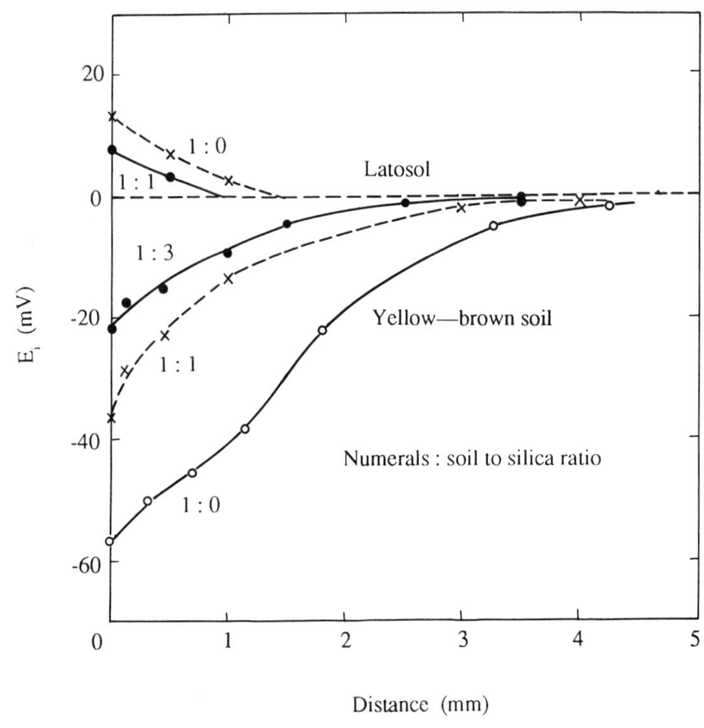

Figure 4. Liquid-junction potential as a function of the distance between salt bridge and two soils with different electric charge properties (from Ji and Yu, 1985, with permission of publisher).

the use of a reference electrode with a salt bridge, particularly in soil science studies. Neglecting this effect may result in measuring errors. Because this potential is time-dependent and is affected by a variety of factors, no method is available to eliminate it entirely. In order to reduce its effect to a minimum, for soil suspensions that can be separated into a sediment phase and a supernatant solution phase it is advisable to put the salt bridge of the reference electrode in the supernatant solution and the ion-selective electrode in the sediment. This is the practice generally followed in soil pH determinations. If the two phases cannot be separated, the salt bridge can be connected to the system through a strip of filter paper previously wetted by the soil solution. In field measurements, several layers of filter paper may be put on the surface of the soil to allow them to be wetted by the soil solution. Then one can place the salt bridge on the paper, if the soil is sufficiently wet.

V. POTENTIOMETRY: DETERMINATION OF ACTIVITIES OF TWO ION SPECIES OR MOLECULES

A. Merits

The determination of an activity ratio for two ion species of the same charge or of an activity product for two ion species of opposite charges using potentiometric techniques has several merits. In principle, from thermodynamic considerations, individual ionic activities are not definable quantities, but the activities of salts are definable functions of partial molar free energies of salt molecules. Thus, determinations in which the difference in potential between two ion-selective electrodes is directly measured avoid troublesome problems associated with the liquid-junction potential. In practice, for soil systems containing more than one ion species, it is frequently desirable to determine the relationship between two ion species. Moreover, because some factors such as temperature, water-to-soil ratio, and salt content may affect the behavior of two ion species in a similar manner, the measured results are less affected by experimental conditions.

In the early literature, some attempts were made to determine the activity ratio or product, such as $a_H/(a_{Al})^{1/3}$ (Russell and Cox, 1950), $a_H \cdot a_{Cl}$ (Schofield and Taylor, 1955a; Xuan and Yu, 1964), or $a_{Na} \cdot a_{Cl}$ (Baker and Low, 1970), in soil or clay systems, using old-fashioned electrodes. In recent years, with the advancement of modern ion-selective electrodes, this research field has greatly advanced. In the following discussions, some examples are given to illustrate the underlying principles and applications of ion activity determinations for two ions or molecules.

B. Potassium–Calcium Activity Ratio

The interrelationships between potassium and calcium, as affected by soil surfaces, have been investigated by a number of soil chemists. Beckett (1964), based on Schofield's ratio law, proposed the use of the activity ratio $a_K/(a_{Ca})^{1/2}$ as a measure of the availability of soil potassium (Sparks and Huang, 1985). Some soil scientists (Kysnezov and Snakin, 1987) determined potassium activities and calcium activities separately with respective ion-selective electrodes, and then calculated the pK − 0.5pCa value. Wang *et al.* (1988) determined the potassium Q/I relationship in soils with a potassium ion-selective electrode and a calcium ion-selective electrode.

Yu *et al.* (1989a) devised a method for the direct determination of the pK − 0.5pCa value.

According to the Nernst equation, the measured electrode potentials E_{Ca} and E_K are

$$E_{Ca} = E_{Ca}{}^0 + S_{Ca}\log a_{Ca} + E_j \tag{6}$$

and

$$E_K = E_K{}^0 + S_K\log a_K + E_j \tag{7}$$

where $E_{Ca}{}^0$ and $E_K{}^0$ are "standard potentials", S_{Ca} and S_K are the slopes of the respective E–$\log a$ relation, and E_j is the liquid-junction potential. It follows that

$$\begin{aligned} E_K - E_{Ca} &= (E_K{}^0 - E_{Ca}{}^0) + S_K\log a_K - S_{Ca}\log a_{Ca} \\ &= (E_K{}^0 - E_{Ca}{}^0) - S_K\mathrm{pK} + S_{Ca}\mathrm{pCa}. \end{aligned} \tag{8}$$

After rearrangement, the pK–0.5pCa would be

$$\mathrm{pK} - 0.5\mathrm{pCa} = \mathrm{pK}(1 - S_K/2S_{Ca}) - [(E_K - E_{Ca}) - (E_K{}^0 - E_{Ca}{}^0)]/2S_{Ca} \tag{9}$$

or

$$\begin{aligned} \mathrm{pK} - 0.5\mathrm{pCa} = 0.5\mathrm{pCa}(2S_{Ca}/S_K - 1) - [(E_K - E_{Ca}) \\ - (E_K{}^0 - E_{Ca}{}^0)]/S_K. \end{aligned} \tag{10}$$

In Eqs. (9) and (10), $(E_K - E_{Ca})$ is the potential difference between the potassium ion-selective electrode and the calcium ion-selective electrode, and it can be measured. If conditions are chosen so that slight deviations of $2S_{Ca}/S_K$ from unity do not affect the measured results significantly so that the first term of the right side of Eqs. (9) and (10) can be neglected, the pK − 0.5pCa can be directly calculated from the measured $E_K - E_{Ca}$ value and the known value of S_K or S_{Ca} after calibration with a standard mixed solution to find the $E_K{}^0 - E_{Ca}{}^0$ value. Thus, only one potential measurement is required for a test system.

Yu *et al.* (1989a) presented data showing that the difference in potential between the two electrodes measured in a soil paste was identical to that measured in the supernatant clear solution. Within a certain concentration range neutral salts did not affect the pK − 0.5pCa value. The pK − 0.5pCa values were nearly constant within a certain range of water to soil ratio, although the pK values varied greatly (Table IV). They interpreted these results in terms of Donnan equilibrium theory.

This technique has been applied to the study of the interactions of potassium and calcium ions with soils (Yu *et al.*, 1989b).

Table IV

Effect of Water-to-Soil Ratio on pK and pK−0.5pCa in Some Soils[a]

	pK					pK−0.5pCa				
Water:soil	A15[b]	A24	A25	A26	H45	A15	A24	A25	A26	H45
0.5:1	3.38	4.65	4.57	4.34	4.39	2.21	3.27	3.25	3.09	3.03
1:1	3.51	4.68	4.71	4.48	4.52	2.25	3.33	3.25	3.11	3.08
2:1	3.62	4.93	4.86	4.59	4.67	2.28	3.34	3.30	3.11	3.08
4:1	3.75	5.05	4.96	4.71	4.79	2.33	3.37	3.30	3.14	3.08
8:1	3.75	5.17	5.12	4.84	4.88	2.33	3.39	3.33	3.15	3.10

[a] From Yu *et al.* (1989a) by courtesy of the publisher.
[b] Soil sample number.

C. Lime Potential

The concept of lime potential was introduced by Aslyng (1954) and Schofield and Taylor (1955b). Lime potential is defined as pH−0.5pCa. It is related to the chemical potential of lime as

$$\frac{(a_{\mathrm{Ca}})^{1/2}}{a_{\mathrm{H}}} = \frac{(a_{\mathrm{Ca}})^{1/2} \cdot a_{\mathrm{OH}}}{a_{\mathrm{H}} \cdot a_{\mathrm{OH}}}. \tag{11}$$

At 293 K,

$$\begin{aligned} \mathrm{pH} - 0.5\mathrm{pCa} &= 14.2 - 0.5\mathrm{pCa} - \mathrm{pOH} \\ &= 0.5 \log a_{\mathrm{Ca(OH)_2}} + 14.2. \end{aligned} \tag{12}$$

Thus, pH − 0.5pCa is a single function of the chemical potential of lime, $Ca(OH)_2$.

Ordinarily, lime potential is determined by measuring the pH and analyzing for the calcium concentration and then correcting for the activity coefficient of the Ca^{2+} ions. Some authors (Yu, 1985) determined the lime potential by measuring the pH and pCa separately with respective ion-selective electrodes. Wang and Yu (1981) devised a method for the direct determination of lime potential with two ion-selective electrodes.

For pH − 0.5pCa, an equation similar to that for pK − 0.5pCa can be derived:

$$\mathrm{pH} - 0.5\mathrm{pCa} = \frac{(E_{\mathrm{Ca}} - E_{\mathrm{H}}) - (E_{\mathrm{Ca}}{}^{0} - E_{\mathrm{H}}{}^{0})}{S_{\mathrm{H}}} + 0.5\mathrm{pCa}\left(\frac{2S_{\mathrm{Ca}}}{S_{\mathrm{H}}} - 1\right) \tag{13}$$

In this case, because it is difficult to prepare a standard solution with known pH and pCa, it is necessary to find the $E_{\mathrm{Ca}}{}^{0}$ and $E_{\mathrm{H}}{}^{0}$ values separately. A simple way is to measure the electrode potential in a series of solutions with known pH or pCa, and then to extrapolate to pH = 0 or pCa = 0 graphically or mathematically.

Because the pH − 0.5pCa is practically independent of the soil-to-water ratio within a wide ratio range, in field measurements it is convenient to simply prepare a soil paste. Then, one can measure the difference in potential between a calcium ion-selective electrode and a flat-shaped glass pH electrode.

Wang and Yu (1981, 1983, 1986) conducted a series of studies on lime potential of soils and concluded that lime potential is a more distinct and meaningful index than pH for characterizing the acidity status of soils, because it is determined by both the pH and the exchangeable calcium status of a soil, which are closely interrelated. An example of the variation of lime potential with depth for agricultural soils is shown in Fig. 5. Wang and Yu suggested that three ranges of lime potential could be distinguished for characterizing soil acidity, namely 1.5–3.0, 3.0–5.0, and 5.0–7.0. These ranges correspond to the pH ranges of 4.0–5.2, 5.2–7.0 and 7.0–8.2, respectively.

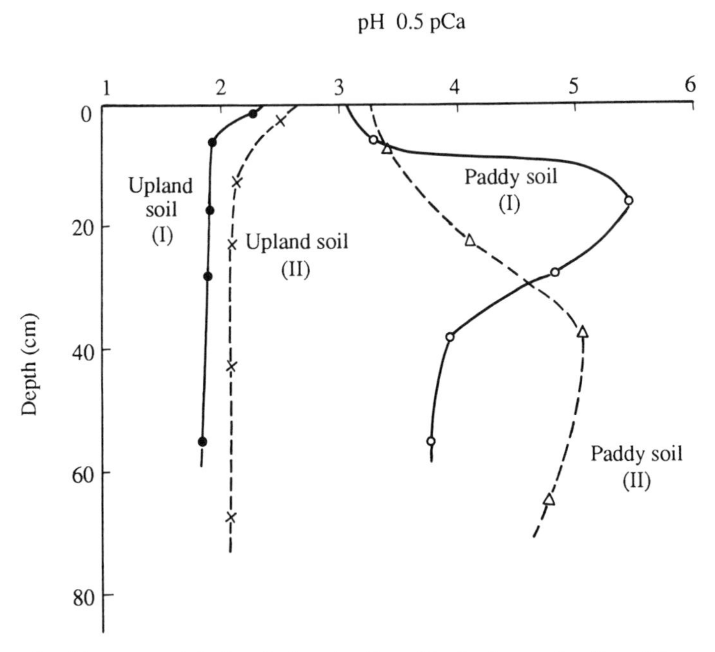

Figure 5. Lime potential of agricultural soil profiles in the tropics (from Wang and Yu, 1983, with permission of publisher).

D. Chloride–Nitrate Activity Ratio

Nitrate and chloride are two anion species commonly found in soils. Their chemical properties are similar in some aspects but different in others. In order to study the interactions of these ions with variable charge soils, Wang *et al.* (1986, 1987) used a chloride ion-selective electrode and a nitrate ion-selective electrode placed in the soil suspension so that the activity ratio a_{Cl}/a_{NO_3} could be determined directly. They found that such factors as the iron oxide content of the soil, the pH of the suspension, the concentration of the respective anion, the kind of accompanying cations, and the dielectric constant of the medium could affect the ratio of the two adsorbed anions. An example of the effect of pH on the activity ratio in three variable charge soils is shown in Fig. 6. Wang *et al.* (1987) interpreted these results in terms of the difference in behavior of the two ion species when they react with soil surfaces. They suggested that, in addition to the coulombic force, a covalent force between the anions, particularly Cl^-, and the metal atom on the surface of soil particles is involved.

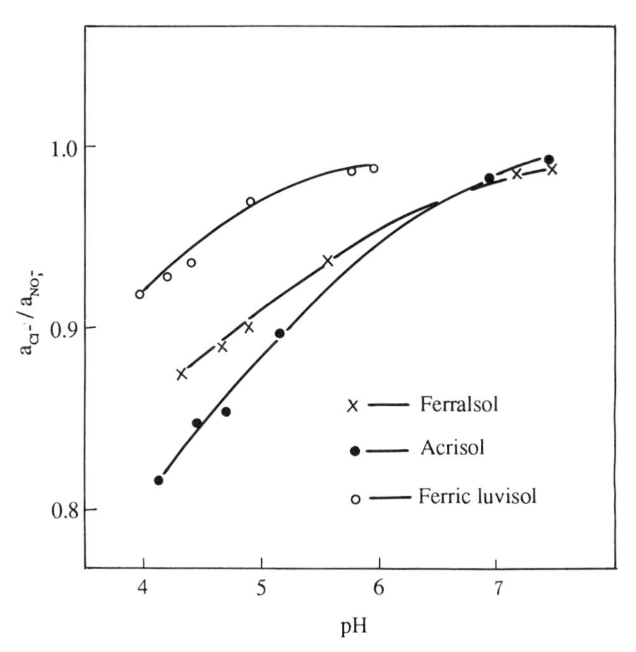

Figure 6. Effect of pH on chloride–nitrate activity ratio after reaction of equal concentrations of the two ions with variable charge soils (from Wang *et al.*, 1987, with permission of publisher).

E. Mean Activity of NaCl

For a solution containing sodium chloride, the difference in potential between a sodium ion-selective electrode and a chloride ion-selective electrode would be

$$E = (E_{Na}{}^{0} - E_{Cl}{}^{0}) + \frac{RT}{F} \ln(a_{Na} \cdot a_{Cl}). \tag{14}$$

By definition, the mean activity of NaCl is

$$a_{NaCl} = (a_{Na} \cdot a_{Cl})^{1/2} \tag{15}$$

or

$$\text{pNaCl} = 0.5(\text{pNa} + \text{pCl}). \tag{16}$$

Thus, in principle, the mean activity of NaCl can be calculated from the measured potential difference. However, in soils the activities of sodium ions and chloride ions are generally not equal to each other. Therefore, the calculated pNaCl value should be termed "apparent mean activity," in short, the mean activity.

Zhang *et al.* (1979) used a spear-shaped sodium ion-selective electrode and a spear-shaped chloride ion-selective electrode to determine the mean activity of NaCl in saline soils. The two electrodes were inserted directly into the soil. In order to examine the distribution of NaCl within a soil profile, a hole was dug. They investigated the distribution of NaCl in soil profiles, the differentiation of chloride ions and sodium ions within a profile, the heterogeneity in NaCl content of surface soils, and the correlation between pNaCl and plant growth among different fields. An example of the distribution of NaCl in three salinized soil profiles is shown in Fig. 7.

One main advantage of the determination of the mean activity of NaCl *in situ* is that the measured value represents the effective concentration of the ions in solution in equilibrium with the solid phase under natural water content conditions, which is directly related to the effect of NaCl on plant growth. Thus, it would be a better parameter than the conventional total NaCl content of the soil. This would be particularly useful in soil survey work. Zhang *et al.* (1979) found a fairly good correlation between the pNaCl value and plant growth for salinized soils.

F. Hydrogen Sulfide

In submerged soils toxicity of hydrogen sulfide to plant growth is frequently encountered. Hydrogen sulfide is conventionally determined by

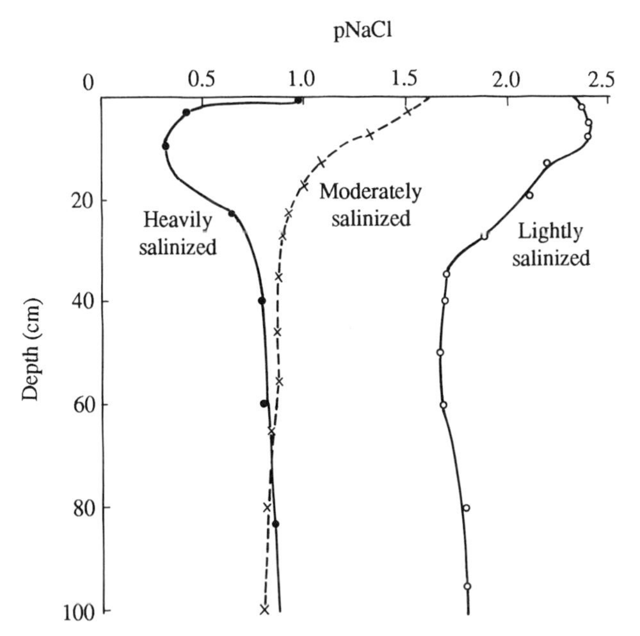

Figure 7. Distribution of pNaCl in salinized soil profiles measured *in situ* (from Zhang *et al.*, 1979, with permission of publisher).

chemical methods. Actually it is not possible to determine true concentration of hydrogen sulfide of a soil by any chemical method, because hydrogen sulfide is in chemical equilibria with several other forms of sulfide, and these equilibria would invariably be disturbed if a chemical treatment is applied. Allam *et al.* (1972) and Ayotade (1977) determined sulfide ions with a sulfide ion-selective electrode and then converted the measured value to H_2S concentration through an equation derived from the chemical equilibrium between H_sS and S^{2-}. Pan *et al.* (1982), based on the principle described by Ross *et al.* (1973), constructed a H_2S sensor for the study of chemical equilibria among various forms of sulfide in soils.

The sensor consists of two ion-selective electrodes. One is the sulfide ion-selective electrode, which responds to S^{2-} activity in the inner solution formed due to the dissociation of hydrogen sulfide from the test solution by difffusion through a gas-permeable membrane. The other is a chloride ion-selective electrode. Because the Cl^- ion activity and the pH in the inner solution are constant, the measured difference in potential between the two electrodes is directly related to the concentration of hydrogen sulfide.

Results showed that the chemical equilibria of hydrogen sulfide in soils were affected by a variety of factors, including the total amount of sulfides,

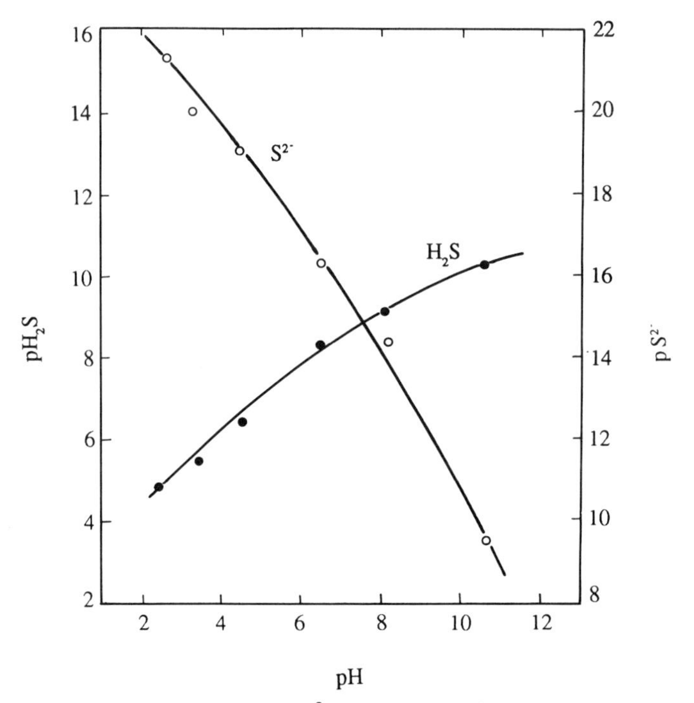

Figure 8. Effect of pH on pH_2S and pS^{2-} in a paddy soil (from Pan, 1985, with permission of publisher).

some metal ions that affect the precipitation–dissolution of sulfides, e.g., Fe^{2+}, Mn^{2+}, and Zn^{2+}, and particularly the pH of the medium. An example for a paddy soil is shown in Fig. 8; the effect of pH on pS^{2-}, as determined with a sulfide ion-selective electrode, is also included. Pan (1985) concluded that the most likely incidence of hydrogen sulfide toxicity to plants is in strongly acid soils having low contents of easily reducible iron and manganese.

VI. VOLTAMMETRY

A. Principles

Voltammetry is a technique for the determination of reducible or oxidizable substances based on the current–voltage relationship at a polarized electrode. The basic difference between this technique and potentiometry is that in the latter technique the relationship between electrode potential and solution composition is evaluated under zero-current conditions, whereas with voltammetry the parameter that is of interest is the relationship between electrode current and solution composition during the flow of a current through the electrode. If the polarized electrode is a dropping mercury electrode, the technique is called polarography. Polarography has been used for the determination of minor elements such as Cu, Cd, Pb, and Zn in soils. The term voltammetry is generally confined to the technique where solid electrodes such as carbon electrodes and platinum electrodes are used as the polarized electrode.

When an external positive or negative voltage is applied to two electrodes in a solution so that the potential of the small working electrode is changed (polarized) and that of the large reference electrode remains practically constant, a current may be produced as a result of an electrochemical reaction occurring at the surface of the working electrode. If the voltage is positive, the electrode reaction is oxidation, and the current is anodic. Conversely, if the voltage is negative the electrode reaction is reduction and the current is cathodic. A curve characterizing the relationship between the current and the applied voltage is called a current–voltage curve, or an I–V curve. The pattern of the curve differs with the manner of change in applied voltage. Two parameters may be measured from the curve to characterize the electrochemical reaction. One is the diffusion current, the current produced by the oxidation or reduction of an electroactive substance diffusing from the bulk solution to the electrode surface. Under specified conditions the diffusion current is proportional to the concentration of the electroactive substance and can be utilized as the basis for quantitative analysis. On the I–V curve, this current is the current at the plateau region. Another parameter is the half-wave potential, the electrode potential at which the current is equal to half of the diffusion current, and it can be utilized as the basis for qualitative characterization. If the change in applied voltage is sufficiently rapid so that the I–V curve is the shape of a peak, the height of the peak is a measure of the concentration, and the half-peak potential, the potential at which the current is equal to half of the peak current, is a measure of the nature of the electroactive substance.

A variety of solids have been used as materials for the fabrication of the working electrode in voltammetry. The primary consideration in choosing electrode material is the cathodic and anodic working potential range. In soil science research, the convenience in handling is also important. Carbon (graphite, glassy carbon) (both cathodic and anodic) and platinum (cathodic) have been commonly used.

B. Determination of Reducing Substances

In soils there are a variety of redox systems. Under submerged conditions most of them are in their reduced state. Even in drained forest soils a considerable amount of reducing substances may be present, if the organic matter content is high (Ding *et al.*, 1984). These substances comprise ferrous and manganous ions, sulfides, and organic compounds. Each of them can produce a characteristic anodic current at the carbon electrode under an appropriate positive voltage. Figure 9 shows the change in reducing substances of different reduction intensities as reflected by the *I–V* curve when a paddy soil is incubated. At the stage of intensive decomposition of organic matter there are several groups of reducing substances. In

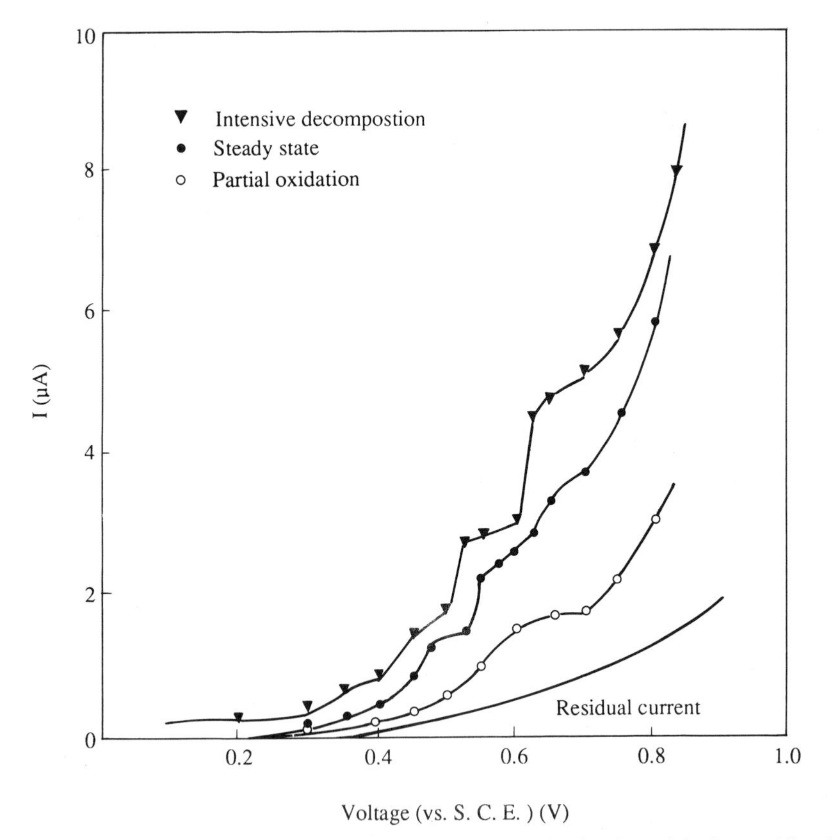

Figure 9. Changes in reducing substances of different reduction intensities in a paddy soil during incubation under submerged conditions (from Ding and Liu, 1985, with permission of publisher).

order to determine these substance in soils *in situ*, Ding *et al.* (1982, 1984) and Ding and Liu (1985) developed a voltammetric technique, using a graphite electrode as the working electrode. The reference electrode was a large-area silver–silver chloride electrode, which was practically unpolarized. Figure 10 shows the distribution of reducing substances within soil profiles in the tropics under different forests. As can be seen, the amount of reducing substances, as a capacity factor of redox status, is closely related to the intensity factor, the oxidation–reduction potential. They designated those substances oxidizable at an applied voltage of +0.35 V vs M Ag–AgCl electrode as strongly reducing substances, and those at +0.70 V as weakly reducing substances. Besides organic substances, the former portion includes sulfide and ferrous ions, and the latter portion includes manganous ions. If required, ferrous ions and manganous ions can be extracted and then determined by sweeping from +0.3 to −0.3 V and from 0 to +0.6 V on an oscilloscopic polarograph, respectively, using a glassy carbon electrode as the working electrode.

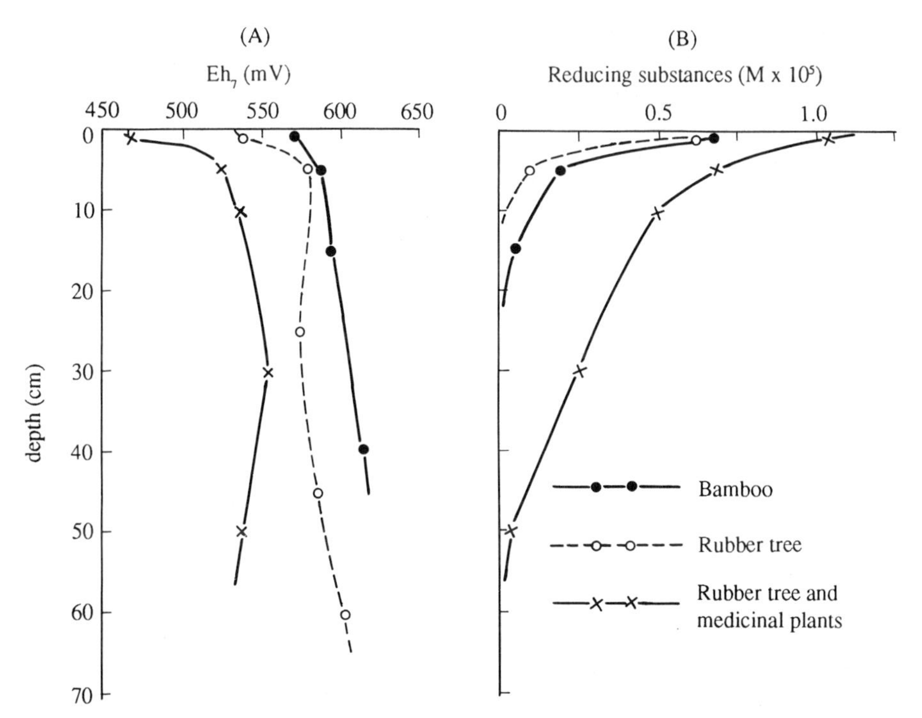

Figure 10. Distribution of oxidation–reduction potential and reducing substances in soil profiles under forest in the tropics (from Ding *et al.*, 1984, with permission of publisher).

In water analysis, these reducing substances are in reality equivalent to COD. Ding and Wang (1991) discussed the superiority of the determination of COD using a voltammetric technique versus conventional chemical methods.

C. Determination of Stability Constants of Fe^{2+}- and Mn^{2+}-Complexes

In soil solution, ferrous and manganous ions can be complexed by organic ligands. It has been found (Bao, 1985) that in submerged paddy soils 10–40% of the ferrous iron was in the complexed form. In order to evaluate the stability of these complexes, Bao *et al.* (1983) and Bao and Yu (1987) developed a voltammetric technique for the determination of their stability constants.

The principle of the technique is based on the shift in peak potential on the I–V curve for the electrochemical oxidation of Mn^{2+} ions or the electrochemical reduction of Fe^{2+} ions in the presence of organic ligands. A plot is made for the shift in peak potential against the logarithm of the concentration of uncomplexed ligands (L), and the stability constant ($\log K$) is found by extrapolating to $\log L = 0$. The mean ligand number of the complex can also be calculated from the slope of the straight line.

The $\log K$ of Fe^{2+}-complexes ranged from 2.6 to 4.5. A change in pH by 1 unit induced a change in $\log K$ of 0.92 units. The $\log K$ of Fe^{2+}-complexes was about 0.8 units larger than that of Mn^{2+}-complexes.

D. Determination of Oxygen

Perhaps the most extensive use of voltammetry in soil science is the determination of oxygen. In this case, a negative voltage of about 0.65 V is applied to a working electrode, usually a microplatinum electrode, and oxygen is reduced at the electrode surface. Two types of platinum electrodes have been used. One is the bare electrode in which the platinum wire is in direct contact with the soil. Another type is the membrane-covered electrode, in which the flat platinum surface is in contact with a thin layer of 0.5 M KCl solution, which is in turn covered by a gas-permeable membrane. For *in situ* determinations the bare-type electrode is convenient to handle. However, because of the deposition of electrolysis products on the electrode surface, the sensitivity of the electrode decreases gradually after repeated use. Therefore, the electrode must be cleaned after each determination. With the latter type of electrode the gas-permeable membrane has a function of "filtration," and thus can prevent the poisoning of the electrode. One disadvantage of this type of electrode is that it is difficult to use in water-unsaturated soils.

E. Voltammetric Titration

Just as in potentiometric titration, in voltammetry the change in electrode current can be utilized to judge the end-point in a titration. This technique is called voltammetric titration, or amperometric titration. Several shapes of titration curves can be distinguished, depending on whether the substance to be determined or the titrant is electroactive.

In soil science research, chloride ions can be determined by titration with silver nitrate, using a carbon electrode or a platinum electrode as the working electrode. After the equivalent point, the excess Ag^+ ions are reduced, resulting in an abrupt increase in electrode current. At an applied voltage of −0.1 V (vs M Ag–AgCl reference electrode) oxygen will not be reduced, and therefore need not be removed from the solution.

Sulfate can be precipitated with barium chloride, and the remaining barium ions titrated by sodium tetraphenylboron (NaTPB) in the presence of polyethylene glycol (PEG) (Zhang *et al.*, 1987). The NaTPB can form a complex of the composition 2TPB · Ba · *n*PEG. After the equivalent point, the excess TPB is oxidized at a carbon electrode under an applied voltage of +0.6 V, producing an anodic current.

VII. CONDUCTOMETRY

A. Principles

Under the influence of an applied electric field, cations and anions in an electrolyte solution move in opposite directions. If the solution contains charged colloidal particles, they migrate in a similar manner. These result in an electrical conductance. Conductometry is a technique for the study of the electrical conductance in relation to the composition of the medium. The electrical conductance of an electrolyte solution is determined by the quantity, charge, and mobility of ions present in the solution. For a solution composed of different ions, the total conductance is the sum of the conductances of individual ions. Therefore, in most cases conductometry cannot be used to discriminate individual ion species. Nevertheless, because of its simplicity and rapidity, conductometry has been used widely in chemical analyses.

The conductivity of an ion is proportional to its mobility. Mobility is the rate of migration under unity potential gradient and is expressed as cm/sec/volt/cm. At infinite dilution, because there are no interactions among ions during their migration, the mobility attains a limiting value, the absolute mobility. For most cations and anions encountered in soils the absolute mobility at room temperature lies within the range of 4×10^{-4}–8×10^{-4} cm/s. Hydrogen ions and hydroxyl ions are two exceptions. At 298 K, the mobilities of these ions are 36.2×10^{-4} and 20.5×10^{-4} cm/s, respectively. The peculiar behavior of these two ions can be utilized in conductometric titrations.

The electrode material commonly used in conductometry includes platinum, stainless steel, and carbon. Traditionally, an alternating current is applied to the electrodes. The main purpose of using alternating current is to avoid electrode polarization, because in this case the net polarity at the electrodes is assumed to be zero, and therefore no electrolysis would occur. Actually, however, for most commercially available conductometers the situation is not so simple, particularly if miniature electrodes and microelectrodes are used. Li (1991) discussed this question in detail. With the advancement in instrumentation, it becomes apparent that the direct-current technique has some advantages. In this technique, four electrodes are employed. An external voltage is applied to the two outer electrodes so that a current flows through the medium. This current produces a potential drop across the two inner electrodes. This potential drop is equal to the product of the current times the resistance between the two inner electrodes, which is in turn a measure of the conductance of the solution. The current can be measured by determining the potential drop across a standard resistance connected in series in the circuit. With the use of a properly constructed constant-current source, electrode polarization at the outer electrodes does not affect the measured results. Thus, because the principle is based simply on Ohms's law and only two potentials are measured, the technique is very accurate. The direct-current technique has been successfully applied in soil science research (Sun *et al.*, 1983).

B. Electrical Conductance of Soil Systems

A soil system consists of solid, liquid, and air phases. Even a water-saturated soil may differ from a solution in that the ions migrate within a heterogeneous medium. In particular, in an electric field charged clay particles together with their counter-ions migrate in a complicated manner. Here conductometry provides a means for studying the interactions between clay particles and ions. If conditions are chosen so that the clay does not migrate, the difference in electrical conductance between the whole system and the free solution is called surface conductance, the conductance contributed by ions of the electrical double layer. It is generally assumed that unhydrated ions close to the clay surface do not contribute to surface conductance, and that ions in the diffuse layer migrate to different degrees in an electric field, with the mobility increasing outward until it approaches that in free solution. Utilization of this phenomenon has been made to evaluate the apparent mobility of counterions and coions. Some authors have estimated the diffusion coefficients of ions in soils.

Apparently, the surface conductance of a given soil is closely related to the nature of adsorbed ions. Generally, the "apparent degree of dissociation" for different cations as estimated by the conductometric method follows the lyotropic series. This phenomenon together with the fact that the mobilities of hydrogen ions and hydroxyl ions are particularly large can be utilized to characterize the cation status of soils by titration with a base and to estimate the cation exchange capacity.

C. Estimation of Salinity in Soils

The utilization of the water extract or soil paste to estimate the salinity of soils has a history of one century (Rhoades *et al.*, 1989). At present this technique is still widely used. In order to estimate salinity in the field, various sensors have been constructed.

For soil survey work, the four-electrode technique is particularly useful. Shea and Luthin (1961) first applied this technique to the measurement of soil salinity. Afterward, Rhoades and co-workers of the U.S. Soil Salinity Laboratory conducted a series of research on this topic. In their technique, four electrodes are vertically inserted into the surface soil on a straight line with equal distances A. The specific conductance of the soil to a depth of A can be calculated from the measured resistance. The conductances of different depths can be determined by varying A. Thus, it is possible to estimate the salinity of the whole profile without the necessity of digging a hole.

For nonsaline soils, most of the ions are plant nutrients. This is particularly true for strongly leached acid soils. Results from a large number of determinations obtained on China soils show that the electrical conductance may be used as a comprehensive index for evaluating the fertility level of a soil. Figure 11, for paddy soils derived from red earths, shows such an example.

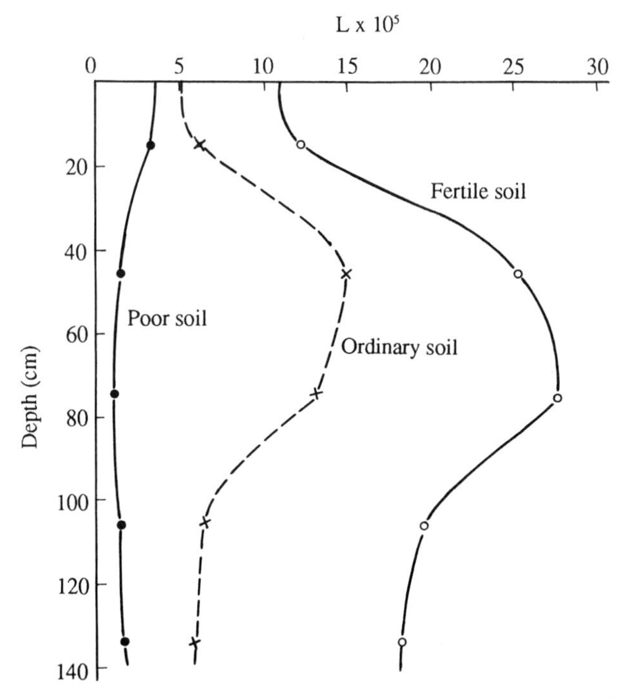

Figure 11. Electrical conductance of paddy soil profiles with different fertility levels (from Sun *et al.*, 1983, with permission of publisher).

D. Conductivity Dispersion

The electrical conductance of electrolyte solutions increases with the frequency of alternating current at sufficiently high frequencies. This phenomenon, called conductivity dispersion, is caused by a decrease in the relaxation effect of the ion atmosphere around a central ion. Soil clay may be regarded as a polyvalent weak electrolyte. Because the adsorbed ions can dissociate only to a limited degree, it may be expected that the conductivity dispersion would be more pronounced than that for electrolyte solutions. Deshpande and Marshall (1961) found that the electrical conductivity of montmorillonitic suspensions was higher when the applied frequency was 10,000 Hz than that when it was 60 Hz. Arulanandan and Mitchell (1968) and Mehran and Arulanandan (1977) observed conductivity dispersion at still higher frequencies with kaolinite and illite. Li and Yu (1987, 1990) studied the conductivity dispersion of variable charge soils at frequencies up to 1.7 MHz.

Variable charge soils carrying positive charges and negative charges simultaneously can adsorb anions as well as cations, and in adsorption, a variety of mechanisms, including specific adsorption, may be involved. Thus, the magnitude of conductivity dispersion would be related to the soil type, particularly to the nature of the ions present. It was found that up to 200 Hz the ability of adsorbed cations to create conductivity dispersion was of the order Na>K>Ca, and for anions, chloride> sulfate>phosphate. At 1.7 MHz, the order for the anions reversed.

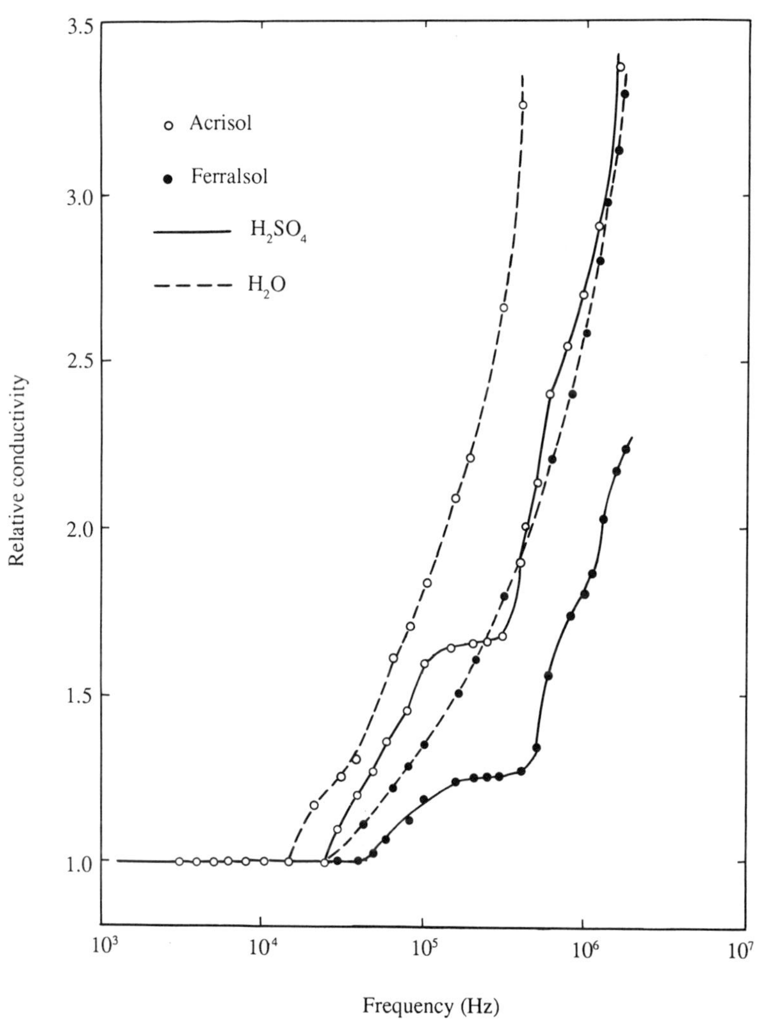

Figure 12. Electrical conductivities of electrodialyzed soils in water and in H_2SO_4 as a function of applied frequency (redrawn from Li and Yu, 1991, with permission of publisher).

When the relative conductivity, the conductivity of the suspension minus that of the equilibrium solution at a given frequency relative to that at 300 Hz, is plotted against the applied frequency, a *c–f* curve characteristic of the system can be obtained. Figure 12 shows typical examples for two electrodialyzed soils in water and in H_2SO_4. At first the relative conductivity does not change. Beyond a threshold frequency it increases rather sharply with an increase in frequency and then reaches a plateau region. Finally, it increases sharply again. It may be assumed that the threshold frequency reflects the level of the energy barrier that must be overcome for an anion associated with the clay to migrate in an electric field, and that the increased conductance at the plateau region is caused by anions at a lower energy level, while that at 1.7 MHz is created by anions at a higher energy level.

VIII. CONCLUDING REMARKS

Electrochemical techniques have helped soil scientists to characterize soil chemical properties more rapidly and more accurately. With the use of these techniques some new research fields have been exploited, which were difficult to study using traditional chemical methods. However, if compared with other sciences such as biology and medicine, the application of electrochemical techniques to soil science is far behind what it should be. There are great potentials for research in this field.

Soil is one of the most complex systems in nature. Some electrochemical techniques, which look straightforward in chemistry, are complicated when applied to soils. Further development in the application of electrochemical techniques to soils will require not only progress in electroanalytical chemistry but also more research by soil scientists.

REFERENCES

Allam, A. I., Pitts, G., and Hollis, J. P. (1972). *Soil Sci.* **114,** 456–467.
Arulanandan, K., and Mitchell, J. K. (1968). *Clays Clay Miner.* **16,** 337–351.
Aslyng, H. C. (1954). "Roy. Vet. Agric. College Copenhagen Year Book 1954," pp. 1–50.
Ayotade, K. A. (1977). *Plant Soil* **46,** 381–389.
Bailey, E. H. (1943). *Soil Sci.* **55,** 143–146.
Baker, D. E., and Low, P. F. (1970). *Soil Sci. Soc. Am. Proc.* **34,** 49–56.
Bao, X. M. (1985). *In* "Physical Chemistry of Paddy Soils" (T. R. Yu, ed.), pp. 69–91. Science Press–Springer Verlag, Beijing/Berlin.
Bao, X. M., Ding, C. P., and Yu, T. R. (1983). *Z. Pflanzenernahr. Bodenk.* **146,** 285–294.
Bao, X. M., and Yu, T. R. (1987). *Biol. Fert. Soils* **5,** 88–92.

Beckett, P. H. T. (1964). *J. Soil Sci.* **15,** 9–23.
Bockris, J. O., and Reddy, A. K. N. (1970). "Modern Electrochemistry," Vol. 1. MacDonald, London.
Bower, C. A. (1961). *Soil Sci. Soc. Am. Proc.* **25,** 18–21.
Chapman, H. D., Axley, J. H., and Curtis, D. S. (1940). *Soil Sci. Soc. Am. Proc.* **5,** 191–200.
Clark, J. S. (1966). *Soil Sci. Soc. Am. Proc.* **30,** 11–14.
Deshpande, K. B., and Marshall, C. E. (1961). *J. Phys. Chem.* **65,** 33–36.
Ding, C. P., and Liu, Z. G. (1985). *In* "Physical Chemistry of Paddy Soils" (T. R. Yu, ed.), pp. 27–46. Science Press–Springer Verlag, Beijing/Berlin.
Ding, C. P., and Wang, J. H. (1991). *In* "Electrochemical Methods in Soil and Water Research" (T. R. Yu, ed.), Chap. 13. Science Press, Beijing.
Ding, C. P., Liu, Z. G., and Yu, T. R. (1982). *Soil Sci.* **134,** 252–257.
Ding, C. P., Liu, Z. G., and Yu, T. R. (1984). *Geoderma* **32,** 287–295.
Haas, A. R. C. (1941). *Soil Sci.* **51,** 17–39.
Huberty, M. R., and Haas, A. R. C. (1940). *Soil Sci.* **49,** 455–478.
Jenny, H., Nielsen, T. R., Coleman, N. T., and Williams, D. E. (1950). *Science* **112,** 164–167.
Ji, G. L. (1991). *In* "Electrochemical Methods in Soil And Water Research" (T. R. Yu ed.), Chap. 8. Science Press, Beijing.
Ji, G. L., and Wang, J. H. (1978). *Acta Pedol. Sinica* **15,** 182–186.
Ji, G. L., and Yu, T. R. (1985). *Soil Sci.* **139,** 166–171.
Kysnezov, M. C., and Snakin, B. B. (1987). "Ionometry in Soil Science." Pusheno, Moskow.
Li, C. B. (1991). *In* "Electrochemical Methods in Soil And Water Research" (T. R. Yu, ed.), Chap. 12. Science Press, Beijing.
Li, C. B., and Yu, T. R. (1987). *Soil Sci.* **144,** 403–407.
Li, C. B., and Yu, T. R. (1990). *Soil Sci.* **150,** 831–835.
Low, P. F. (1981). *In* "Chemistry in the Soil Environment" (R. H. Dowdy, ed.), pp. 34–45. Amer. Soc. Agron., Madison, WI.
Mahendrappa, M. K. (1969). *Soil Sci.* **108,** 132–136.
Malcolm, R. L., and Kennedy, V. C. (1969). *Soil Sci. Soc. Am. Proc.* **33,** 247–253.
Marshall, C. E. (1942). *Soil Sci. Soc. Am. Proc.* **7,** 182–186.
Marshall, C. E. (1953). *In* "Mineral Nutrition of Plants" (E. Truog, ed.), pp. 57–77. Oxford & I.B.M., New Delhi.
Marshall, C. E. (1964). *In* "The Physical Chemistry of Soils," pp. 211–259. Wiley, New York.
Mehran, M., and Arulandandan, K. (1977). *Clays Clay Miner.* **25,** 39–48.
Olsen, R. A., and Robbins, J. E. (1971). *Soil Sci. Soc. Am. Proc.* **35,** 260–265.
Onken, A. B., and Sunderman, H. D. (1970). *Commun. Soil Sci. Plant Anal.* **1,** 155–161.
Overbeek, J. Th. G. (1953). *J. Colloid Sci.* **8,** 593–605.
Overbeek, J. Th. G. (1956). *Prog. Biophys. Biophys. Chem.* **6,** 57–64.
Pallman, H. (1930). *Koll. Chem. Beih.* **30,** 334–405.
Pan, S. Z. (1985). *In* "Physical Chemistry of Paddy Soils" (T. R. Yu, ed.), pp. 93–110. Science Press–Springer Verlag, Beijing/Berlin.
Pan, S. Z., Liu, Z. G., and Yu, T. R. (1982). *Soil Sci.* **134,** 171–175.
Panicolau, E. P. (1970). *Z. Pflanzenernahr. Bodenk.* **126,** 33–42.
Parra, M. A., and Torrent, J. (1983). *Soil Sci. Soc. Am. J.* **47,** 335–337.
Peinemann, N., and Helmy, A. K. (1973). *Soil Sci.* **115,** 331–335.
Rhoades, J. D., Manteghi, N. A., Shouse, P. J., and Alves, W. J. (1989). *Soil Sci. Soc. Amer.* **53,** 428–433.

Ross, J. W., Reiseman, J. H., and Krueger, J. A. (1973). *Pure Appl. Chem.* **36,** 473–489.
Russell, E. W., and Cox, G. A. (1950). *In* "Trans. 4th Int. Congr. Soil Sci.", Vol. 1, pp. 138–141.
Schofield, R. K., and Taylor, A. W. (1955a). *J. Soil Sci.* **6,** 137–140.
Schofield, R. K., and Taylor, A. W. (1955b). *Soil Sci. Soc. Am. Proc.* **19,** 164–167.
Shea, P. F., and Luthin, J. N. (1961). *Soil Sci.* **92,** 331–339.
Sparks, D. L., and Huang, P. M. (1985). *In* "Potassium in Agriculture" (R. D. Munson, ed.), pp. 201–276. Amer. Soc. Agron., Madison, WI.
Su, Y. S. (1991). *In* "Electrochemical Methods in Soil and Water Research" (T. R. Yu, ed.), Chap. 6. Science Press, Beijing.
Sun, X. Z., Wu, J., and Yu, T. R. (1983). *Acta Pedol. Sinica* **20,** 69–78.
Tschapek, M. (1961). *Agrochimica* **6,** 12–20.
Wang, E. J., and Yu, T. R. (1989a). *Soil Sci.* **147,** 34–39.
Wang, E. J., and Yu, T. R. (1989b). *Soil Sci.* **147,** 91–96.
Wang, E. J., and Yu, T. R. (1989c). *Soil Sci.* **147,** 174–178.
Wang, J., Farrell, R. E., and Scott, A. D. (1988). *Soil Sci. Soc. Am.* **52,** 657–662.
Wang, J. H., and Yu, T. R. (1981). *Z. Pflanzenernahr. Bodenk.* **144,** 514–523.
Wang, J. H., and Yu, T. R. (1983). *Acta Pedol. Sinica* **20,** 286–294.
Wang, J. H., and Yu, T. R. (1986). *Z. Pflanzenernahr. Bodenk.* **149,** 598–607.
Wang, P. G., Ji, G. L., and Yu, T. R. (1986). *In* "Current Progresses in Soil Research in People's Republic of China" (Soil Sci. Soc. China, ed.), pp. 85–91. Soil Sci. Soc. China, Nanjing.
Wang, P. G., Ji, G. L., and Yu, T. R. (1987). *Z. Pflanzenernahr. Bodenk.* **150,** 17–23.
Wiegner, G., and Pallman, H. (1930). *Z. Pflanzenernahr. Dung. Bodenk.* **A16,** 1–57.
Xuan, J. X., and Yu, T. R. (1964). *Acta Pedol. Sinica* **12,** 307–319.
Yu, T. R. (1985). *Ion select. Elect. Rev.* **7,** 165–202.
Yu, T. R., Beyme, B., and Richter, J. (1989a). *Z. Pflanzenernahr. Bodenk.* **152,** 353–358.
Yu, T. R., Beyme, B., and Richter, J. (1989b). *Z. Pflanzenernahr. Bodenk.* **152,** 359–365.
Zhang, D. M., Zhang, X. N., Wu, J., and Yu, T. R. (1979). *Acta Pedol. Sinica* **16,** 362–371.
Zhang, G. Y., Zhang, X. N., and Yu, T. R. (1987). *J. Soil Sci.* **38,** 29–38.

History and Status of Host Plant Resistance in Cotton to Insects in the United States[1]

C. Wayne Smith

Department of Soil and Crop Sciences,
Texas A & M University,
College Station, Texas 77843

I. INTRODUCTION

Prior to 1892, only the cotton leafworm, *Alabama argillacea* (Hbu.); the bollworm, principally *Helicoverpa zea* (Boddie); and the aphid or plant louse, principally *Aphis gossypii* (Glover) were considered pest of cotton, and then only occasionally (Brown, 1938; Paddock, 1919; Newsom and

Advances in Agronomy, Volume 48

Brazzel, 1968). This situation apparently meant that very little natural selection for resistance to the myriad of insect pests that attack cotton today took place from the time that cotton was introduced as a commercial crop in the United States through the 1800s. But the situation changed dramatically in 1892 when the Mexican boll weevil (*Anthonomus grandis* Boh.) migrated into south Texas. Thus began a continuing battle between the American cotton producer and insects that would deprive the producer of the fruits of his labor. By 1965 over 100 species of insects and mites were known to attack cotton in the United States.

To understand the present status of host plant resistance in cotton to insects, one must follow the evolution of cotton production in the United States since its introduction in 1607 at Jamestown, Virginia, and its continual, and expanded, production in this country since 1621.

The first cottons grown along the eastern seaboard were undoubtedly the Old World species, *Gossypium herbaceum*, probably originating from the Levant, and *G. arboreum*, variety "Chinese Nankeen" and/or "White Siam," being introduced into the Louisiana settlement, which included parts of Alabama, the Mississippi River Valley and parts of Texas (Ware, 1951).

Cotton was well adapted to the southeastern United States, where rainfall is plentiful, about 60 in. per year with reasonable distribution most years, and there are nearly 200 frost-free days per year. Cotton was accepted quickly by the colonists settling inland from the coast as a source of fiber for clothing, bedding, etc. Along the Atlantic coast, the colonists looked upon cotton as a secondary crop since they could purchase necessary woven items for their personal use as a consequence of their proximity to sea trade routes.

Although the scientific principles of plant breeding awaited the rediscovery of Mendel's research, selection of desirable plant types of crops such as maize (*Zea mays* L.) and cotton had been occurring for centuries. Nonshattering panicles of grasses such as sorghum (*Sorghum bicolor* L. Moench) and wheat (*Triticum aestivum* L.) were selected by early civilized man in the Old World. In the Western Hemisphere, the natural mutation and selection by Western man within populations of teosinte (*Z. mexicana* [Schrad] Kuntze) resulted in the development of modern corn at least 8000 years ago (Galinat, 1988). Hybridization of maize by the Indians of Central and South America has been reported to have occurred long before the 1800s and "varietal" hybrids were being developed as early as 1805 in Virginia.

When man began to develop an agrarian society, his livelihood depended upon a successful "crop"; surely he took measures to ensure the production of that crop. Producers of corn and other grains selected plants for desirable height, resistance to lodging, grain size, etc., as the progenitors of the next season's crop. Cotton farmers in the United States undoubtedly did likewise.

II. EVOLUTION OF COTTON TYPES

A terse review of the germplasm grown and available for hybridization and selection for host plant resistance to combat insect pests in this century is necessary to understand the rationale, success, and failures in developing insect resistance and tolerance in U.S. cotton cultivars.

The tetraploid, or New World, types were introduced into the United States prior to 1730 and quickly replaced the diploid species. It is not obvious from the literature whether the diploid species were ever grown to any extent. Both *G. hirsutum* and *G. barbadense* soon became established, with *G. barbadense* being grown on the coastal islands and lowlands along the Atlantic coast—thereby referred to as "sea island" cottons—whereas the *hirsutum* types were better adapted inland or upland from the coast—thereby referred to as "upland" types. Today, both types are produced in the United States. *Gossypium barbadense* is no longer grown along the Atlantic and Gulf of Mexico coasts, but it is grown on limited acreage in the western United States and referred to as "Pima," so named for a county in Arizona, or "ELS" (extra long staple).

The upland types that were to become dominant by 1800 were "Georgia Green Seed" and "Creole Black." Georgia Green was introduced to the southeastern coastal states from the West Indies by Philip Miller, while Creole Black was introduced by the French around 1730. Creole Black was grown primarily in the lower Mississippi River Valley, that being parts of the present states of Mississippi and Louisiana, and perhaps portions of Alabama and Texas. Sea island cotton was, and Pima cotton is today, more vegetative than upland and required a much longer growing season, a point of much importance relative to cotton insects.

In 1806, Walter Burling, a cotton producer from Natchez, Mississippi, journeyed to Mexico to discuss a boundary dispute between the Spanish territory of Mexico and the Louisiana territory, newly purchased from France by the United States. Burling's real purpose was to acquire seed of a cotton grown by the native Indians of the Mexican Plateau. This cotton was reported to be earlier and more productive and to possess better quality fibers than either the Georgia Green or the Creole Black cottons. Burling was officially denied permission to export cottonseed of this material, but did, with the help of the Viceroy of Mexico, smuggle some seed into the United States hidden in "Mexican Dolls" (Wailes, 1854).

The following year Burling gave the seed to a friend, William Dunbar, who apparently received favorable reports on the fiber from textile experts in England. Between 1807 and 1810, Dunbar increased the contraband seeds to over 3000 pounds of ginned cottonseed. By 1820, this Mexican introduction probably had outcrossed with both Georgia Green Seed and Creole Black Seed. Apparently the 1806 introduction was also known as Mexican Hybrid, Mexican Highland Stock, and probably by several other names. History suggests that other introductions of the Mexican cottons were made in the early part of the 1880s; however, definite proof is lacking since the appearance of the Mexican phenotype in Georgia and the Carolinas around 1825 could have originated from seed of the 1806 introduction. Aside from seed brought by returning U.S. soldiers from the Mexican War of 1847–1848, the historical record notes one other introduction of the stock, this by the Wyche brothers around 1857. When the brothers emigrated from Germany in 1853, one went to Algeria and the other settled in Georgia. In 1857, the brother in Algeria sent a package of cottonseed, apparently of Mexican descent, to the brother in Georgia (Ware, 1951).

Although the events surrounding the introductions of the Mexican cotton had nothing to do with host plant resistance in 1806, this stock is credited with providing the necessary genetic variability for the survival of the U.S. cotton industry following the migration of the Mexican boll weevil. This insect has caused more loss, altered more production practices, and influenced more decisions than perhaps any other insect in modern agricultural history.

Although there were essentially two types of cotton or "varieties" in 1806, the natural outcrossing with the Mexican stock, selection, movement, and reselection as cotton production expanded westward resulted in 58 cultivars by 1880, 118 by 1895, and almost 400 by 1907 (Brown, 1938). Tyler (1910), on the other hand, identified over 600 cultivars in 1907. Obviously, there were numerous producers making selections and reselections in cotton many years prior to the rediscovery of Mendel's principles. None of these cultivars were selected for host resistance to insects as far as the author has determined. However, Ware (1951) notes that Dickson cultivar was grown in Texas in 1880 as an early-maturing cultivar to escape the late-season effects of caterpillars, probably cabbage loopers. Dickson was selected in Georgia from Boyd Prolific, a semicluster type commonly grown in Mississippi by 1847.

III. THE BOLL WEEVIL INVADES TEXAS

Because the boll weevil was the first major pest to consistently impact U.S. cotton production, and continues to do so, it is reasonable to address that insect first. We will then proceed to review the host plant resistance arena for the myriad of other cotton insect pests.

Cotton producers in the United States had grown cotton with essentially no production restrictions other than the exhaustion of soil nutrients that occurred under continuous monocropping. Production spread westward from Virginia as the country and its population grew. With cotton being an introduced species, one would perhaps not expect to find insects adapted to feed and survive on this plant. However, cotton is native to Mexico, and as cotton production in that country spread northward, and as production in the United States moved southward into Texas, the natural barrier of vast areas of land between U.S. cotton and sympatric insect pests of cotton in Mexico shrank. The boll weevil, a near obligate feeder of cotton pollen, approached U.S. cotton with each year of expansion and was within range of U.S. cotton by the late 1800s. In 1880, a journeying American noted its damage to the cotton crop around Monclova, Mexico, a town about 120 miles south of the U.S.–Mexican border. A strong flyer, the weevil was within range of U.S. cotton and was first reported in Texas in 1892.

A. Preinsecticide Era

The U.S. cotton producer was unprepared for the boll weevil, and by 1895, accounts of 30 to 90% crop loss were reported (Townsend, 1895). Early recommendations to quarantine and arrest its movements in Texas were not heeded; cultural control recommendations such as early stalk destruction and removal of infested flower buds met limited success. The only weapon against this pest that was quickly accepted was the use of early-maturing cultivars, first recommended in 1896, 4 short years after the first reports of the weevil's presence in Texas (Howard, 1896). Early maturity provided an escape mechanism from the weevil by producing bolls relatively early in the growing season rather than in the latter part of the growing season when weevil numbers could become extremely high. This early maturity was without doubt the result of the importation of the Mexican stock in 1806 by Burling and by others during the first half of the nineteenth century.

The producers in the southern regions of the U.S. Cotton Belt first turned to early-maturing cultivars that had been selected for the short growing seasons further north. These cultivars required fewer days to maturity, but sacrificed fiber quality and gin turnout. Gradually, however, producers and seedsmen began to select for earlier maturity within material having better fiber length (staple) and improved lint production (Bennett, 1904). Bennett noted several factors about short-season cultivars, such as lower node of first fruiting branch and shortened sympodial internodes, that would be quantified years later by a number of scientists (McNamara *et al.*, 1940; Ray and Richmond, 1966; Smith, 1984).

Although it is obvious that farmer selections and reselections of desirable plant types among the genetically variable cultivars that resulted from the intermating of Burling's Mexican stock, Georgia Green Seed, and Creole Black Seed types helped to combat the boll weevil, scientific plant breeding did not begin in the United States until 1898 when Dr. H. J. Webber was hired by the USDA to develop improved cotton cultivars. By 1920, several scientifically based breeding programs were in place: Coker Pedigreed Seed Co. in Hartsville, South Carolina; the USDA program at Stoneville, Mississippi; the Stoneville Pedigreed Seed Company located in Stoneville, Mississippi; and the Delta and Pine Land Co. in Scott, Mississippi. The Stoneville Pedigreed Seed Co., the Delta and Pine Land Co., and the USDA program at Stoneville, Mississippi, continue today, but the Coker program was purchased by Stoneville in 1990.

B. Postinsecticide Era

Prior to the boll weevil, and later the bollworm (*Helicoverpa zea*), earliness was a term used by producers to mean the production of a reasonable number of bolls prior to frost. Post-boll-weevil connotation is the production that occurs before the late season build-up of boll weevil populations, and later, before outbreaks of bollworm, tobacco budworm (*Heliothis virescens*), and pink bollworm (*Pectinophora gossypiella*) (Walker and Smith, 1993).

The emphasis on earliness and escape/resistance to the boll weevil remained strong until about 1920 when the USDA recommended calcium arsenate as a control measure for this devestating pest (Walker and Smith, 1993). With the discovery of calcium arsenate as a poison for the weevil, the pressure to develop earlier-maturing cultivars slowed. And with the discovery of the chlorinated hydrocarbons, such as DDT after World War II, emphasis on earliness came to a near stop. Breeders turned their attention to yield potential. Genetic gains in yield potential in the midsouth region of the U.S. Cotton Belt from 1910 through the 1960s averaged

9.46 kg/ha/yr (Bridge and Meredith, 1983). Similar results could be documented elsewhere.

Calcium arsenate had provided the producer with a means of significantly reducing losses to the boll weevil, but its use was often linked with outbreaks of secondary pests such as bollworms and aphids (Ballou, 1919; Sherman, 1930; Paddock, 1919). However, the shortcomings of this chemical were soon forgotten with the introduction of the synthetic chlorinated hydrocarbon insecticides, and the exodus away from early-maturing cultivars became complete (Walker, 1984). Breeding efforts in the rainbelt and irrigated West were directed almost totally to yield potential of full-season types, to take advantage of the botanical indeterminacy of cotton. Better cultivars, along with commercial fertilizer, resulted in tremendous yield gains, by historical standards.

1. Boll Weevil Resistance to Insecticides

By the mid-1950s, the constant selection pressure on the boll weevil population effected a shift in the creature's gene pool to one that contained a large percentage of individuals resistant to the chlorinated hydrocarbons (Brazzel, 1961; Roussel and Clower, 1955). The agricultural chemical industry responded with the organophosphate methyl parathion, and it gave almost complete control of weevils; chlorinated hydrocarbons, such as DDT, were added to the mixture to control bollworms and tobacco budworms. The U.S. cotton producer was mesmerized, for the moment, into thinking that all insect problems could be corrected with the right chemical(s). However, within a few years, increased dosages of the chlorinated hydrocarbon compounds were often required to control bollworms and budworms, and by 1965 these chemicals were deemed ineffective in certain areas of the Cotton Belt (Adkisson, 1964; Adkisson and Nemec, 1966; Brazzel, 1963, 1964; Harris *et al.*, 1972; Nemec and Adkisson, 1969). For a brief period, bollworms and budworms were handled by larger rates of methyl parathion, but resistance was soon detected in *H. virescens* F., the tobacco budworm (Brazzel, 1963).

2. Earliness Rediscovered

Researchers began turning their attention toward the advice of Mally and Bennett of nearly a century ago. New interest in developing earlier-maturing cultivars began in the 1950s and accelerated in the 1960s and 1970s. Producers and entomologists began to look, again, for earlier maturity as the cornerstone of cotton insect management. Shortening the growing season reduced exposure time to insects, thereby reducing the number of insecticide applications. Less insecticide was lauded by all, as it placed less selection pressure on insect populations, was cheaper, and was good for the environment. In regions where outbreaks of bollworms and budworms were often attendant with late-season insecticide applications for control of weevils, earliness became the structure around which schemes were designed to manage the boll weevil, schemes with less dependence on insecticides.

With the rediscovery of earliness and cultural control, equipped now with new knowledge of the biology of weevils and other pests, integrated pest management (IPM), systems approach, short-season concept, and community-wide approach became the slogans of the day, and early-maturing, more agronomically determinate cultivars were once again the cornerstone of these concepts. Walker and Niles (1971), working to understand economic thresholds of the weevil, found that fast-fruiting genotypes could "set" an acceptable crop of bolls that could escape first-generation weevil damage if fields were infested with 20 or fewer overwintering females per acre. But if 60 or more females were found, then the first generation population of weevils would be sufficient to cause economic loss. This work led to the conclusion that it was important to have 30 days of blooming before weevils built to damaging levels. Thirty days of blooming would result in a sufficient number of bolls, of sufficient age, to escape major damage (Walker and Niles, 1971). Therefore, genotypes setting the greatest number of bolls in the first 30 days of blooming would have a production advantage under reduced insecticide production schemes.

This understanding of overwintering weevils and population dynamics was refined by Parker *et al.* (1980), who reported that 5000 or more punctured squares per acre before bloom suggested that a destructive population would build by the twentieth day of bloom, whereas 1500 or fewer punctures indicated that damaging levels would not occur until the thirtieth day of blooming, or later.

From 1970 through 1973, Sterling and Haney (1973) effected an increase in yield and a decrease in insecticide use by using the systems approach to insect management on the farms of the Texas Department of Corrections. Other researchers reported on the economic advantage of integrated pest management (Carruth and Moore, 1973; Frisbie *et al.*, 1976; Larson *et al.*, 1975; Collins *et al.*, 1979).

Another series of events occurred that sparked interest in short-season cultivars in more modern times. The energy crises of the 1970s and the resulting inflation hammered home a startling point: chemicals were no longer cheap and irrigation water would become more and more expensive (Schaunak *et al.*, 1982).

3. Road to Early Maturity

We must now digress to the mid-1950s to document the rapid move toward earlier-maturing cultivars. Carl Mooseburg, USDA cotton breeder headquartered at Marianna, Arkansas, in 1957, released a cultivar called Rex, which was meaningfully faster fruiting than then currently available cultivars (Waddle, 1957). Developed for mechanical picking and rainbelt production, Rex was 10 to 14 days earlier than other commercially available cultivars in Arkansas, and in one comparison at Marianna, Rex produced 1112 pounds of seedcotton at first harvest while a "popular cultivar" produced only 409 pounds. Rex outyielded the check cultivar by approximately 350 pounds of seedcotton (Mooseberg and Waddle, 1958).

With the development of Rex, Mooseburg, as had Bennett (1904), demonstrated that earliness could be obtained without sacrifice of yield or quality in picker-type cottons. Ironically, Rex's phenotype was similar to that advocated by Bennett 52 years before; it had short sympodial internodes and apparently much shortened, for 1956, vertical and horizontal fruiting intervals.

With the cumulative problems of resistance in populations of weevils and bollworms and/or budworms; with harvest problems of late-maturing cultivars, especially in years when the effective growing season is reduced by the early onset of low temperatures; with harvest problems that arose with excessive rates of fertilizer, especially nitrogen; and with the possible delays in maturity associated with organophosphate insecticides, breeders began to follow Mooseburg's lead, giving consideration to earlier-maturing genotypes. The move to earlier-maturing picker types picked up steam with the release of the DES cultivars in Mississippi in the early 1970s, quickly followed by privately developed early-maturing cultivars, and today all cultivars grown in the Midsouth are considered early-maturing. In fact, Bridge and McDonald (1987) found that 34 fewer days were required from planting to final harvest of the Mississippi Cotton Cultivar Trials at Sumner and Stoneville in 1986 than in 1968.

As breeders in the Southeast, Midsouth, Southwest, and Far West proceeded cautiously toward short-season cultivars, attempting to maintain agronomic indeterminacy, breeders in Texas moved quickly to determinate, ultra-short-season (for their day) cultivars. The work of Walker and Niles (1971) had demonstrated the wisdom of these types and L. S. Bird and others put the concepts into practice. Tamcot SP21, SP23, and SP37 were released in 1973 and fit the short-season requirements for early, determinate cultivars (Bird, 1975). These cultivars were not only early, but also productive, especially in the Coastal Bend area of Texas. Insect control, or the lack thereof, had almost driven cotton production out of the five-county area surrounding Corpus Christi, with only 50,000 acres grown by the early 1970s. Acreage increased dramatically during the years following the release of the determinate Tamcot germplasm, and by 1979 near 300,000 acres were grown in this region. This adoption of the short-season cotton cultivars and the attendant cultural control of the boll weevil through early harvest and stalk destruction resulted in an estimated increase of $11,000,000 in producer profits in 1979 (Lacewell and Taylor, 1980; Masud *et al.*, 1980).

In other areas of Texas, and the nation, short-season technology, as that used in the Coastal Bend of Texas, could not be utilized. Normal rain patterns and amounts, less than 40 in., supported the use of determinate types in the drier Coastal Bend of Texas. In production areas that receive 50 in. of rain per year or more, the new determinate Texas cultivars were poorly adapted. Rainfall amounts and distribution in those areas dictate a less determinate and larger plant type for optimum economic yields. In the irrigated areas of New Mexico, Arizona, and California, agronomically determinate types are currently not acceptable because of the availability of irrigation water and an accommodating long production season that allows the use of indeterminate cultivars for maximum yields of superior quality lint. However, all producers in the United States recognize the value of earlier cotton production; they recognize the dollar savings associated with the reduced inputs and are more comfortable with the lower risks that earliness carries.

C. Morphological Traits

Although earliness is not strictly a host plant resistance (HPR) mechanism to the boll weevil, it is an effective escape (also referred to as klenducity) mechanism that has been exploited since the weevil moved into Texas. Other HPR mechanisms, both morphological and biochemical, have been studied: (*a*) frego bract; (*b*) pubescence levels; (*c*) reduced androecium; (*d*) red plant color; (*e*) gossypol content; and (*f*) oviposition suppression factor.

1. Frego Bract

About 1940, a mutant plant growing in a field of Stoneville 2B was identified by Mr. George Frego. Whereas normal plants have bracts that are essentially flat and surround the flower bud and later the developing boll, frego bracts are more narrow with margins rolled inward and the entire bract extended away from the flower bud, resulting in complete exposure of the bud. Frego bract reduces the desirability of cotton as an oviposition site for the boll weevil.

Several workers (Pieters and Bird, 1976; Lincoln *et al.*, 1971; Lincoln and Waddle, 1966; Jenkins and Parrott, 1971; Maxwell *et al.*, 1969; Jenkins *et al.*, 1969; Jones *et al.*, 1987) have reported that frego bract reduces weevil oviposition. Reduction in egg-lay ranged from 60 to 100%, compared with normal bracted cotton under small plot research conditions. In larger plots, Jenkins and Parrott (1971) reported reductions of 66, 71, 75, and 94% when compared with commercial, normal bracted cultivars grown under producer conditions with each of the four tracts of frego being 10 to 20 AC in size. On two of the four tracts, insecticide treatment was not required for suppression of weevil below economic thresholds.

Schuster *et al.* (1981) reported that frego bract did not carry resistance to the boll weevil. This conclusion may be consistent with others in that previous reports measured preference and not antibiosis. Schuster *et al.* (1981) placed female weevils in cages of frego or nonfrego and found no difference in oviposition punctures, suggesting that weevils oviposit at comparable rates on frego bract and normal bract cottons when given no choice.

There is some controversy over the accurate classification of oviposition and feeding punctures in frego bract cotton. There is general agreement that punctures sealed by a substance secreted by the female weevil after oviposition are definitely oviposition punctures and that feeding punctures are not sealed (Buford *et al.*, 1967). However, Jones *et al.* (1980) reported that two frego bract lines had significantly lower hatchout from obvious oviposition-punctured squares than did their normal bract control, but twice as many weevils emerged as had been indicated by punctures classified as oviposition punctures. This suggests either that multiple eggs are placed in some oviposition sites, which has not been documented, or that not all oviposition punctures in frego bract types are sealed. Given the fact that frego bract acts as a behavioral modifier, the latter explanation seems most reasonable.

It is the experience of the author that frego bract is not preferred by the boll weevil when given a choice and that fewer punctured squares will result under small plot conditions; however, weevil pressure can become sufficiently high to overwhelm this type of "resistance." This observation, and reports previously noted raise questions about the use of frego bract planted community-wide.

There are some data (Bates, 1989) that indicate a degree of antibiosis is associated with frego bract in some genetic backgrounds. Square size and anther density have been shown to contribute to variability in weevil size, and these factors could contribute to observations of antibiosis. Data obtained to date do not clearly indicate that reduced viability is the result of reduced emerging weevil size.

No cultivar released for production in the United States has frego bracts. Although this character is potentially useful as a resistance/escape mechanism from weevils, it makes these genotypes more susceptible to the "plant bug" complex, *Lygus lineolaris* P. de B. (tarnished plant bug), *Neurocolpus nubilus* (clouded plant bug), *Pseudatomoscelis seriatus* (fleahopper), and *Adelphucoris rapidus* (rapid plant bug). These insects attack cotton throughout the fruiting season, but are particularly disastrous early when plants are beginning to fruit. Treatment of frego bract cotton during the early part of the fruiting season could set in motion a scenario where the cotton bollworm; tobacco budworm; and pink bollworm, *Pectinophora gossypiella* (Saunders), would be unchecked by natural predators. Jones *et al.* (1987, 1989) have reported progress is overcoming this sensitivity to the plant bug complex. The success of these germplasms remains to be seen.

2. Plant Pubescense

Wannamaker (1957) evaluated plant hairiness in relation to feeding and egg deposition by the boll weevil, concluding that extreme hairiness, or pilose, significantly reduced damage. Others (Stephens, 1957; Stephens and Lee, 1961; Hunter *et al.*, 1965; Wessling 1958a) reported similar results and suggested that resistance was mechanical because the hairs along the margins of the bracts were interlaced, thus temporarily enclosing the flower buds.

Heavy plant pubescence confers a high degree of resistance to the boll weevil and also to the plant bug complex (the United States does not have jassid, which also is deterred from feeding by plant hairs). However, this trait is of limited value because it results in an increased amount of plant trash in mechanically harvested cotton, which is difficult and expensive to remove. Second, plants with increased levels of hairiness are preferred by the bollworm complex, suggesting it unwise to incorporate this trait into U.S. cultivars. And finally, pilose is associated with short, coarse fibers that are unacceptable to the U.S. cotton industry.

3. Reduced Androecium

Meyer (1965, 1971, 1972) reported that anther numbers were reduced when the cytoplasm of *G. anomalum* Wawr. and Peyr. (DES-ANOM-16), *G. herbaceum* L. (DES-HERB-16), or *G. arboreum* L. (DES-ARB-16), all diploid species, was transferred into Deltapine 16, a U.S. commercial tetraploid. McCarty (1974) found that these three lines were less attacked by weevils than was Deltapine 16, by as much as 38%. He noted fewer anthers in the three experimentals than in the commercial Deltapine 16 and suggested that the resistance was a result of reduced androecium, a logical conclusion as weevils are primarily pollen feeders. Weaver and Graham (1977) substantiated this line of reasoning by evaluating cytoplasmic male sterile lines for feeding and oviposition preference in Georgia. There are very few data to suggest that this trait would result in increased pressure from other cotton insects; however, Glover *et al.* (1975) reported increased bollworm damage on DES-ARB-16 and DES-ANOM-16. Whereas there is a paucity of information on anther density and weevil resistance, there appears to be no reason not to incorporate this trait into commercial cotton cultivars, although the level of resistance is too low to be of practical value except under extremely light infestations. Reduced androecium might enhance the resistance conferred by frego bract or red plant color if problems associated with these HPR characters are solved.

4. Plant Pigmentation

Isley (1928) referenced the work of C. R. Jones in 1907, who determined that red color was less preferred by the boll weevil than were other colors, including green. In 1925 and 1926, Isley studied the relationship of plant color, green versus red, and leaf size. Winesap cultivar, having red foliage, was compared with a green-colored, early-maturing Acala cultivar of similar size and maturity. In both years, fewer squares of the red plants were infested than of the green Acala control, results that have been substantiated by several researchers since (Wessling, 1958b; Stephens, 1957; Hunter *et al.*, 1965; Jenkins *et al.*, 1969; Jones *et al.*, 1980). Hunter *et al.* (1965) documented that the resistance is nonpreference and that it is not expressed as strongly in large blocks of red leaf types as in small research plots adjacent to green foliage cotton. Jones (1972) stated that the degree of nonpreference was related to the intensity of the red color: the darkest red type, R1, being the least preferred; Ak Djura (also called red stem) red with a medium red foliage and stems being intermediate; and red margin (or N. C. margin), which has leaves almost green with light red leaf margins, having the least degree of nonpreference. However, Jones *et al.* (1970) and Jones (1972) reported that the R1 locus (or red leaf), the darkest red color, was associated with an unacceptable reduction in yield of 7 to 30%, when compared with its green isoline. This reduction in yield potential has been observed by others (Kohel *et al.*, 1967; Karami and

Weaver, 1972), with less yield reduction reported for the AK Djura red and red margin reds (So. Coop Res. Bull., 1981).

5. Gossypol

There is little evidence that gossypol gland density or total gossypol content has a significant impact on the boll weevil. Singh and Weaver (1971) found that weevils emerging from high gossypol cotton lines were smaller than those from normal gossypol or gossypol-free strains. Maxwell *et al.* (1966) studied 12 pairs of isogenic or near isogenic lines for susceptibility of the glandless lines to boll weevil with mixed results. More eggs were deposited in the glandless isoline of three pairs with no differences in the other nine. Similar results were obtained in antibiosis tests where no differences were found in weevil development times.

There appears to be no solid evidence that gossypol impacts the biology of the boll weevil in any fashion that might be beneficial in developing resistant cottons.

6. Oviposition Suppression Factor(s)

There is appreciable amount of evidence that there are some biochemical resistance factors that make certain genotypes of cotton less desirable for oviposition. Evidence dates to Buford *et al.* (1967) who reported that oviposition was reduced in 26 of 252 cotton lines, both *G. hirsutum* and *G. barbadense.* Sea Island Seaberry (*G. barbadense*) had the lowest percentage egg deposition, which is noteworthy since the boll weevil was the primary factor in the demise of the sea island cotton industry in the United States. The resistance in Seaberry was subsequently substantiated by Jenkins *et al.* (1969). The primary reasons that sea island genotypes were so vulnerable to the boll weevil were (*a*) sea island genotypes are slow fruiting and maturing and (*b*) the carpel walls remain soft and vulnerable to attack, whereas the carpel walls of *G. hirsutum* become sufficiently hardened to resist attack from the boll weevil about 10 days after anthesis (Walker *et al.*, 1976; Jenkins, 1942).

Earnheart (1973) found 49 *G. hirsutum* race stocks that showed some resistance to weevil oviposition under laboratory conditions. Earnheart noted that most of the identified resistant stocks were photoperiodic, were late-maturing, had small flower buds, and were not very productive, relative to commercial uplands. The USDA is converting all race stock accessions that are photoperiodic to day neutrality. Several of these have been released and others are in the backcrossing and testing phases prior to release. This program has and will continue to make the germplasm collections much easier to work with and hasten the development of HPR cultivars.

Other researchers (Bates *et al.*, 1988; Lukefahr and Vieiera, 1986; McCarty *et al.*, 1977, 1979a,b, 1986, 1987; Jenkins *et al.*, 1978; Walker *et al.*, 1987; Jones *et al.*, 1988) have reported results substantiating Earnheart's findings. Bates *et al.* (1988) reported that resistance to oviposition and feeding damage could be detected on individual plants within converted race stocks T277 and MT1180. Prior to that report, resistance had been reported on a population basis only. The significance of the report by Bates *et al.* (1989) is that it offers the possibility of selecting for weevil resistance in individual plants within breeding nurseries, thereby speeding the development of resistant cultivars.

Another avenue along this line of thought that has received little attention is the development of doubled haploids from these resistant race stocks through the use of semigamy. Although the success rate is very low, 12% or fewer F_1's will be haploid and therefore suitable to be doubled, a concentrated effort could yield tremendous returns. This system would involve the crossing of a semigametic strain, homozygous dominant for the semigametic locus and a marker locus, such as virescent foliage, with a number of plants of each converted race stock; race stocks are not homogeneous. Mahill *et al.* (1983) reported 3 to 5% success in deriving plants with haploid sectors whose chromosome complement could be doubled by the use of colchicine. This method would circumvent all segregating generations and avoid the necessity of evaluating single plants. Stelly *et al.* (1988) proposed the development of hybrid-eliminating, haploid-inducing (HEHP) lines by incorporation of a lethality allele from *G. davidsonii*. In this proposed system, true hybrids would die in the seedling stage and therefore only haploid seedlings would survive; these would then be treated en masse with colchicine to recover the normal chromosome compliment. To date this system has not been developed.

7. Klenducity

The desirability of earliness as an escape mechanism has been discussed in detail. However, cultural practices that enhance and supplement this genetic escape have evolved in the United States. These include proper plant nutrition, early-season insect control through uniform delayed planting, community-wide insecticide management, termination of crop growth with plant growth regulators, prompt stalk destruction, and general field sanitation (Walker and Smith, 1993; Smith *et al.*, 1986; Kittock *et al.*, 1973).

IV. PINK BOLLWORM

A. Movement to and Occurrence in the United States

The pink bollworm, *Pectinophora gossypiella* (Saunders), is native to India where it was reported to be destructive to cotton in 1842 (Brown, 1927). This insect is considered one of the most destructive insects to cotton and probably ranks second only to the boll weevil in sheer destructive power.

The life cycle of the pink bollworm is such that the insect pupates, or enters a "resting stage," which can last from 3 to 18 months. This stage, the last larval instar, can occur in the soil, in bolls left in the field, or inside seeds. Occasionally, two cottonseeds will be hollowed out by the larvae and fastened together to make a rather sizable cell. Survival within seeds accounts for the movement of the insect from India to practically everywhere cotton is grown today.

The pink bollworm was established in Egypt through a large shipment of cottonseed from India around 1906 and was found in the Philippines prior to that time. During the years 1911, 1912, and 1913, the Brazilian government imported 9 tons of Egyptian cottonseed that had not been fumigated, thus introducing the pink bollworm into the Western Hemisphere. Two importations of cottonseed from Egypt were made to Mexico around 1913, both probably containing contaminated seed. The pink hollworm was reportedly found alive on rail freight cars at the Mexican–U.S. border before 1917.

The pink hollworm was first cited in Texas near the town of Hearne, about 350 miles north of the Mexican border, on 10 September 1917. The infestation was traceable to the importation of Mexican-grown cottonseed to an oil mill at Hearne. The immediate response was a massive clean-up effort to destroy all cotton and cotton plants remaining in nearby fields. The clean-up took 500 laborers working about 5 months around Hearne and other areas in Texas where contaminated cottonseed had been shipped. The Texas government took immediate steps to prevent the growing of cotton in infested areas and to prevent the importation of cottonseed or its products from northern Mexico. This latter provision was designed to prevent the establishment of a cotton industry just south of the U.S.–Mexican border, which would have created a natural bridge between U.S. and Mexican cotton production, thus providing a continual source of pink bollworm.

From this beginning, the pink bollworm has been found in several states of the U.S. Cotton Belt. The pink bollworm was identified in Louisiana and in New Mexico by 1918. Today, it is the primary pest in parts of southwestern United States, especially in the state of Arizona.

B. Cultural Control

Control of the early infestations in U.S. cotton was achieved through the use of strict quarantines, cultural practices, and improved farm and gin equipment. Following the first serious outbreak of pink bollworm in the United States in 1952, a state and federal cooperative effort identified such things as stalk-shredding machinery that would kill a large percentage of larvae in seedcotton remaining after harvest and trash conveyor systems that would do likewise during ginning as vital elements in managing the pest. When the pink bollworm is found in any area in the United States today, that production area is quarantined and all cottonseed and unginned seedcotton moved out of the quarantined area must be fumigated and certified so by state and/or federal agencies. Other cultural practices that yield a measure of control are uniform late planting dates and soil tillage (Martin, 1962).

C. Host Plant Resistance

Host plant resistance in cotton to the pink bollworm was first documented by several workers outside the United States. Wolcott (1927) reported that the pink bollworm preferred tetraploid *G. hirsutum* over *G. arboreum* and the native cottons grown in Haiti. According to Brazzel and Martin (1956), resistance to pink bollworm has been reported in *G. thurberi, G. armourianum, G. somalense*, and *G. anomalum*.

Whereas quarantines and cultural control measures provided reasonable control of the pink bollworm in most of the United States, considerable effort has been devoted to identifying and enhancing HPR in cotton to this most destructive pest.

The vast majority of HPR work in the United States has been conducted in Texas and Arizona. Fenton (1928) reported that bolls of *G. thurberi* (diploid) were essentially ignored by the pink bollworm when grown near commercial cultivars of *G. hirsutum*. No resistance among available commercial cultivars of upland cotton was found by Chapman (1938) in studies conducted in Texas. Brazzel and Martin (1956) reported that of 107 diploid, tetraploid, and their hexaploid F_1's, only one, *G. hirsutum* CV MW-147, from Guatemala, had fewer larvae recovered per gram of boll weight than from Deltapine 15, a commercial upland cotton cultivar of their day. Sixty-eight of the 107 genotypes were found to have more ($P \leq 0.05$) larvae recovered per gram of boll weight, including *G. thurberi* and *G. arboreum*. However, in other experiments, significantly ($P \leq 0.05$) fewer larvae were recovered from *G. thurberi* than from Deltapine 15. The apparent reason for these conflicting data is the extremely small bolls of *G. thurberi*, so small in fact that increased mortality from lack of food supply and extremely slower growth rates would be obvious. On a more positive note, Brazzel and Martin found a high degree of antibiosis, measured as the ratio of the number of larvae that enter the bolls to the number that survive or exit, in an F_1 of Stoneville 2B × *G. tomentosum*; but again, boll size may have created an artifact, as the boll size reported for Deltapine 15 was three times that of the resistant F_1. *Gossypium thurberi*, MW-298, and *G. stocksii* showed similar resistance but had bolls only about 7% of the size of Deltapine 15. The authors suggested that the resistance in *G. thurberi* was actually escape or nonpreference because of the morphology of vegetative and fruiting structures and not the result of extremely small bolls. These structures included a tight fitting calyx without triangular lobes and/or flared bracts. On the other hand, they observed that moths were attracted to genotypes having heavy, coarse pubescence and leaves having heavy veins, with oviposition taking place along the veination to the exclusion of fruiting forms, thus also making the newly hatched larvae more susceptible to weather, predators, parasites and chemical control. Smith *et al.* (1975) concluded 20 years later that plant hairs are attractive to the pink bollworm for oviposition, but also may impede movement of first-instar larvae, making it a resistance mechanism also. As noted above, quarantine and cultural control methods have worked well in Texas, the Midsouth, and southeastern United States, but these measures have been only partially successful in the irrigated cotton-producing regions of the far West. Producers in these regions prefer full-season cultivars and are reluctant to harvest and destroy residue before the induction of diapause or resting phase in pink bollworms. This creates an environment conducive to large-scale overwintering of the pink bollworm and therefore makes control or eradication much more difficult (Ridgeway, 1984).

Records indicate that this pest was eradicated twice in Arizona before 1958 (Reynolds and Leigh, 1967). That year the pink bollworm again was found in Arizona but federal and state agencies were not able to bring the pest under control. By 1965 it had spread northward and westward into southeastern California. Yield losses to the pink bollworm in that state were reported for the first time in 1966.

The desirability for the production of full-season, highly productive cottons of excellent quality in Arizona and California has prompted considerable research into HPR to the pink bollworm, primarily by the USDA.

Wilson and Wilson (1975) tested 304 race stock lines using a larval diet of carpel walls and 258 lines using a diet consisting of boll contents. Of these, 42 lines significantly affected growth and development rates of larvae fed the carpel wall diet and 23 lines adversely affected pink bollworm larvae fed boll content diets. Further analysis of these race stocks and hybrids with commercial cultivars failed to support resistance (Wilson and Wilson, 1977).

Several morphological characters have been evaluated extensively since 1977 (Wilson *et al.*, 1980a; 1981, 1987; George and Wilson, 1982; Wilson, 1989; Wilson and George, 1984). These include nectarilessness (Wilson and Wilson, 1976a; Wilson and George, 1982), glabrousness (Wilson and Wilson, 1976b; Wilson *et al.*, 1980a), and okra leaf (Wilson *et al.*, 1987). Nectarilessness, the absence of nectaries on leaves, bracts and flowers, imparts a low but consistent level of resistance to the pink bollworm (Wilson, 1982; Niles, 1980). Conflicting reports on the glabrous, or smooth plant, character suggest that, while it may be less preferred for oviposition, its nonpreference is not sufficient to categorize glabrous cottons as resistant to the pink bollworm (Wilson and George, 1982). Wilson *et al.* (1987) suspected that resistance imparted by the okra leaf trait was caused by a change in the microclimate around the boll that affected the behavior of the pink bollworm. However, mixed results were obtained when this character was evaluated in different genetic backgrounds. Early maturity will provide an escape mechanism from pink bollworm damage but extreme earliness resulted in a significant decrease in yield potential under Arizona's long-season production system (Wilson *et al.*, 1980a; 1981). Combinations of these traits have, in some instances, shown an additive effect for resistance (Wilson, 1987, 1989).

Wilson (1982) identified six stocks of upland cotton that apparently possess an unidentified mechanism(s) of resistance. These were Coker Foster 300, an obsolete American cultivar; Laxmi, an Indian cultivar; T-86 mut × Deltapine G24-2 (a race stock × cultivar hybrid); AET-Br-2-1; AET-Br-2-8; and 7203-14-104. In addition to these six, one breeding line, AET-5, has shown consistent antibiotic resistance (Wilson *et al.*, 1980a,b, 1979; Wilson, 1987). AET-5 has a complex pedigree that includes Triple Hybrid material, (*G. arboreum* × *G. thurberi*) × *G. hirsutum*, and its resistance reaction supports the findings of early workers that *G. arboreum* and *G. thurberi* are resistant to pink bollworm (Wolcott, 1927; Fenton, 1928). Antibiotic resistance has also been documented in two okra leaf breeding lines (George and Wilson, 1982; Wilson *et al.*, 1987). In addition to these unknown antibiotic factors, hemigossypolone, heliocides H1 and

H2, and gossypol have been shown to reduce larval weight gain and to slow development (Lukefahr *et al.*, 1977).

Two cultivars that have okra-leaf-shaped leaves have been released in the United States and several nectariless cultivars have been released as commercial cultivars. However, these were released and grown in the Midsouth or southeastern United States where pink bollworm is not a problem; no cultivar of cotton has been released specifically for resistance to the pink bollworm in the United States.

V. COTTON BOLLWORM AND TOBACCO BUDWORM

The cotton bollworm, *Helicoverpa zea* (formally *Heliothis zea, H. obsolete*, or *Chloridea obsoleta*) has been known to attack cotton in the United States since the 1800s. By 1905, Quaintance and Brues (1905) recorded that the bollworm was recognized as a serious pest that occasionally caused severe damage and loss of yield. Bishop (1929) noted that this pest was responsible for the loss of $8.5 million damage annually to the U.S. cotton crop by the 1920s. The most severe damage occurred in East Texas, Oklahoma, and Arkansas. In some seasons, the pest caused serious damage in Louisiana, Mississippi, and Alabama.

In more recent years we have learned that the tobacco budworm, *Heliothis virescens* (F.), also attacks cotton, inflicting the same type of damage as the bollworm. It is reasonable to assume that the tobacco budworm was categorized as a cotton bollworm in older reports of outbreaks, as they are difficult to distinguish. At any rate, these two species have become known as the bollworm–budworm complex (BBC).

A. The Old South—Minor Pests

The history of this complex in the United States is much less understood than that of the boll weevil or pink bollworm. The bollworm attacks a variety of host crops grown in this country, including corn, cotton, tomato, alfalfa, cowpeas, and a number of garden vegetables. On the other hand, the tobacco budworm attacks fewer crops, and therefore, populations build much more slowly during the growing season than populations of cotton bollworm. Prior to mechanization, farmers in the southeastern and southern United States grew corn as feed for mules and cotton as a cash crop. The mules, of course, were a necessary source of power and cotton evolved as the only cash crop available to these producers. Farmers were self-sufficient in vegetables, poultry, and meats; movement of these commodities over great distances were prohibitive, resulting in only local markets.

This pastorial life created a situation where the bollworm was essentially confined to corn and vegetables, to the exclusion of cotton, until mid-July or early August, at which time it would be forced to seek more succulent hosts. During years of plentiful rain, cotton was a most suitable host, being tall and succulent, with a large crop of floral buds and young bolls for developing larvae, and an adequate nectary flow for adults, such that significant losses could occur. During dry years, cotton was not a very desirable host, as the plants would be small, with few floral buds and tough, dry carpel walls on reduced numbers of bolls, and little nectary flow for adult moths. This situation probably accounts, to a great extent, for the bollworm being a sporadic pest of cotton in the early days of U.S. production. Small, drought-stressed cotton, to this day, is often not attacked by the bollworm to any great extent.

B. Machine and Pesticide Era—Major Pests

Two events occurred that thrust the bollworm into major pest status in the United States. First, and probably the least important of the two, was the movement from animal power to machine power. As the number of mules declined, the land area in the U.S. Cotton Belt devoted to corn declined. This meant that bollworm moths moving into cotton from maturing corn on reduced land area, or from wild hosts withered under midsummer temperatures, or that moths migrating from corn and other hosts further south had fewer choices of hosts, thereby becoming more concentrated in cotton fields.

The second, and more documentable, was the destruction of natural predators through efforts to chemically control the boll weevil. Within a few years after its recommendation by the USDA for boll weevil control, calcium arsenate use was recognized as causing build-ups of aphids, *Aphis gossypii*, and, in many cases, BBC (Folsom, 1928; Sherman, 1930; Ewing and Ivy, 1943). As noted previously, calcium arsenate gave way to the more powerful chlorinated hydrocarbons just after World War II, and these were replaced shortly by the more powerful organophosphates in the 1950s. During these years, producers looked to insecticides as a way to annihilate insect pests, and increased resistance within populations of both *H. zea* and *H. virescens* was met first with higher dosages and then with innovative mixtures of chemicals.

C. Host Plant Resistance

It became painfully clear by the 1960s that unabated use of pesticides for the control of bollworms and budworms was unacceptable and unworkable. With resistant populations of these insects causing control failures with increasing regularity, scientists again turned their attention to host plant resistance.

1. Early Maturity

The "Texas system" of extremely short-season production, coupled with early stalk destruction and autumn tillage (to reduce boll weevil, bollworm, and budworm pupal survival), has reduced losses to the BBC and dramatically reduced the amount of insecticides needed for their control. By reducing the amount of insecticide needed to control boll weevils, producers can avoid or delay the build-up of boll worms. This delay results from maximum use of the complex of natural predators to the BBC that exist in a cotton ecosystem.

East of Texas, where rainfall amounts create a very different ecosystem, genetic earliness of maturity is still valuable, allowing producers to "set" an acceptable crop in a shorter time frame and before numbers of bollworms and budworms build to devastating proportions. Producers in these regions will treat more often for BBC than the Texas producers, but treatment will be delayed. This delay helps prevent the selection of resistant populations. Earlier maturity has also become a cornerstone of cotton culture in these states. So, while early maturity is an escape mechanism, it is a genetic means of producing cotton without or with reduced chemical control of the BBC.

2. Allelochemics

In the early 1960s, scientists turned their attention to possible biochemical sources of host plant resistance to bollworms and budworms. Cook (1906) had suggested that the multitude of lysigenous glands found in cotton could be a defence mechanism against insects. Bottger *et al.* (1964) were the first to report that cotton lacking gossypol glands suffered more damage by a number of insects, including bollworms, than did their glanded counterparts. By 1966, according to Lukefahr and Martin (1966), scientists had determined that these glands contained gossypol, querimeritrin, gossypurpurin, gossypectin, gossycaerulin, gossyfulrin, gossypitrin, and gossyrerdurin. Evidence had already surfaced that these glands, when intact, were more toxic than an equivalent amount of pure gossypol (Eagle *et al.*, 1948).

a. Gossypol

The toxicity of gossypol and related compounds to the BBC has been confirmed many times since Cook's observation of 1906 (Lukefahr and Martin, 1966; Lukefahr *et al.*, 1966a, 1977; Lukefahr and Houghtaling, 1969; Shaver and Lukefahr, 1971; Wilson and Lee, 1971; Stipanovic *et al.*, 1976; Sappenfield and Dilday, 1980; Shaver *et al.*, 1980; Jenkins *et al.*, 1982, 1983; Zummo *et al.*, 1984). Quaintance and Brues (1905) and Cook (1906) noted the near immunity of certain strains of *G. barbadense* to the bollworm to be the result of it's "oil glands" and Smith (1962) confirmed that the gossypol content of Pima (*G. barbadense*) was about triple that found in cultivated upland cottons. Although Smith (1962) could not attribute resistance directly to gossypol, he did compare the gossypol content of Pima, a Coker upland cultivar, and a glandless experimental line and found that bollworm larvae growth rates decreased as gossypol levels increased. Armed with this information, Lukefahr and Martin (1966) and Shaver and Lukefahr (1969) confirmed that gossypol was toxic to both the cotton bollworm and the tobacco budworm when mixed in artificial diet.

The next step in the development of elevated levels of gossypol in upland cotton involved the introgression of this trait through a plant introduction from Socorro Island, Mexico. Originally identified in 1897, this biotype, according to Fryxell and Moran (1963), was reported again around 1900 and again in 1931. This feral cotton was collected by Moran in 1957 and classified as *G. hirsutum* var. *punctatum* sensu Hutchinson. Fryxell and Moran (1963) described this cotton as having prominent pigment glands and foliar nectaries, but lacking bract nectaries.

Lukefahr and Houghtaling (1969) reported on the successful recovery of high gossypol from this Socorro Island collection when crossed with Deltapine 15 and then crossed and backcrossed once to the experimental nectariless strain M-11, obtained from J. R. Meyer. Experimental line XG-15 was selected from the F_5BC_2 generation and had a total square gossypol content of 1.7%, whereas buds of Deltapine 15 had only 0.44%. This high gossypol, experimental line sustained 59% fewer damaged squares and 47% fewer damaged bolls than did M-8, a stable upland experimental genotype.

Chemical determination of gossypol content is, as with most chemical analyses, a slow process and requires breeders to evaluate fewer progeny than desirable, often dictating selection of poor agronomic performers for the sake of maintaining high levels of gossypol. This limitation was first circumvented by Wilson and Lee (1970) who reported positive and significant association between seed gossypol content and gland numbers in cotyledonary leaf petioles. However, these authors pointed out that they could occasionally get a high number of glands in upland plants that were homozygous $Gl_2Gl_2Gl_3raiGl_3rai$ for gossypol that would not have the gossypol level of some "heavily glanded" cottons. Wilson and Smith (1976) surmised that a high potency Gl_3 allele was responsible for flower gossypol of 2.63% compared with a content of 0.95% in the normal gossypol line, Empire Glanded. Lee (1974) had substantiated the existence of the Gl_3 high-potency allele and proposed that it controlled ovary or carpel wall glanding and therefore gossypol content. He found an association between the allelic configuration at the Gl_3 locus and pitted or rugate fruit surface such that selection for rugate bolls would result in selection for very high levels of gossypol.

Sappenfield *et al.* (1974) discussed a selection procedure for high gossypol that has enjoyed the widest acceptance among cotton breeders in the United States. This system involves the selection of progeny having an increased number of glands in the sepal tips; normal glanded upland has none or only a few glands in this area. Genotypes having varying levels of gossypol can be selected with this system by varying the selection pressure for number and size of glands in sepal tips. Tamcot HQ95, released by the Texas Agricultural Experiment Station, has about three glands per sepal tip while Tamcot 2111 averages just over one per tip. Part of the selection procedure for Tamcot HQ95 involved evaluation for bollworm resistance, but not in the selection of Tamcot 2111. The overall gossypol level of Tamcot HQ95 is similar to that of other normal gossypol upland cultivars.

Since first-instar larvae of the BBC feed along the calyx margins, development of genotypes with gossypol glands densely distributed in this area could improve resistance to BBC (personal communications, Dr. Jack Jones, Louisiana Agric. Exp. Sta.) without increasing overall gossypol content. The thrust behind this kind of reasoning is that cotton seed crushers and the cattle feed industries in the United States do not want cottonseed with elevated levels of gossypol. Such an event would decrease the value of cottonseed and cottonseed meal as a poultry and cattle feed, gossypol being toxic to nonruminants and young cattle (Bailey *et al.*, 1990), and increase the cost of removal of gossypol from cottonseed oil for human consumption.

The other research that deserves special mention here is that of Jenkins *et al.* (1982). Methods to achieve uniform field infestations of tobacco budworm were developed by these USDA scientists at Mississippi State, Mississippi. These techniques involve laboratory rearing, handling, and distribution of first-instar larvae to field plots. Using this technique, Parrott *et al.* (1981) reported five lines having resistance to the tobacco budworm: BW-76-31 DH, AET-5, NC-177-16-30, HG 469, and HGT-216.

As indicated in the review of Lukefahr and Martin (1966), the 1960s saw increasing interest in exploring and understanding the complex allelochemicals conditioning host plant resistance in cotton to insects. Eight compounds were reference by Lukefahr and Martin (1966) as being compartmentalized in gossypol glands. Bell and Stipanovic (1977) reviewed the chemical composition of cotton pigment glands and reported 23 compounds (reduced to 18 in a later publication (Bell, 1986)) found in gossypol glands, although 4 predominated in upland cotton and 10 predominated in Pima. Chan and Waiss (1981) reported to the Beltwide Cotton Physiology Conference that gossypol glands not only contained gossypol and related terpene aldehydes, but also contained several flavonoids and condensed tannins, suggesting that selection for elevated levels of gossypol may increase the levels of flavonoids or other compounds that dictate resistance to the BBC. Obviously, reports of gossypol content in different genotypes and genetic backgrounds can mean profoundly different things and may be the reason for much of the variability in toxicity reports. Bell and Stipanovic (unpublished data) noted that heliocide H1 is four times more toxic to the tobacco budworm than heliocide H2 and that gossypol is 12 times more toxic than hemigossypolone, all of which occur in gossypol glands. Because of the expense as well as the selection difficulty, no program in the United States devotes a concentrated effort to breeding for specific terpenoid aldehydes found in the lysigenous glands of cotton.

As noted previously, levels of seed gossypol and associated compounds substantially higher than those found in modern cultivars are unacceptable in the United States today. Fryxell (1965) reported that plants of the wild Australian diploid, *Gossypium sturtianum* Willis, have glands distributed throughout its foliage but not in its seeds. Dilday (1986) reported on finding a hexaploid [chromosome-doubled progeny of *G. hirsutum* ($4\times = 2n = 52$) × *G. sturtianum* ($2\times = 2n = 26$)] having a glanding pattern like the *G. sturtianum* parent. Altman (personal communication), USDA geneticist at College Station, has worked without complete success with this material for several years in efforts to recover a 52-chromosome upland with glanded plant-glandless seed. However, he has recovered types having reduced seed gossypol and normal or elevated foliage gossypol.

b. Condensed Tannin

In addition to the traditionally accepted allelochemicals reported as residing in gossypol glands, researchers have evaluated condensed tannins, cyanidin, delphinidin, catechin, chrysanthemin, isoquercetin, quercetin, and unidentified compounds as sources of resistance. Mixtures of polyphenolics called condensed tannins have received considerable attention over the past decade (Chan *et al.*, 1978; Schuster, 1979; Zummon *et al.*, 1983, 1984; Altamarino *et al.*, 1988, 1989; Schuster *et al.*, 1990; Lege *et al.*, 1991; Smith *et al.*, 1991). Tannins are believed to occur universally in higher plants as a protection mechanism from insects and diseases.

Chan *et al.* (1978) reported that growth of tobacco budworm larvae was retarded when diets containing freeze-dried and ground squares of race accession Texas 254, which contained 3.4% condensed tannins, were fed. Zummo *et al.* (1983) reported a negative relationship between tannin concentration in cotton plants and the number of bollworm-damaged fruit. Similar associations were reported by Schuster (1979) and Schuster and Lane (1980). However, Smith *et al.* (1991) reported that growth and survival rates of *Heliothis virescens* were not affected by level of condensed tannins in several breeding lines selected for elevated condensed tannin levels.

These discrepancies might result from different laboratory procedures for determining condensed tannins and different bioassays. Chan *et al.* (1978) used freeze-dried flower buds to adjust an artificial diet for tannin level; Zummo *et al.* (1984) reported that the best correlation was between reduced bollworm damage and relative astringency or tannin quality. Smith *et al.* (1991) evaluated survival, growth rates, and plant damage in breeding lines selected from crosses of upland cotton cultivars and wild accessions identified as having high levels of condensed tannins. Condensed tannins in these materials were 50 to 100% higher than those in Stoneville 213. The prospects for the usefulness of condensed tannins as a HPR allelochemical against the BBC appear dim at this time.

c. Unknown Sources

Lambert *et al.* (1982) identified five genotypes, three foreign and two U.S. breeding lines, exhibiting resistance to the BBC. The two U.S. breeding lines were high gossypol lines, but the resistance mechanism(s) in the three foreign lines was not identified. Culp *et al.* (1979) reported a slight level of resistance to the BBC in germplasm lines Pee Dee 695, a frego bract line; Pee Dee 875; and Pee Dee 8619. Although all of these lines were superior to the commercial controls, resistance measured as percentage BBC-injured squares, egg numbers or number of live larvae was minimal. This source of resistance was later designated as "Q-factor" resistance. Jenkins *et al.* (1984, 1985), utilizing the techniques reported above (Jenkins *et al.*, 1982) supported the work of Culp *et al.* (1979). Pee Dee 875 and 8619, normal bract cottons, yielded a significantly greater percentage of their potential yield under heavy tobacco budworm pressure than did commercial cultivars.

The 1985 report of Jenkins *et al.* also showed that Stoneville 506 and Tamcot CAMD-E cultivars were more resistant than five other commercial cultivars. Unpublished data of the author and associates support the finding that commercial cultivars differ in their degree of resistance or susccptibility.

3. Morphological HPR Traits

a. Nectarilessness

Myer and Meyer (1961) initiated a program in the 1950s to transfer recessive alleles (ne_1ne_2) for the absence of leaf and extrafloral nectaries found in *G. tomentosum* Nuttall. Their original motive was simply to use the nectariless trait as a genetic marker. C. L. Rhyne, Jr. (apparently in a personal communique), suggested to Meyer and Meyer that lack of nectaries could be important as a host plant resistance mechanism by depriving certain insects of an important source of food. D. F. Martin (personal communication to Meyer and Meyer) reported that plots of nectariless cottons interspersed with nectaried cottons had "about" half as many pink bollworms. Lukefahr and Rhyne (1960) subsequently reported reductions in populations of cabbage loopers and cotton leafworms in nectariless cotton versus a nectaried, commercial cultivar.

Lukefahr *et al.* (1965) found reductions in BBC eggs in nectariless phenotypes ranging from 29 to 46% compared with nectried cottons. In 1966, Lukefahr *et al.* reported significantly fewer bollworms on a nectariless and smooth experimental strain than on nectaried and hirsute cultivars. Shaver *et al.* (1970) noted that nectariless strains reduced fecundity of "*Heliothis* spp.," probably *Heliothis virescens* and *Helicoverpa zea*, by 50% and that reduction of trichomes from 500 to 50 or less (per unit not given) would reduce egg population by 60%. Maxwell *et al.* (1976) reported reductions of 16 to 58% in BBC-damaged squares in nectariless cotton; Schuster and Maxwell (1974) found a 58% reduction in damaged squares and 46% fewer bollworm eggs. There are reports that nectarilessness has no effect on certain components of the bollworm or budworm life cycle (Ha *et al.*, 1987; Lukefahr, 1977; Sosa *et al.*, 1981), but the preponderance of evidence suggests that the absence of leaf and bract nectaries reduces losses to the cotton bollworm and tobacco budworm.

b. Plant Pubescence

According to Gilliam (1963), villous surfaces were preferred by the bollworm as oviposition sites (Callahan, 1957), spurring speculation that smooth leaves would provide a measure of resistance. By 1963, several smooth cottons had been developed by transfer of the D_2 gene from *G. armourianum*, which resulted in completely smooth leaves, petioles, and stems. Other smooth-leaf cottons, i.e., having trichomes on petioles and stems only or having reduced leaf trichomes, have been developed through selection within *G. hirsutum* germplasm. Gilliam confirmed that the cotton bollworm preferred extremely pubescent types and that smooth types effected increased egg losses and increased larvae mortality.

Lukefahr *et al.* (1965) reported 61 to 72% reduction in cotton bollworm and tobacco budworm eggs on glabrous cotton compared with hirsute types, and Lukefahr *et al.* (1971) found consistent and similar results from 1965 through 1969. Jones *et al.* (1977) found significant reductions in Louisiana; Robinson *et al.* (1980) reported similar results in Texas. However, Culp *et al.* (1979) found no differences in percentage BBC-injured squares, number of live larvae, or number of eggs on a glabrous–nectariless line, La. 17801, a glabrous cultivar, Coker 420, and the commercial hirsute control, Stoneville 213. Sosa *et al.* (1981) reported reductions of over 40% of eggs and larvae on LA 17801 when compared with standard cultivars. Robinson *et al.* (1981) reported data suggesting that the resistance in the Socorro Island collection could have resulted from smooth leaves and not from elevated levels of gossypol.

There appears to be little doubt that smooth leaves are not preferred by BBC females for oviposition sites. As will be pointed out later, other insects show a definite preference for smooth(er) genotypes.

c. Plant Pigmentation

Two other morphological characters slightly affect the cotton bollworm and the tobacco budworm. Bailey (1981) reported that growth of *H. virescens* larvae was retarded when yellow or orange pollen was fed instead of the more normal cream pollen. Yellow pollen depressed growth rates more than orange. Bhardwaj and Weaver (1983) observed more bollworm eggs on red leaf cotton than on normal green but significantly less damage occurred on the red leaf line.

VI. OTHER INSECT PESTS OF COTTON

While the boll weevil, pink bollworm, cotton bollworm, and tobacco budworm are considered the major insect pests of cotton in the United States, several other injurious insects are a part of the cotton ecosystem in this country. These include the cotton fleahopper (*Pseudatomoscelis seriatus* (Reuter)); the clouded plant bug (*Neurocolpus nubilus* (Say)); the tarnished plant bug (*Lygus lineolaris* (Palisot de Beaurosis)); lygus bugs (often referred to as plant bug, *Lygus hesperus* (Knight)); thrips (*Frankliniella tritici* (Fitch), *F. fusca* (Hinds), *F. exigna* (Hood), *F. occidentalis* (Pergande), *Thrips tabaci* (Lindeman), *Sericothrips variabilis* (Beach), *Caliothrips fasciatus* (Pergande)); aphids (*Aphis gossypii* (Glover), *A. caraccivora (Khochi)); the leaf miner (Liriomyza munda* (Frick)); the cotton leaf perforator (*Bucculatrix thurberiella* (Busck)); and the whitefly (*Trialenrodes abutilonea* (Haldeman), *Bemisia tabaci* (Gennadius)). Of this group, those referred to as "plant bugs" (family Miridae), i.e., clouded plant bug, tarnished plant bug, lygus, and fleahopper, are usually the most destructive.

Historical perspective on damage caused by these insect pests is somewhat clouded. Thrips and plant bugs have been indicted as pests that delay maturity in cotton, but even on this issue there is considerable disagreement. As recently as the 1960s and early 1970s, entomologists often denied that these insects were pests. For example, when the author began work as a breeder in Arkansas in 1974, there were a number of agronomists and entomologists that did not believe that plant bugs, predominately the clouded and tarnished, caused economic losses in cotton. Another commonly expressed sentiment was that thrips occurred so early (damaged plants would recover) or so late (too late to cause yield losses) in the growing season that insecticidal control was not necessary. Common advice during early-season infestations was to treat with insecticide or go fishing for 2 weeks; either way, the problem would be solved.

However, renewed interest in reducing the length of growing season required for cotton production in all regions of the U.S. Cotton Belt focused scientists on any pest that delayed maturity. Getting a stand of cotton and having rapid, early-season growth gained priority, as was noted earlier in this chapter, and breeders began looking at HPR against early-season pests that could delay maturity by destroying the first squares produced on young plants.

A. Plant Bugs

1. Historical Review

Lygus elisus, referred to as the tarnished plant bug at the time, was reported to damage cotton in the Imperial Valley of California as early as 1913 (McGregor, 1927). The insect was not a problem in 1916 but caused considerable damage again in 1917 and 1918. McGregor reported that *L. elisus* was probably native to western America.

Benedict and Leigh (1976) reported that *Lygus hesperus* (Knight) was the most important insect pest of cotton in the San Joaquin Valley of California and was a primary pest of cotton in much of the Western United States. However, since plant bugs were sporadic in appearance and damage in Texas, the midsouth, and southeastern United States, and since insecticides that controlled the boll weevil and bollworm complex in these areas probably controlled plant bugs also, there was little interest in the United States in HPR to plant bugs before the 1960s. As noted elsewhere, this interest was spurred by resistance to insecticides in populations of the boll weevil and the BBC. Hanny *et al.* (1979) in Mississippi and Peck and Tugwell (1976) in Arkansas suggested that the plant bug was not a serious pest of cotton in the Midsouth before about 1970.

2. Morphological Traits

The search for HPR characters has identified four possibilities: plant bug suppression factor(s); nectarilessness; plant pubescence; and, possibly, high gossypol.

a. Plant Bug Supression Factor

Tingey *et al.* (1975) identified 19 cotton genotypes that adversely affected *Lygus hesperus*. Breeding line 247-1-6 (a high-gossypol, smooth, nectariless upland), four *G. barbadense* lines, and five upland stocks with cytoplasms from *G. harknessi* or *G. longicalyx* displayed the greatest resistance, as identified by reduced growth rates, reduced survival, or reduction in the number of nymphs hatched per plant. Meredith *et al.* (1979) reported that cytoplasms from *G. herbaceum, G. arboreum, G. anomalum, G. barbadense*, or *G. tomentosum* had no effect or provided no resistance to the tarnished plant bug.

b. Nectarilessness

Nectarilessness is the character most often noted as effecting a level of resistance to plant bugs (Schuster and Maxwell, 1974; Meredith and Laster, 1975; Schuster *et al.*, 1975; Benedict and Leigh, 1976; Schuster and Frazier, 1976; Lukefahr, 1977; Benedict *et al.*, 1981). Reductions of nymphs and adults on nectariless lines ranged as high as 60% and egg-lay was reduced as much as 90%, relative to nectaried cotton. Frego bract cottons have been noted to reduce boll weevil damage but these types are extremely susceptible to plant bugs (Milam *et al.*, 1982) and fleahopper (Young *et al.*, 1986). Several scentists are attempting to combine frego bract for resistance to the boll weevil with the nectariless trait to reduce sensitivity to plant bugs. A measure of success along this avenue was reported by Jones *et al.* (1989).

c. Plant Pubescence

A common assumption is that trichomes on cotton provide mechanical resistance to the movement of and damage caused by plant bugs. A logical continuance of this line of thought is that pilose, or extremely hairy, types would be even more resistant. Work of Meredith and Schuster (1979) supports this reasoning. However, Tingey *et al.* (1975) found that *L. hesperus* actually preferred pilose over normal hirsute types in a free-choice test. In no-choice studies, no difference in ovipositioning was found.

Lukefahr *et al.* (1968, 1970) reported that glabrousness effects a reduction in fleahopper populations. Walker and Niles (1973) and Walker *et al.* (1974) later reported that glabrous lines reduced the populations of fleahopper but suffered as much or more damage than their hairy counterparts. This same effect (Laser and Meredith, 1974; Meredith and Schuster, 1979) is seen in the case of plant bugs, limiting to date the use of the glabrous trait as a source of resistance.

Jenkins *et al.* (1977) reported resistance to the tarnished plant bug in a smooth plant accession called TIMOK 811. However, Meredith and Schuster (1979) reported that crosses of TIMOK 811 with smooth parents could result in hirsute F_1's because the dominance of glabrous alleles is not complete in all germplasms (Lee, 1971). Additional work with TIMOK 811 and other smooth germplasms is necessary to develop resistance in glabrous types.

d. Gossypol

Increased levels of gossypol that effect resistance to the BBC have no effect on plant bugs. However, glandlessness, or the absence of gossypol in seeds and plant foliage, increases susceptibility or sensitivity (Tingey *et al.*, 1975; Benedict and Leigh, 1976). Early reports suggested that *Lygus hesperus* developed twice as fast on glandless cotton as on normal glanded types, but glandless lines developed in the last 10 years show less sensitivity to this insect complex, spurring hopes that resistance can be developed within glandless types.

B. Fleahopper

1. Historical Review

The cotton fleahopper is usually catorgized as a part of the "plant bug complex" along with *Lygus* spp. and *Neurocolpus nubilus*, since these insects all attack the cotton plant in the same general way, i.e., feeding in very young squares and meristematic tissue early in the growing season. If these insects delay moving into cotton until after flowering, the producer will often experience little loss of yield or delayed maturity. The comments relative to plant bugs are applicable to the fleahopper; however, they are separated here to review the large amount of research that has been dedicated to "cotton fleas."

Hunter (1926) noted that the cotton fleahopper, then classified as *Psallus seriatus* Reuti, had been credited for causing shedding of small squares and abnormal, excessive growth of the main stem, later to be called "cabbage" or "crazy top." Still later it would be noted that this was nothing more than the destruction of the apical meristem as a result of feeding by plant bugs, which would release apical dominance, thus allowing the growth of several adventitious buds located at compressed upper main stem nodes. This insect had been associated with this type of damage for 25 years in South Texas, being first identified in 1896 by L. O. Howard. In 1923, the same plant symptoms occurred in much of East Texas as far as the northern border with Oklahoma, and chemical control measures were under evaluation by 1927 (Reinhard, 1927). Around this same time, Eddy (1927) reported that this insect damaged cotton in South Carolina and that there were reports that it had damaged cotton in other southeastern states for several years.

2. Morphological Traits

a. Canopy Density

Traditionally the cotton fleahopper has been of greater importance to Texas than to other cotton producing states, and the literature reflects this. Lukefahr *et al.* (1966b) reported that experimental strain 1514, a strain both nectariless and glabrous, might have a low degree of resistance to the fleahopper. The authors of this research did not suggest that either glabrousness or the absence of nectaries was the mechanism, rather they attributed it to the observation of Mr. R. L. Garr in a personal communication that 1514 had a more open or less compact terminal than other cottons evaluated.

b. Plant Pubescence

By 1968, Lukefahr *et al.* (1968) had documented that smooth cottons suppressed fleahopper populations, although the documentation would soon surface that smooth cottons are hypersensitive to plant bugs, including fleahoppers (Walker *et al.*, 1974; Niles *et al.*, 1974; Laster and Meredith, 1974; Jones *et al.*, 1977; Meredith and Schuster, 1979). Pilosity, on the other hand, may impede the movement of fleahopper nymphs and/or adversely affect oviposition (Meredith and Schuster, 1979; Lidell *et al.*, 1986).

c. Nectarilessness

Essentially the same comments concerning nectarilessness as a HPR character against plant bugs can be made for the fleahopper. Pest populations are reduced up to 60% (Laster and Meredith, 1974; Schuster *et al.*, 1975; Meredith, 1976; Lidell *et al.*, 1986) on nectariless lines compared to nectaried cottons. The genetic background of nectariless strains apparently plays a role in resistance to fleahoppers. Lidell *et al.* (1986) found one nectariless strain, 7Ane × 6M, that had moderate resistance but reported that another seven nectariless lines suffered significant loss of yield and/or earliness components under high fleahopper populations.

d. Gossypol

Increased levels of gossypol have been shown to be the most consistent HPR mechanism to the BBC and, on a few occasions, it has been noted to effect resistance to fleahoppers (Cowan and Lukefahr, 1970; Lukefahr and Houghtaling, 1975; Schuster and Frazier, 1976). However, inconsistent results indicate caution in attributing resistance to this character.

C. Minor Pests

As noted above, there are a number of minor insect pests of cotton in the United States and a paucity of research aimed at host plant resistance has been conducted on these insects. Brown, in his 1917 text, did not reference plant bugs (other than the fleahopper), thrips, leaf miner, leaf perforator, or whitefly. Christidis and Harrison (1955) essentially ignored all cotton insects, other than jassid, and made only passing reference to host plant resistance. Obviously then, these insects have not impacted U.S. cotton production to the same extent that the boll weevil, the bollworm-budworm complex, and the pink bollworm have. When there is such a small amount of research relative to any scientific question, contradictions often appear in the literature. This is particularly true in the area of host plant resistance where genetic background and cytoplasm impact the expression of resistance.

1. Whitefly

In 1967, Butler and Muramoto (1967) referred to a report by Mound (1965) that banded-winged whitefly populations were larger on pubescent cotton types than on glabrous cultivars. Butler and Muramato, however, found no significant association between banded-winged whitefly populations and leaf pubescence except for progeny of a cross of germplasm line Dwarf A × Lankart.

Banded-winged whitefly population rarely establish on okra- or super-okra-leafed cottons and infestations seem to be less severe on smooth-leaf genotypes relative to a normal levels or pubescence (Jones *et al.*, 1975). Lukefahr (1977) reported a personal communication with Clower and Jones of Louisiana State University stating that whitefly and thrips are attracted to high gossypol genotypes. The author has made this same observation. The whitefly deposits a sticky "honewdew" on cotton lint that lowers the value of raw cotton because of spinning difficulties (Perkins, 1983).

The sweet potato whitefly (SPW) was reported on cotton in Arizona and California in the 1920s (Russell, 1975). Butler *et al.* (1986) reported that *Bemisia tabaci* was only a sporadic pest of cotton prior to 1981, but had been a serious pest since that date. This pest can build to extremely high populations and is known to transmit several infectious diseases from lettuce, (*Lactucia sativa* L.), melons (*Cuncumis melo* L.), cucurbits (*Cucurbita* spp), and sugar beets (*Beta vulgaris* L.) (Duffus and Flock, 1982). The SPW was found in overwhelming numbers in the Lower Rio Grande Valley of Texas in 1991. Numbers were apparently sufficient to cause premature leaf dessication and to cause significant yield losses. The current wisdom is that *B. tabaci* will become more of an established pest of cotton in this geographical area because of the diversity of alternate hosts, including vegetables, melons, and wild hosts.

Limited evidence suggests that smoother leaves confer some degree of resistance/nonpreference to the SPW (Butler *et al.*, 1986; Butler and Wilson, 1984; Butler and Henneberry, 1984). Butler and Henneberry (1984) found that eight smooth-leaf isolines harbored significantly fewer adults of *B. tabaci* than did their original, pubescent parents. Butler and Wilson (1984) reported that same year that among pink bollworm-resistant AET-5 isolines with and without leaf hairs, significantly fewer adults of banded-winged and sweet potato whiteflies were found in the smooth-leaf lines. However, the DES-24 smooth-leaf isoline had significantly more whiteflies than did the original, hairy parent. Okra-leaf and/or nectariless-ness actually conferred some susceptibility in certain genetic backgrounds. Mixed results for the effectiveness of the smooth-leaf character in different genetic backgrounds were also reported by Butler *et al.* in 1986.

2. Aphid

The cotton aphid was recognized as a pest of cotton before 1854 according to a reference found in a review of the subject by Paddock (1919). The economic damage caused by the cotton aphid today is in dispute; yield losses are rarely definable. But in the days of hand harvest in the United

States, a large infestation of aphids could make hand-picking even less of a desirable chore. The sticky exudate, or honeydew, from aphids is the major objection to this insect because it causes considerable problems in the spinning of raw cotton. This problem continues today with no appropriate means of removing the honeydew from raw stocks. In addition to the stickiness problem, bacteria and fungi grow on the honeydew, discoloring the fiber. The problem with aphids has been exacerbated in recent years because season-long use of pyrethroid insecticides destroys aphid's natural enemies and the insect goes unchecked, being unaffected by this class of insecticides. The ability of aphids to soar in numbers after insecticide use was noted as far back as 1930 by Folsom and Bondy (1930). Dunnam and Clark (1939) examined the possibility that pilose cotton cultivars would be desirable because more calcium arsenate would be held between the leaf trichomes, thereby improving insecticidal control. However, in 1937, they reported that the number of aphids "increased in direction proportion to the number of hairs on the lower leaf surface." There are other reports that pilose cottons are favored by the cotton aphid (Harding and Cowan, 1971).

VII. CONCLUDING REMARKS

Host plant resistance is universally accepted as the most desirable means of achieving control of insect pests. We often review the seemingly endless list of insects that cause economic damage to cotton and question why no insect-resistant cultivars are available, especially considering the large expenditures in human and monetary terms that have been applied to the cotton insect problem. We usually fail to consider the increase in yield potential as a mechanism of tolerance to insects, even though insecticides must be applied in many or most cases. We also fail to understand that today's cotton cultivars do carry a measurable degree of resistance to many of cotton's insect pests. For example, Jenkins (1982) noted that tobacco budworms reared on Stoneville 213 or Deltapine 61 cotton cultivars for 5 days weighed 83% less than those grown for the same period of time on artificial diet.

The plant breeding research teams in the United States have made giant strides toward controlling cotton's insect pests through HPR. Another benchmark is on the horizon. Genetically engineered plants await the identification of genes in almost any organism for transplant into cotton's genome. Genes from bacteria, *Bacillus thuringiensis*, that code for the production of a deadly toxin that dissolves the gut membrane of the tobacco budworm, cotton bollworm, and the pink bollworm have already been transferred into cotton. Commercial cultivars with the *Bt* gene are very close to release, probably within the next 5 years.

Continued success in traditional plant breeding and molecular biology dictate that exciting times are ahead as scientists continue to strive to feed and clothe an expanding world population.

References

Adkisson, P. L. (1964). "Comparative Effectiveness of Several Insecticides for Controlling Bollworms and Tobacco Budworms," Texas Agric. Exp. Sta. Bull. MP-709.

Adkisson, P. L., and Nemec, S. J., (1966). "Comparative Effectiveness of Certain Insecticides for Killing Bollworms and Tobacco Budworms," Texas Agric Exp. Sta. Bull. 1048.

Altamarino, T. P., Smith, C. W., and Lege, K. E. (1989). Combining ability for condensed tannins in cotton. *In* "Proceedings, Beltwide Cotton Prod. Res. Conf. 1989," p. 138.

Altamarino, T. P., Smith, C. W., Love, J., Bell, A. A., and Stipanovic, R. D. (1988). Progress in developing high tannin cotton for *Heliothis* resistance. *In* "Proceedings, Beltwide Cotton Prod. Res. Conf. 1988," p. 553.

Anonymous (1965). "Cotton Insects: A Report of a Panel of the President's Science Advisory Committee." U.S. Gov't. Printing Office. Washington, D.C.

Bailey, E. M., Jr., Smith, C. W., and Coppock, C. E. (1990). "Gossypol: Status in Cotton and Impacts on Livestock," Texas Agric. Exp. Sta. Bull. MP-167.

Bailey, J. C. (1981). Growth comparison of *Heliothis virescens* (F.) larvae fed white, yellow and orange cotton pollen. *In* "Proceedings, Beltwide Cotton Prod. Res. Conf. 1981," p. 79.

Ballou, H. A. (1919). The poisoning of the boll weevil. *Agric. News* **18,** 122,123.

Bates, S. L. (1989). "Evaluation of Two Cotton Race Stocks for Resistance to the Cotton Boll Weevil, *Anthonomus grandis* Boh." M. S. thesis, Texas A&M University, College Station, Texas.

Bates, S. L., Walker, J. K., and Smith, C. W. (1988). Detecting boll weevil resistance in converted race stocks by sampling single plants. *In* "Proceedings, Beltwide Cotton Prod. Res. Conf. 1988," pp. 552,553.

Bell, A. A. (1986). Physiology of secondary products. *In* "Cotton Physiology" (Mauney and Stewart, eds.). The Cotton Foundation, Memphis, Tennessee.

Bell, A. A., and Stipanovic, R. D. (1977). The chemical composition, biological activity, and genetics of pigment glands in cotton. *In* "Proceedings, Beltwide Cotton Prod. Res. Conf. 1977," pp. 244–258.

Benedict, J. H., and Leigh, T. F. (1976). Host plant resistance in cotton to lygus bugs and other insect pests. *In* "Proceedings, Western Cotton Prod. Conf. 1976," pp. 80–83.

Benedict, J. H., Leigh, T. F., Hyer, A. H., and Wynholds, P. F. (1981). Nectariless cotton: Effect on growth, survival, and fecundity of lygus bugs. *Crop Sci.* **21,** 28–30.

Bennett, R. L. (1904). "Early Cotton," Texas Agric. Exp. Sta. Bull. 75.

Bhardwaj, H. L., and Weaver, J. B., Jr. (1983). Bollworm resistance in red leaf cotton. *In* Proceedings, Beltwide Cotton Prod. Res. Conf. 1983," pp. 117–119.

Birds, L. S. (1975). "Tamcot SP21, SP23, SP37 Cotton Varieties," Texas Agric. Exp. Sta. Bull. L.-1351.

Bishop, F. C. (1929). "The Bollworm or Corn Earworm as a Cotton Pest," USDA Farmer's Bull. 1595.

Bottger, G. T., Sheehan, E. T., and Lukefahr, M. J. (1964). Relation of gossypol content of cotton plants to insect resistance. *J. Econ. Entomol.* **57,** 283–285.

Brazzel, J. R. (1961). "Boll Weevil Resistance to Insecticides in Texas in 1960," Texas Agric. Exp. Sta. Bull. PR-2171.

Brazzel, J. R. (1963). Resistance to DDT in *Heliothis virescens. J. Econ. Entomol.* **56,** 561.

Brazzel, J. R. (1964). DDT resistance in *Heliothis zea. J. Econ. Entomol.* **57,** 455.

Brazzel, J. R., and Martin, D. F. (1956). "Resistance of Cotton to Pink Bollworm Damage," Texas Agric. Exp. Sta. Bull. 843.

Bridge, R. R., and McDonald, L. D. (1987). Beltwide efforts and trends in development of varieties for short season production system. *In* "Proceedings, Beltwide Cotton Prod. Conf. 1987," pp. 44–48.

Bridge, R. R., and Meredith, W. R., Jr. (1983). Comparative performance of obsolete and current cotton cultivars. *Crop Sci.* **23,** 949–952.

Brown, H. B. (1927). "Cotton." McGraw–Hill, New York/London.

Brown, H. B. (1938). "Cotton," 2nd ed. McGraw–Hill, New York/London.

Buford, W. T., Jenkins, J. N., and Maxwell, F. G. (1967). A laboratory technique to evaluate boll weevil oviposition preference among cotton lines. *Crop Sci.* **7,** 579–581.

Buford, W. T., Jenkins, J. N., and Maxwell, F. G. (1968). A boll weevil oviposition suppression factor in cotton. *Crop Sci.* **8,** 647–649.

Butler, G. D., and Henneberry, T. J. (1984). *Bemisia tabaci*: Effect of cotton leaf pubescence on abundance. *Southwest Entomol.* **9**(1), 91–94.

Butler, G. D., Henneberry, T. J., and Wilson, F. D. (1986). *Bemisia tabaci* (Homoptera: Aleyrodidae) on cotton: Adult activity and cultivar oviposition preference. *J. Econ. Entomol.* **79,** 350–354.

Butler, G. D., and Wilson, F. D. (1984). Activity of adult whiteflies (Homoptera: Aleyrodidae) within plantings of different strains and cultivars as determined by sticky-trap catches. *J. Econ. Entomol.* **77,** 1137–1140.

Butler, G. D., Jr., and Muramoto, H. (1967). Banded-wing whitefly abundance and cotton leaf pubescence in Arizona. *J. Econ. Entomol.* **60,** 1176,1177.

Callahan, P. S. (1957). Oviposition response of the corn earworm to differences in surface texture. *J. Kansas Entomol. Soc.* **30,** 59–63.

Carruth, L. A., and Moore, L. (1973). Cotton scounting and pesticide use in Earn Arizona. *J. Econ. Entomol.* **66,** 187–190.

Chan, B. C., and Waiss, A. C., Jr. (1981). Evidence for acetogenic and shikimic pathways in cotton glands. *In* "Proceedings, Beltwide Cotton Prod. Res. Conf. 1981," pp. 49–51.

Chan, B. G., Waiss, A. C., and Lukefahr, M. (1978). Condensed tannins, an antiobiotic chemical from *Gossypium hirsutum. Insect Physiol.* **24,** 113–118.

Chapman, A. J. (1938). Pink bollworm. *Texas Agric. Expt. Sta. Annu. Rep.* **51,** 44.

Christidis, B. G., and Harrison, G. J. (1955). "Cotton Growing Problems." McGraw–Hill, New York.

Collins, G. S., Lacewell, R. D., and Norman, J. (1979). Economic comparison of alternative cotton production practices: Texas Lower Rio Grande Valley, *South J. Agric. Econ.* **11,** 79–82.

Cook, O. F. (1906). "Weevil Resisting Adaptations of the Cotton Plant," USDA Bull. 88.

Cowan, C. B., and Lukefahr, M. J. (1970). Characters of cotton plants that affect infestations of cotton fleahoppers. *In* "Proceedings, Beltwide Cotton Prod. Res. Conf. 1970," pp. 781–80.

Culp, T. W., Hopkins, A. R., and Taft, H. M. (1979). "Breeding Insect-Resistant Cottons in South Carolina," South Carolina Agric. Exp. Sta. Tech. Bull. 1074.

Dilday, R. H. (1986). Development of a cotton plant with glandless seeds and glanded foliage and fruiting forms. *Crop Sci.* **26,** 639–641.
Duffus, J. E., and Flock, R. A. (1982). Whitefly transmitted disease complex of the desert southwest. *Calif. Agric.* **36,**(11,12), 4–6.
Dunnam, E. W., and Clark, J. C. (1939). The cotton aphid in relation to the pilosity of cotton leaves. *J. Econ. Entomol.* **31,** 663–666.
Eagle, E. L., Castillon, E., Hall, C. M., and Boatner, C. H. (1948). Acute oral toxicity of gossypol in cottonseed pigment glands for rats, mice, rabbits, and guinea pigs. *Arch. Biochem.* **18,** 271–277.
Earnheart, A. T., Jr. (1973). "Evaluation of Cotton, *Gossypium hirsutum* L., Race Stocks for Resistance to Boll Weevil Attack." M.S. thesis, Mississippi State University, State College, Mississippi.
Eddy, C. O. (1927). "The Cotton Fleahopper," South Carolina Agric. Exp. Sta. Bull. 235.
Ewing, K. P., and Ivy, E. E. (1943). Some factors influencing bollworm populations and damage. *J. Econ. Entomol.* **36,** 602–606.
Fenton, F. A. (1928). Biological notes on the pink bollworm (*Pectinophora gossypiella* Saunders) in Texas. *Fourth Int. Cong. Entomol.* **2,** 439–447.
Folsom, J. W. (1928). Calcium arsenate as a cause for aphid infestation. *J. Econ. Entomol.* **21,** 174.
Folsom, J. W., and Bondy, F. F. (1930). "Calcium Arsenate Dusting as a Cause of Aphid Infestation," USDA Circ. 116.
Frisbie, R. E., Sprott, J. M., Lacewell, R. D., Parker, R. D., Buxkemper, W. E., Baghley, W. E., and Norman, J. W. (1976). A practical method of economically evaluating an operational cotton pest management program in Texas. *J. Econ. Entomol.* **69,** 211–214.
Fryxell, P. A. (1965). A revision of the Australian species of *Gossypium* with observations on the occurrence of *Thespesia* in Australia (Malvaceae). *Aust. J. Bot.* **13,** 71–102.
Fryxell, P. A., and Moran, R. (1963). Neglected form of *Gossypium hirsutum* on Socorro Island, Mexico. *Empire Cotton Grow. Rev.* **40,** 289–291.
Galinat, W. C. (1988). The origin of corn. *In* "Corn and Corn Improvement" (G. F. Sprague and J. W. Dudley, eds.), pp. 1–31. Am. Soc. Agron., Madison, Wisconsin.
George, B. W., and Wilson, F. D. (1982). Infestation of okra-leaf and normal-leaf near isolines by pink bollworm. *In* "Proceedings, Beltwide Cotton Prod. Res. Conf. 1982," p. 133.
Gilliam, F. E. M. (1963). A study in the response of bollworm, (*Heliothis zea* (Boddie)), to different genotypes of upland cotton. *In* "Proceedings, Beltwide Cotton Prod. Res. Conf. (formally "Proceedings of Fifteenth Annu. Cotton Imp. Conf.") 1963," pp. 80–88.
Glover, D., Glower, D. F., and Jones, J. E. (1975). Boll weevil and bollworm damage as affected by upland cotton strains with different cytoplasms. *In* "Proceedings, Beltwide Cotton Prod. Res. Conf. 1975," pp. 99–102.
Ha, S. B., Young, J. H., Wilson, L. J., and Verhalen, L. M. (1987). Effects of morphological traits in cotton on natural infestations of the cotton fleahopper and bollworm. *In* "Proceedings, Cotton Beltwide Prod. Res. Conf. 1987," p. 104.
Hanny, B. W., Cleveland, T. C., and Meredith, W. R., Jr. (1979). Effects of tarnished plant bug (*Lygus lineolaris*), infestation on presquaring cotton (*Gossypium hirsutum*). *Environ. Entomol.* **6,** 460–462.
Harding, J. A., and Cowan, C. B., Jr. (1971). "Infestations of Seven Cotton Insects on Pilose, Glanded, Frego Bract and Colored Cotton in 1969," Texas Agri. Exp. Sta. Prog. Rep. 2862.
Harland, C. S. (1919). A suggested method for the control of certain bollworms in cotton. *Empire Cotton Grow. Rev.* **6,** 333–334.

Harris, F. A., Graves, J. B., Nemec, S. J., Vinson, S. B., and Wolfenbarger, D. A. (1972). "Distribution, Abundance, and Control of *Heliothis* spp. in Cotton and Other Host Plants." So. Coop. Ser. Bull. 169.

Howard, L. O. (1896). Some Miscellaneous Results of the Work of the Division of Entomology," USDA Div. Entomol. Bull. 18.

Hunter, R. C., Leigh, T. F., Lincoln, C., Waddle, B. A., and Bariola, L. A. (1965). "Evaluation of a Selected Cross-Section of Cottons for Resistance to the Boll Weevil," Ark. Agric. Exp. Sta. Bull. 700.

Hunter, W. D. (1926). "The Cotton Hopper, or So-Called "Cotton Fleas," USDA Dep. Cir. 361.

Isley, D. (1928). The relation of leaf color and leaf size to boll weevil infestation. *J. Econ. Entomol.* **21,** 553–559.

Jenkins, J. G. (1942). "The Growing of Sea Island Cotton in the Coastal Plain of Georgia," Georgia Coastal Plain Exp. Sta. Bull. 33, University System of Georgia.

Jenkins, J. N. (1982). Present state of the art and science of cotton breeding for insect resistance in the southeast. *In* "Proceedings, Beltwide Cotton Prod. Res. Conf. 1982," pp. 117–125.

Jenkins, J. N., Hedin, P. A., Parrott, W. L., McCarty, J. C., Jr., and White, H. H. (1983). Cotton allelochemics and growth of tobacco budworm larvae. *Crop Sci.* **23,** 1195–1198.

Jenkins, J. N., Maxwell, F. G., Parrott, W. L., and Buford, W. T. (1969). Resistance to the boll weevil (*Anthonomus grandis* Boh.) oviposition in cotton. *Crop Sci.* **9,** 369–372.

Jenkins, J. N., McCarty, J. C., and Parrot, W. L. (1977). Inheritance of resistance to tarnished plant bugs in a cross of Stoneville 213 by TIMOK 811. *In* "Proceedings, Beltwide Cotton Prod. Res. Conf. 1977," p. 97.

Jenkins, J. N., and Parrott, W. L. (1971). Effectiveness of frego bract as a boll weevil resistance character in cotton. *Crop Sci.* **11,** 739–743.

Jenkins, J. N., Parrott, W. L., McCarty, J. C., and Dearing, L. (1985). Performance of cotton when infested with tobacco budworm. *Crop Sci.* **26,** 93–95.

Jenkins, J. N., Parrott, W. L., McCarty, J. C., and Earnheart, A. T. (1978). "Evaluation of Primitive Races of *Gossypium hirsutum* L. for Resistance to Boll Weevil," Miss. Agric. and Forestry Exp. Sta. Tech. Bull. 91.

Jenkins, J. N., Parrott, W. L., and McCarty, J. C., Jr. (1984). *Heliothis virescens* F. resistance in cotton. *In* "Proceedings, Beltwide Cotton Prod. Res. Conf. 1984," pp. 371,372.

Jenkins, J. N., Parrott, W. L., McCarty, J. C. Jr., and White, W. H. (1982). Breeding cotton for resistance to the tobacco budworm: Techniques to achieve uniform field infestations. *Crop Sci.* **22,** 400–404.

Jones, J. E. (1972). Effect of morphological characters of cotton on insects and pathogens. *In* "Proceedings, Beltwide Cotton Prod. Res. Conf. 1972," 88–92.

Jones, J. E., Bowman, D. T., Brand, J. W., and Playloff, M. A. (1980). Effects of species cytoplasm and morphological traits on resistance to the boll weevil in upland cotton. *In* "Proceedings, Beltwide and Cotton Prod. Res. Conf. 1980," pp. 286,287.

Jones, J. E., Clower, D. F., Milam, M. R., Caldwell, W. D., and Melville, D. R. (1975). Resistance in upland cotton to the banded-wing whitefly. *Trialeurodes abutilonea* (Haldeman). *In* "Proceedings, Beltwide Cotton Prod. Res. Conf. 1975," pp. 98,99.

Jones, J. E., Clower, D. F., Williams, B. R., Brand, J. W., Quebedeaux, K. L., and Milam, M. R. (1977). Isogenic evaluation of different sources of glabrousness for agronomic performance and pest resistance. *In* "Proceedings, Beltwide Cotton Prod. Res. Conf. 1977," pp. 110–112.

Jones, J. E., Dickson, J. I., and Beasley, J. P. (1987). Preference and nonpreference of boll weevils to selected cotton. *In* "Proceedings, Beltwide Cotton Prod. Res. Conf. 1987," pp. 98–101.

Jones, J. E., Dickson, J. I., Graves, J. B., Pavloff, A. M., Leonard, B. R., Burris, E.,

Caldwell, W. D., Micinski, S., and Moore, S. H. (1989). Agronomically enhanced insect-resistant cotton. *In* "Proceedings, Beltwide Cotton Prod. Res. Conf. 1989," pp. 135–137.

Jones, J. E., Novick, R. G., and Dickson, J. I. (1988). Boll weevil resistance in day-neutral converted primitive race stocks of *Gossypium hirsutum* L. *In* "Proceedings, Beltwide Cotton Prod. Res. Conf. 1988," pp. 99,100.

Jones, J. E., Sloane, L. W., and Phillips, S. A. (1970). Isogenic studies of red plant color in upland cotton. *Agron. Abstr.* 13.

Karami, E., and Weaver, J. B., Jr. (1972). Growth analysis of American upland cotton, *Gossypium hirsutum* L., with different leaf shapes and colors. *Crop Sci.* **12,** 317–320.

Kittock, D. L., Mauney, J. R., Arle, H. F., and Bariola, L. A., (1973). Termination of late season cotton fruiting with growth regulators as an insect-control technique. *J. Environ. Qual.* **2,** 405–408.

Kohel, R. J., Lewis, C. F., and Richmond, T. R. (1967). Isogenic lines of American upland cotton, *Gossypium hirsutum* L.: Preliminary evaluation of lint measurements. *Crop Sci.* **7,** 67–70.

Lambert, L., Jenkins, J. N., Parrott, W. L., and McCarty, J. C. (1982). Effect of 43 foreign and domestic cotton cultivars and strains on growth of tobacco budworm larvae. *Crop Sci.* **22,** 543–545.

Lacewell, R. D., and Taylor, C. R. (1980). "Economic Analysis of Cotton Pest Management Programs," Texas Agric. Exp. Sta. T.A. 15972.

Larson, J. L., Lacewell, R. D., Casey, J. D., Namkin, L. N., Heilman, M. D., and Parker, R. D. (1975). "Impact of Short Season Cotton Production in Texas Lower Rio Grande Valley on Producer Returns, Insecticide Use and Energy Consumption," Texas Agric. Exp. Sta. Bull. MP-1204.

Laster, M. L., and Meredith, W. R., Jr. (1974). Evaluating the response of cotton cultivars to tarnished plant bug injury. *J. Econ. Entomol.* **67,** 686–688.

Lee, J. A. (1971). Some problems in breeding smooth-leaved cottons. *Crop Sci.* **11,** 448–451.

Lee, J. A. (1972). Allele determining rugate fruit surface in cotton. *Crop Sci.* **18,** 251–254.

Lee, J. A. (1974). Some genetic relationships between seed and square gossypol in *Gossypium hirsutum* L. *In* "Proceedings, Beltwide Cotton Prod. Res. Conf. 1974," p. 87.

Lege, K. E., Smith, C. W., and Cothren, J. T. (1991). Condensed tannins in cotton as affected by planting date, planting density, stage of growth and plant part. *In* "Proceedings, Beltwide Cotton Prod. Res. Conf. 1991," pp. 579.

Lidell, M. C., Niles, G. A., and Walker, J. K. (1986). Response of nectariless cotton genotypes to cotton fleahopper (Heteroptera: Miridae) infestation. *J. Econ. Entomol.* **79,** 1372–1376.

Lincoln, C., Dean, G., Waddle, B. A., Yearian, W. C., Phillips, J. R., and Roberts, L. (1971). Resistance of frego-type cotton to boll weevil and bollworm. *J. Econ. Entomol.* **64,** 1326,1327.

Lincoln, C., and Waddle, B. A. (1966). Insect resistance in frego-type cotton. *Ark. Farm Res.* **15,** 5.

Lukefahr, M. J. (1977). Varietal resistance to cotton insects. *In* "Proceedings, Beltwide Cotton Prod. Res. Conf. 1977," pp. 236,237.

Lukefahr, M. J., Bottger, G. T., and Maxwell, F. (1966a). Utilization of gossypol as a source of insect resistance. *In* "Proceedings, Beltwide Cotton Prod. Res. Conf. 1966," pp. 215–219.

Lukefahr, M. J., Cowan, C. B., Pfrimmer, T. R., and Noble, L. W. (1966b). Resistance of experimental cotton strain 1514 to the bollworm and cotton fleahopper. *J. Econ. Entomol.* **59,** 393–395.

Lukefahr, M. J., Cowan, C. B., Bariola, E. A., and Houghtaling, J. E. (1968). Cotton strains resistant to the cotton fleahopper. *J. Econ. Entomol.* **61,** 661–664.

Lukefahr, M. J., Cowan, C. B., and Houghtaling, J. E. (1970). Field evaluations of improved cotton strains resistant to the cotton fleahopper. *J. Econ. Entomol.* **63,** 1101–1103.

Lukefahr, M. J., and Houghtaling, J. E. (1969). Resistance oof cotton strains with high gossypol content to *Heliothis* spp. *J. Econ. Entomol.* **62,** 588–591.

Lukefahr, M. J., and Houghtaling, J. E. (1975). High gossypol cottons as a source of resistance to the cotton fleahopper. *In* "Proceedings, Beltwide Cotton Prod. Res. Conf. 1975," pp. 93,94.

Lukefahr, M. J., Houghtaling, J. E., and Graham, H. M. (1971). Suppression of *Heliothis* populations with glabrous cotton strains. *J. Econ. Entomol.* **64,** 486–488.

Lukefahr, M. J., and Martin, D. F. (1966). Cotton-plant pigments as a source of resistance to the bollworm and tobacco budworm. *J. Econ. Entomol.* **59,** 176–179.

Lukefahr, M. J., Martin, D. F., and Meyer, J. R. (1965). Plant resistance to five lepidoptera attacking cotton. *J. Econ. Entomol.* **58,** 516–518.

Lukefahr, M. J., and Rhyne, C. (1960). Effects of nectariless cottons on populations of three lepidopterous insects. *J. Econ. Entomol.* **53,** 242–244.

Lukefahr, M. J., Stipanovic, R. D., Bell, A. A., and Gray, J. R. (1977). Biological activity of new terpenoid compounds from *Gossypium hirsutum* against the tobacco budworm and the pink bollworm. *In* "Proceedings, Beltwide Cotton Prod. Res. Conf. 1977," pp. 97–100.

Lukefahr, M. J., and Vieiera, R. M. (1986). New sources of boll weevil resistance in primitive race stocks of *Gossypium hirsutum. In* "Proceedings, Beltwide Cotton Prod. Res. Conf. 1986," pp. 493–495.

Mahill, J. F., Jenkins, J. N., and McCarty, J. C. (1983). Registration of eight germplasm lines of cotton. *Crop Sci.* **23,** 403,404.

Martin, D. F. (1962). "A Summary of Recent Research Basic to the Cultural Control of Pink Bollworm," Texas Agric. Exp. Sta. MP-579.

Masud, S. M., Lacewell, R. D., Taylor, C. R., Benedit, J. H., and Lippke, L. A. (1980). "An Economic Analysis of Integrated Pest Management Strategies for Cotton Production in the Coastal Bend Region of Texas," Texas Agric. Exp. Sta. Bull. MP-1467.

Maxwell, F. G., Jenkins, J. N., Parrott, W. L., and Buford, W. T. (1969). Factors contributing to resistance and susceptibility of cotton and other hosts to the boll wevil, *Anthonomos grandis* Boh. *Entomol. Exp. Appl.* **12,** 801–810.

Maxwell, F. G., Lafever, H. N., and Jenkins, J. N. (1966). Influence of the glandless genes in cotton on feeding, oviposition, and development of the boll weevil in the laboratory. *J. Econ. Entomol.* **59,** 585–588.

Maxwell, F. G., Schuster, M. F., Meredith, W. R., and Laster, M. L. (1976). Influence of the nectariless character in cotton on harmful and beneficial insects. *Symp. Biol. Hung.* **16,** 157–161.

McCarty, J. C. (1974). "Evaluation of Species and Primitive Races of Cotton for Boll Weevil Resistance and Agronomic Qualities." Ph.D. dissertation, Mississippi State University. State College, Mississippi.

McCarty, J. C., Creech, R. G., and Parrott, W. L. (1979a). Registration of 77 cotton germplasm. *Crop Sci.* **19,** 933,934.

McCarty, J. C., Jr., Jenkins, J. N., Parrott, W. L. (1979b). The Conversion of Photoperiodic Primitive Race Stocks of Cotton to Day-Neutral Stocks," Miss. Agric. and Forestry Exp. Sta. Res. Rep. 4, 819.

McCarty, J. C., Jenkins, J. N., and Parrott W. L. (1977). Boll weevil resistance, agronomic characteristics, and fiber quality in progenies of a cotton cultivar crossed with 20 primitive stocks. *Crop Sci.* **17,** 5–7.

McCarty, J. C., Jr., Jenkins, J. N., and Parrott, W. L. (1986). Registration of two cotton germplasm lines with resistance to boll weevil. *Crop Sci.* **26,** 1088.

McCarty, J. C., Jr., Jenkins, J. N., and Parrott, W. L. (1987). Genetic resistance to boll weevil oviposition in primitive cotton. *Crop Sci.* **27,** 263,264.

McGregor, E. A. (1927). "*Lygus elisus*: A Pest of the Cotton Regions in Arizona and California," USDA Tech. Bull. 4.

McNamara, H. C., Hooton, D. H., and Porter, D. D. (1940). "Differential Cotton Growth Rates in Cotton Varieties and Their Response to Seasonal Conditions at Greenville, Texas," USDA Tech. Bull. 710.

Meredith, W. R. (1976). Nectariless cottons. *In* "Proceedings, Beltwide Cotton Prod. Res. Conf. 1976," pp. 34–37.

Meredith, W. R., and Laster, M. L. (1975). Agronomic and genetic analyses of tarnished plant bug tolerance in cotton. *Crop Sci.* **15,** 535–538.

Meredith, W. R., Jr., Meyer, V. G., Hanney, B. W., and Bailey, J. C. (1970). Influence of five *Gossypium* species cytoplasms on yield, yield components, fiber properties, and insect resistance in upland cotton. *Crop Sci.* **19,** 647–650.

Meredith, W. R., Jr., and Schuster, M. F. (1979). Tolerance of glabrous and pubescent cottons to tarnished plant bugs. *Crop Sci.* **19,** 484–488.

Meyer, J. R., and Meyer, V. G. (1961). Origin and inheritance of nectariless cotton. *Crop Sci.* **1,** 167–169.

Meyer, V. G. (1965). Cytoplasmic effects of anther numbers in interspecific hybrids of cotton. *J. Hered.* **56,** 292–294.

Meyer, V. G. (1971). Cytoplasmic effects of anther numbers in interspecific hybrids of cotton. II. *Gossypium herbaceum* and *G. harknessii*. *J. Hered.* **62,** 77,78.

Meyer, V. G. (1972). Cytoplasmic effects of anther numbers in interspecific hybrids of cotton. III. *Gossypium longicalyx*. *J. Hered.* **63,** 33,34.

Milam, M. R., Jenkins, J. N., Parrott, W. L., and McCarty, J. C., Jr. (1982). Influence of earliness and nectariless traits on reducing the sensitivity to the tarnished plant bug in frego bract cotton. *In* "Proceedings, Beltwide Cotton Prod. Res. Conf. 1982," pp. 132,133.

Mooseberg, C. A., and Waddle, B. A. (1958). Rex cotton . . . A progress report. *Ark. Farm Res.* **7,** 9.

Mound, L. A. (1965). Effect of leaf hairs on cotton whitefly populations in the Sudan Gezira. *Empire Cotton Grow. Rev.* **42,** 33–40.

Nemec S. J., and Adkisson, P. L. (1969). "Laboratory Tests of Insecticides for Bollworm, Tobacco Budworm, and Boll Weevil Control," Texas Agric. Exp. Sta. Bull. PR-2676.

Newsom, L. D., and Brazzel, J. R. (1968). Pests and their control. *In* "Advances on Production and Utilization of Quality Cotton: Principles And Practices" (F. C. Elliot, M. Hoover, and W. K. Porter, eds.). Iowa State Univ. Press, Ames, Iowa.

Niles, G. A. (1980). Breeding cotton for resistance to cotton insects. *In* "Breeding Plants Resistant to Insects" (F. G. Maxwell and P. R. Jennings, eds.). Wiley-Interscience, New York.

Niles, G. A., Walker, J. K., and Gannaway, J. R. (1974). Breeding for insect resistance. *In* "Proceedings, Beltwide Cotton Prod. Res. Conf. 1974," pp. 84–86.

Paddock, F. B. (1919). "The Cotton or Melon Louse," Texas Agric. Exp. Sta. Bull. 257.

Parker, R. D., Walker, J. K., Niles, G. A., and Mulkey, J. R. (1980). "The "Short-Season Effect" on Cotton and Escape from the Boll Weevil," Texas Agric. Exp. Sta. Bull. 1315.

Parrott, W. L., Jenkins, J. N., and McCarty, J. C., Jr. (1981). Performance of high gossypol

strain tests under artificial infestation of tobacco budworm. *In* "Proceedings, Beltwide Cotton Prod. Res. Conf. 1981," p. 82.
Peck, T. M., and Tugwell, N. P. (1976). "Clouded and Tarnished Plant Bugs on Cotton: A Comparison of Injury Symptoms and Damage on Fruit Parts," Ark. Agric. Exp. Sta. Rep. Ser. 226.
Perkins, H. H., Jr. (1983). Effects of whitefly contamination on lint quality of U.S. cotton. *In* "Proceedings, Beltwide Cotton Prod. Res. Conf. 1983," pp. 102,103.
Pieters, E. P., and Bird, L. S. (1976). Field studies of boll weevil resistant cotton lines possessing the okra leaf-frego bract characters. *Crop Sci.* **17,** 431–433.
Quaintance, A. L., and Brues, C. T. (1905). "The Cotton Bollworm," USDA Bur. Entomol. Bull. 50.
Ray, L. L., and Richmond, T. R. (1966). Morphological measures of earliness of crop maturity in cotton. *Crop Sci.* **6,** 527–531.
Reinhard, H. J. (1927). "Control and Spring Emergence of the Cotton Fleahopper," Texas Agric. Exp. Sta. Bull. 356.
Reynolds, H. T., and Leigh, T. F. (1967). "The Pink Bollworm—A Threat to California Cotton," Calif. Agric. Exp. Sta. Ext. Serv. Cir. 544.
Ridgeway, R. L. (1984). Cotton protection practices in the USA and world-insects. *In* "Cotton" (R. J. Kohel and C. F. Lewis, eds.). Am. Soc. of Agron. Madison, Wisconsin.
Robinson, S. H., Wolfenbarger, D. A., and Dilday, R. H. (1980). Antixenosis of smooth leaf cotton to the ovipositional response of tobacco budworm. *Crop Sci.* **20,** 646–649.
Robinson, S. H., Wolfenbarger, D. A., Salgado, E., Silguero, J. F., and Dilday, R. H. (1981). A reevaluation of high bud gossypol in cotton as a deterrent to damage by *Heliothis* species. *In* "Proceedings, Belltwide Cotton Prod. Res. Conf. 1981," 79–81.
Roussel, J. S., and Clower, D. F. (1955). "Resistance to the Chlorinated Hydrocarbon Insecticides in the Boll Weevil (*Anthonomus grandis* Boh.)," Louisiana Agric. Exp. Sta. C. 41.
Russell, L. M. (1975). Collection records of *Bemisia tabaci* (Gennadius) in the United States. USDA Coop. *Insect. Rep.* **25,** 229,230.
Sappenfield, W. P., and Dilday, R. H. (1980). Breeding high terpenoid cottons: The 1978 regional tests. *In* "Proceedings, Beltwide Cotton Prod. Res. Conf. 1980," pp. 92–94.
Sappenfield, W. P., Stokes, L. G., and Harrendorf, K. (1984). Selecting cotton plants with high square gossypol. *In* "Proceedings, Beltwide Cotton Prod. Res. Conf. 1974," pp. 87–93.
Schaunak, R. K., Lacewell, R. D., and Norman, J. (1982). "Economic Implication of Alternative Cotton Production Strategies in the Lower Rio Grande Valley of Texas, 1973–78," Texas Agric. Exp. Sta. Bull. 1420.
Schuster, M. F. (1979). New sources of high tannin resistance to *Heliothis* in upland cotton resulting in feeding deterrence. *In* "Proceedings, Beltwide Cotton Prod. Res. Conf. 1979," pp. 85–87.
Schuster, M. F., Anderson, R. F., and Cannon, C. E. (1981). Boll weevil oviposition on frego bract cotton. *J. Econ. Entomol.* **74,** 346–349.
Schuster, M. F., Gibbs, M., and Smith, C. W. (1990). High condensed tannin and *Heliothis* resistance; Biological effect on larvae damage and larvae effects on yield. *In* "Proceedings, Beltwide Cotton Prod. Res. Conf. 1990," pp. 84,85.
Schuster, M. F., annd Lane, H. C. (1980). Evaluation of high tannin cotton lines for resistance to bollworms. *In* "Proceedings, Beltwide Cotton Prod. Res. Conf. 1980," pp. 83,84.
Schuster, M. F., Lukefahr, M. J., and Maxwell, F. G. (1975). Impact of nectariless cotton on plant bugs and natural enemies. *J. Econ. Entomol.* **69,** 400–402.

Schuster, M. F., and Maxwell, F. G. (1974). The impact of nectariless cotton on plant bugs, bollworm, and beneficial insects. *In* "Proceedings, Beltwide Cotton Prod. Res. Conf. 1974," pp. 26,27.
Schuster, M. L., and Frazier, J. S. (1976). "Mechanisms of Resistance to *Lygus* spp. in *Gossypium hirsutum* L." Eucarpia/OILB Host Plant Resistance to Insects and Mites. Wageningen Proc.
Shaver, T. N., Dilday, R. H., and Wilson, F. D. (1980). Use of glandless breeding stocks to evaluate unknown *Heliothis* growth inhibitors (X-factor) in cotton. *Crop Sci.* **20,** 545–548.
Shaver, T. N., and Lukefahr, M. J. (1969). Effect of flavonoid pigments and gossypol on growth and development of the bollworm, tobacco budworm, and pink bollworm. *J. Econ. Entomol.* **62,** 643–646.
Shaver, T. N., and Lukefahr, M. J. (1971). A bioassey technique for detecting resistance of cotton strains to tobacco budworms. *J. Econ. Entomol.* **64,** 1274–1277.
Shaver, T. N., Lukefahr, M. J., and Wilson, F. D. (1970). Wild races of *Gossypium hirsutum* L. as potential sources of resistance to *Heliothis* spp. *In* "Proceedings, Beltwide Cotton Prod. Res. Conf. 1970," pp. 77–79.
Sherman, F. (1930). Results of airplane dusting on the control of the cotton bollworm (*Heliothis obsoleta* Fahr). *J. Econ. Entomol.* **23,** 810–813.
Singh, I. D., and Weaver, J. B., Jr. (1971). Growth and infestation of boll weevils on normal-glanded, glandless, and high gossypol strains of cotton. *J. Econ. Entomol.* **65,** 821–824.
Smith, C. W. (1984). Combining ability for earliness traits among five diverse cotton genotypes. *In* "Proceedings, Beltwide Cotton Prod. Res. Conf. 1984," p. 98.
Smith, C. W., Cothren, J. T., and Varvil, J. J. (1986). Yield and fiber quality of cotton following application of 2-chloroethyl phosphoric acid. *Agron. J.* **78,** 814–818.
Smith, C. W., McCarthy, J. C., Altamarino, T. P., and Lege, K. E. (1991). Effect of condensed tannins on *Heliothis virescens* growth rates and survival. *In* "Proceedings, Beltwide Cotton Prod. Res. Conf. 1991," p. 550.
Smith, E. H. (1962). Synthesis and translocation of gossypol of cotton plants. *In* "Proceedings, Cottonseed Quality Res. Conf. 1962," pp. 7–12.
Smith, R. L., Wilson, R. L., and Wilson, F. D. (1975). Resistance of cotton plant hairs to mobility of first-instars of the pink bollworm. *J. Econ. Entomol.* **68,** 679–683.
Sosa, E. S., Silguero, J. F., and Wolfenbarger, D. A. (1981). Resistance of LA 17801 to *Heliothis* spp. in southern Tamaulipas. *In* "Proceedings, Beltwide Cotton Prod. Res. Conf. 1981," pp. 80,81.
Southern Cooperative Series (1978). "The Boll Weevil: Management Strategies," Texas Agric. Exp. Sta. Bull. 228.
Southern Coop Res. Bulletin (1981). "Preservation and Utilization of Germplasm in Cotton, 1968–1980," Texas Agric. Exp. Sta. Bull. 256.
Stelly, D. M., Lee, J. A., and Rooney, W. L. (1988). Proposed schemes for mass-extraction of doubled haploids of cotton. *Crop Sci.* **28,** 885–890.
Stephens, S. G. (1957). Sources of resistance of cotton strains to the boll weevil and their possible utilization. *J. Econ. Entomol.* **50,** 415–418.
Stephens, S. G., and Lee, H. S. (1961). Further studies on the feeding and oviposition preferences of the boll weevil (*Anthonomus grandis*). *J. Econ. Entomol.* **54,** 1085–1090.
Sterling, W., and Haney, R. L. (1973). Cotton yields climb, costs drop through pest management systems. *Texas Agric. Prog.* **19,** 4–7.
Stipanovic, R. D., Bell, A. A., Lukefahr, M. J., Gray, J. R., and Mabry, T. J. (1976).

Characterization of terpenoids from squares of *Heliothis* resistant cottons. *In* "Proceedings, Beltwide Cotton Prod. Res. Conf. 1976," p. 91.

Tingey, W. M., Leigh, T. F., and Hyer, G. H. (1975). *Lygus hesperus*: Growth, survival, and egg laying resistance of cotton genotypes. *J. Econ. Entomol.* **68,** 28–30.

Townsend, C. H. T. (1895). Report on the Mexican cotton boll weevil in Texas. (*Anthonomus grandis* Boh.). *In* "Insect Life," Vol. 4, pp. 295–309.

Tyler, F. J. (1910). "Varieties of American Upland Cotton," USDA Bur. Plt. Ind. Bull. 163.

Waddle, B. A. (1957). Rex, A new Arkansas cotton. *Ark. Farm Res.* **6,** 5.

Wailes, B. L. C. (1854). Agriculture. *In* "Report on the Agriculture and Geology of Mississippi," p. 143.

Waiss, A. C., Jr., Chan, B. G., Elliger, C. A., and Binder, R. G. (1981). Biologically active cotton constituents and their significance in HPR. *In* "Proceedings, Beltwide Cotton Prod. Res. Conf. 1981," pp. 61–63.

Walker, J. K. (1984). The boll weevil in Texas and the cultural strategy. *Southwest. Entomol.* **9,** 444–463.

Walker, J. K., Gannaway, J. R., and Niles, G. A. (1976). Age distribution of cotton bolls and damage from the boll weevil. *J. Econ. Entomol.* **70,** 5–8.

Walker, J. K., and Niles, G. A. (1971). "Population Dynamics of the Boll Weevil and modified Cotton Types." Texas Agric. Exp. Sta. Bull. 1109.

Walker, J. K., and Niles, G. A. (1973). Studies of the effect of bollworms and cotton fleahoppers on yield reduction in different cotton genotypes. *In* "Proceedings, Beltwide Cotton Prod. Res. Conf. 1973," pp. 100–102.

Walker, J. K., Niles, G. A., Gannaway, J. R., Robinson, J. V., Cowan, C. E., and Lukefahr, M. J. (1974). Cotton fleahopper damage to cotton genotypes. *J. Econ. Entomol.* **67,** 537–542.

Walker, J. K., and Smith, C. W. (1993). Cultural Control. *In* "Cotton Insects" (J. R. Philips and E. G. King, eds.). Cotton Foundation Reference Book Series.

Walker, J. K., Smith, C. W., and Reina, J. D. (1987). Evaluation of T277-2-6 for boll weevil resistance in Texas, 1986. *In* "Proceedings, Beltwide Cotton Prod. Res. Conf. 1987," pp. 337–340.

Wannamaker, W. K. (1957). The effect of plant hairiness of cotton strains on boll weevil attack. *J. Econ. Entomol.* **50,** 418–423.

Ware, J. O. (1951). "Origin, Rise and Development of American Upland Cotton Varieties and Their Status at Present," Ark. Agric. Exp. Sta. Agronomy Dep. Misc. Pub. (unnumbered).

Weaver, J. B., and Graham, R. (1977). Behavior of boll weevils on cytoplasmic male-sterile cotton in isolated plots. *In* "Proceedings, Beltwide Cotton Prod. Res. Conf. 1977," pp. 100,101.

Wessling, W. H. (1958a). Resistance to boll weevil in mixed populations of resistant and susceptible cotton plants. *J. Econ. Entomol.* **51,** 502–506.

Wessling, W. H. (1958b). Genotypic reactions to boll weevil attack in upland cotton. *J. Econ. Entomol.* **51,** 508–512.

Wilson, F. D. (1982). Present state of the art and science of cotton breeding for insect resistance in the West. *In* "Proceedings, Beltwide Cotton Prod. Res. Conf. 1982," pp. 111–117.

Wilson, F. D. (1987). Pink bollworm resistance, lint yield, and earliness of cotton isolines in resistant genetic backgrounds. *Crop Sci.* **27,** 957–960.

Wilson, F. D. (1989). Yield, earliness, and fiber properties of cotton carrying combined traits for pink bollworm resistance. *Crop Sci.* **29,** 7–12.

Wilson, F. D., and George, B. W. (1981). Breeding cotton for resistance to pink bollworm. *In* "Proceedings, Beltwide Cotton Prod. Res. Conf. 1981," pp. 63–65.

Wilson, F. D., and George, B. W. (1982). Effects of okra-leaf, frego-bract, and smooth-leaf mutants on pink bollworm damage and agronomic properties of cotton. *Crop Sci.* **22,** 798–801.

Wilson, F. D., and George, B. W. (1984). Pink bollworm (Lepidoptera: Gelechiidae): Selecting for antibiosis in artificially and naturally infested cotton plants. *J. Econ. Entomol.* **77,** 720–724.

Wilson, F. D., George, B. W., and Wilson, R. L. (1981). Lint yield and resistance to pink bollworm in early-maturing cotton. *Crop Sci.* **21,** 213–216.

Wilson, F. D., and Lee, J. A. (1970). Tobacco budworm response to variation in gland density in cotton seedlings. *In* "Proceedings, Beltwide Cotton Prod. Res. Conf. 1970," p. 54.

Wilson, F. D., and Lee, J. A. (1971). Genetic relationship between tobacco budworm feeding response and gland number in cotton seedlings. *Crop Sci.* **11,** 419–421.

Wilson, F. D., and Smith, J. N. (1976). Some genetic relationships between gland density and gossypol content in *Gossypium hirsutum* L. *Crop Sci.* **16,** 830–832.

Wilson, F. D., Szaro, J. L., and Hefner, B. A. (1987). Behavior of pink bollworm larvae on bolls of okra-leaf and normal-leaf cotton. *In* "Proceedings, Beltwide Cotton Prod. Res. Conf. 1987," p. 96.

Wilson, F. D., and Wilson, R. L. (1977). Seed damage by pink bollworm to race stocks, cultivars, and hybrids of cotton. *Crop Sci.* **17,** 465–467.

Wilson, F. D., Wilson, R. L., and George, B. W. (1979). Pink bollworm: Reduced growth and survival of larvae placed on bolls of cotton race stocks. *J. Econ. Entomol.* **72,** 860–864.

Wilson, F. D., Wilson, R. L., and George, B. W. (1980a). "Pink Bollworm: Expected Reduction in Damage to Cottons Carrying Combinations of Resistance Characters," USDA Agric. Res. Results W-12.

Wilson, F. D., Wilson, R. L., and George, B. W. (1980b). Resistance to pink bollworm in breeding stocks of upland cotton. *J. Econ. Entomol.* **73,** 502–505.

Wilson, R. L., and Wilson, F. D. (1975). "A Laboratory Evaluation of Primitive Cotton (*Gossypium hirsutum* L.) Races for Pink Bollworm Resistance," USDA-ARS-W-30.

Wilson, R. L., and Wilson, F. D. (1976a). Effects of nectariless and glabrous cotton on pink bollworm in Arizona. *In* "Proceedings, Beltwide Cotton Prod. Res. Conf. 1976," p. 91.

Wilson, R. L., and Wilson, F. D. (1976b). Nectariless and glabrous cottons: Effect on pink bollworm in Arizona. *J. Econ. Entomol.* **69,** 623,624.

Wolcott, G. R. (1927). Haitian cotton and the pink bollworm. *Bull. Entomol. Res.* **18,** 79–82.

Young, J. H., Willson, L. J., and Price, R. G. (1986). Cotton fleahopper perference of cotton cultivars as oviposition sites. *In* "Proceedings, Beltwide Cotton Prod. Res. Conf. 1986," pp. 488,489.

Zummo, G. R., Benedict, J. H., and Segers, J. C. (1983). No choice-study of plant–insect interactions for *Heliothis zea* (Boddie) (Lepidoptera: Noctuidae) on selected cottons. *Environ. Entomol.* **12,** 1833–1836.

Zummo, G. R., Segers, J. C., and Benedict, J. R. (1984). Seasonal phenology of allelochemicals in cotton and resistance to bollworm (Lepidoptera: Noctuidae). *Environ. Entomol.* **13,** 1287–1290.

Index

A

E

H

I

K

L

N

O

P

ISBN 0-12-000748-7